Human Walking in Virtual Environments

Frank Steinicke · Yon Visell
Jennifer Campos · Anatole Lécuyer
Editors

Human Walking in Virtual Environments

Perception, Technology, and Applications

Springer

Editors
Frank Steinicke
University of Würzburg
Würzburg
Germany

Yon Visell
Electrical and Computer Engineering Department
Drexel University
Philadelphia, PA
USA

Jennifer Campos
Toronto Rehabilitation Institute
University of Toronto
Toronto, ON
Canada

Anatole Lécuyer
National Institute for Research in Computer Science and Control (INRIA)
Rennes
France

ISBN 978-1-4419-8431-9 ISBN 978-1-4419-8432-6 (eBook)
DOI 10.1007/978-1-4419-8432-6
Springer New York Heidelberg Dordrecht London

Library of Congress Control Number: 2013937207

Printed on acid-free paper

Springer is part of Springer Science+Business Media (www.springer.com)

Foreword

Walking, strutting, running, shuffling, tiptoeing, climbing, or pirouetting—people move on foot through an impressive variety of activities and contexts. Not surprisingly, there has been a keen scientific awareness and a growing body of knowledge surrounding ways that people execute tasks involving locomotion and how they perceive their environment and its contents during the course of movement on foot. In parallel, locomotion is increasingly seen as a natural and promising means of moving in virtual environments. A number of important questions pertain to how virtual walking may be afforded in new computational systems and how self-motion is affected and perceived in virtual environments. Finally, there is a growing consensus that locomotion in augmented and virtual reality environments may be relevant to a wide range of emerging applications, from immersive training simulations, to entertainment and video games.

This book concerns the science and engineering of walking in virtual environments. It is an attempt to bring together, for the first time in one volume, contributions from a growing interdisciplinary body of knowledge on human self-motion perception, the multisensory nature of walking, conceptual design approaches, current technologies, and applications. The use of VR and movement simulation systems is becoming popular and more accessible within a variety of research fields and applications. Many of the relevant simulation technologies initially focused on developing realistic, interactive visual environments. However, it is becoming apparent that our everyday interactions are highly multisensory. Therefore, investigators are beginning to understand the critical importance of walking interfaces that can allow for realistic, natural behaviors. This book aims to present an overview of what is currently understood about human perception and performance when moving in virtual environments and to situate it relative to the broader scientific and engineering literature on human locomotion and walking interfaces. The contents include scientific background and recent empirical findings related to biomechanics, selfmotion perception, and physical interactions. The book also discusses conceptual approaches to multimodal sensing, display systems, and interaction for walking in real and virtual environments. Finally, it presents current and emerging applications in areas such as gait and posture rehabilitation, gaming, sports, and architectural design.

The organization of this book largely reflects the level of interdisciplinarity of the topical area it addresses, touching on aspects related to human perception and action, virtual reality technologies, and their applications in human–computer interaction design, immersive simulation, health care, and entertainment.

Walking as Perception and Action

Locomotion can be seen to serve two key tasks: those of movement and of sensory awareness, i.e., of action and perception. On one hand, the most basic function of walking might be said to be that of self-motion. A hallmark of our species is that we travel, stand, and otherwise negotiate our surroundings in a mostly bipedal manner. We do so over a range of different speeds, with different manners, and following different patterns that are dependent on the task at hand and the way it is performed. Beyond self-motion, we walk in order to have a look around, surveying our surroundings as we navigate, and generating a great deal of multisensory information about the world. The act of locomotion is intimately tied to the ways that we perceive the ambient spaces and ground surfaces that we traverse. Stable and efficient locomotion is itself known to require the gathering of information about the ground serving as support, and a large amount of sensory information is likely to be available for this purpose.

A pedestrian receives visual information via the eyes, sound information via the auditory channel, haptic information via the sense of touch, and information about movements of the muscles and joints via the proprioceptive sense. These multiple sensory inputs are integrated in the formation of coherent percepts about the contents and activities of space and the pedestrians own motion in it. Walking thus involves a range of human sensory and motor faculties, and the neural processing apparatus that supports them.

When walkers are enabled to navigate within virtual environments, additional factors come into play, including limitations in the presentation of virtual 3D scenes, or to the coupling of body movements that change with perspective and distance. Several of these questions are addressed in Part I of the book. In Chap. 1 Waller and Hodgson describe how spatial knowledge of one's environment during navigation is informed by external (e.g., visual and auditory), internal (e.g., vestibular and proprioception), and cognitive (e.g., attention) sources and the implications for movement simulation. In Chap. 2 Riecke and Schulte-Pelkum summarize multimodal effects on the illusion of self-motion (i.e., vection) and ways in which various sensory inputs can be exploited to maximize this illusion in the most efficient way possible. In Chap. 3, Multon and Olivier review in detail the most current literature describing the biomechanical characteristics of walking in real and virtual environments. In Chap. 4 Fajen explores locomotion from the ecological perspective by considering one's perception of affordances during tasks such as obstacle avoidance and wayfinding. In Chap. 5, Ruddle considers the role and importance of body-based cues during translational and rotational movements when interacting within virtual environments of various scales (model, small, and large scale). Finally, in Chap. 6, Frissen et al., summarize a collection of research focusing on the biomechanics of natural walking, the interactions of proprioceptive and vestibular inputs during curvilinear walking, and the characteristics of unconstrained large-scale walking, all with the intention of illuminating the development and testing of a unique omnidirectional treadmill (the Cyberwalk).

Technologies for Virtual Walking Experiences

Just as walking is fundamental to our negotiation of natural environments, it is of increasing relevance to interactions with computational systems. From Star Treks holodeck, to William Gibsons cyberspace, the idea that people could move through virtual environments via seamless and natural-seeming body movements has long been a staple of science fiction and futurist thinking. However, the potential of realizing such experiences within real laboratory settings has only recently become feasible, due to advances in multimodal 3D display technologies, sensing, and robotic motion simulators. Other contemporary interactive paradigms have emerged as well, including the superposition or mixing of components of virtual worlds within real environments, or via novel body-scale human interactive devices.

Part II of this book surveys a range of technological challenges that arise when designing virtual walking experiences, and some of the predominant solutions that have emerged in the last few years. In Chap. 7, Steed and Bowman review the displays and interaction devices that can be utilized for virtual travel, ranging from desktop to fully immersive visual displays, and hand-held devices to motion tracking systems. In Chap. 8, Multon describes the most popular methods and algorithms used to evaluate the parameters and main properties of human walking (e.g., step length, joint angles, or ground reaction forces). In Chap. 9, Iwata provides an extensive survey of locomotion interfaces, i.e., mechanical devices for creating artificial sensations of physical walking, categorizing them into four types: sliding shoes, treadmills, foot-pads, and robotic tiles. In Chap. 11, Whitton and Peck focus on stepping-driven locomotion techniques (walking-in-place and real-walking interfaces) which do not include treadmills or other mechanical devices and are driven by the users' actual stepping motions to convert those values into viewpoint movement between frames. In Chap. 10, Bruder and Steinicke explain how to implement virtual walking in virtual environments, via different strategies that allow users to actually move through the real world, using physical displacements that are mapped to motions of the camera in the virtual environment (VE) in order to support unlimited omnidirectional walking. Lastly, in Chap. 12, Marchal et al., address the multimodal rendering of walking over virtual ground surfaces, and how to model, simulate, and incorporate haptic, acoustic, and graphic rendering to enable truly multimodal walking experiences.

Applications of Virtual Walking

Beyond the basic scientific and technological issues addressed in the first two parts of the book lies the basic question of why, and for what purposes, it may be desirable to design interactive experiences of virtual walking. From one standpoint, this remains a nascent field of research and development, and, as has been seen in other domains that have emerged on large scales in recent decades (e.g., mobile computing), many of the applications that ultimately take hold may be difficult to foresee from the present, early state of development. Nonetheless, a number of broad domains of potential application can be identified, related to areas such as human–computer interaction design, immersive simulation, health care, and entertainment.

Thus, Part III of the book presents a number of interactive techniques and application scenarios that have been subjects of recent research. In Chap. 13 of Part III, Kulpa, Bideau, and Brault describe the implementation of techniques for allowing athletes to interact via movements in virtual sports setting, and how these interactions may be useful for understanding sports performance. In Chap. 14, Suma, Krum, and Bolas describe the use of redirected walking techniques in the design of immersive simulation training environments. In Chap. 15, Kiefer, Rhea, and Warren review current applications of VR for clinical assessment and rehabilitation of locomotor behavior. In Chap. 16, Williamson, Wingrave, and LaViola present a number of techniques and issues related to using low-cost video game controllers to design affordances for self-motion in virtual environments. Finally, in Chap. 17, Visell and Cooperstock review the state of the art and future directions in human–computer interaction design for computationally augmented floor surfaces.

It is hoped that this diverse collection, organized under the broad umbrella of virtual walking, a topic that was essentially unaddressed in the research literature just two decades ago, may prove interesting for researchers in related fields of engineering, computing, perception, and the movement sciences, and further, that the many challenges that remain may suggest interesting directions for future research.

Yon Visell
Frank Steinicke
Jennifer Campos
Anatole Lécuyer

Contents

Part II Technologies

Part III Applications and Interactive Techniques

Part I
Perception

Chapter 1
Sensory Contributions to Spatial Knowledge of Real and Virtual Environments

David Waller and Eric Hodgson

Abstract Most sensory systems are able to inform people about the spatial structure of their environment, their place in that environment, and their movement through it. We discuss these various sources of sensory information by dividing them into three general categories: external (vision, audition, somatosensory), internal (vestibular, kinesthetic) and efferent (efference copy, attention). Research on the roles of these sensory systems in the creation of environmental knowledge has shown, with few exceptions, that information from a single sensory modality is often sufficient for acquiring at least rudimentary knowledge of one's immediate environment and one's movement through it. After briefly discussing the ways in which sources of sensory information commonly covary in everyday life, we examine the types and quality of sensory information available from contemporary virtual environments, including desktop, CAVE, and HMD-based systems. Because none of these computer mediated systems is yet able to present a perfectly full and veridical sensory experience to its user, it is important for researchers and VE developers to understand the circumstances, tasks, and goals for which different sensory information sources are most critical. We review research on these topics, as well as research on how the omission, limitation, or distortion of different information sources may affect the perception and behavior of users. Finally, we discuss situations in which various types of virtual environment systems may be more or less useful.

Brian and Sarah set up their campsite at dusk. At dawn the next morning, Sarah decides to venture from the campsite briefly to explore the area. As she walks away from the camp, the sights and sounds of a nearby brook recede into the background. She feels her legs working, climbing uphill from the brook, her feet occasionally slipping backwards slightly, down the muddy trail. As an owl flies by, it captures her attention, and she turns her head quickly to watch it light in a nearby tree. Turning her head to admire the bird does not disorient Sarah; nor does it cause her to change

D. Waller (✉) · E. Hodgson
Department of Psychology, Miami University,
Psychology Building 218, Oxford, OH 45056, USA
e-mail: wallerda@muohio.edu

F. Steinicke et al. (eds.), *Human Walking in Virtual Environments*,
DOI: 10.1007/978-1-4419-8432-6_1,

her generally uphill course. After reaching the top of the hill, Sarah directs her gaze down toward a larger river running below her before deciding to move on back to her campsite. Even before her campsite comes into view, she can smell smoke coming from the campfire that Brian had just built.

This brief vignette illustrates the variety and the complexity of information that is available to people as they navigate through known and unknown places. In the story, Sarah uses a variety of online sensory information—visual, kinesthetic, vestibular, somatosensory, auditory, and even olfactory—to provide information about her environment and her place in it. She is also able to use offline information—internally stored knowledge—to determine her goals and destinations. Because nearly every sensory modality can contribute information about one's spatial disposition or about the spatial properties of one's environment, it is a challenge for scientists to understand how this multimodal influx of information is acquired, interpreted, combined, and acted upon. In this chapter, we will examine what is known about the different ways that people can take-in the spatial information that is available to them as they walk through real and computer-simulated (virtual) environments. In the first part of the chapter, we will briefly discuss what we know about the senses' contribution to spatial knowledge, including the circumstances under which each sense in isolation is necessary or sufficient for acquiring spatial knowledge and for enabling spatial behaviors. The modality-by-modality discussion, however, is only expository; it is important to realize that in most real situations, information flows simultaneously through all of the senses—that spatial information is multimodal, overlapping, and heavily redundant. Thus, in the second part of the chapter, we examine situations in which there are differing combinations of sensory information about space. In particular, we discuss the sensory contributions of today's computer-simulated environments, in which some sensory modalities have access to limited or imperfect information.

Before we can begin a meaningful discussion of incoming sensory information, we will do well to recognize that the issue of sensory contributions to spatial knowledge cannot be disentangled from conceptualizations of the type of knowledge that is produced. In general, the type of knowledge that one acquires about his or her environment depends critically on the goals that one has for interacting with the environment. On one hand, many goals are served exclusively by transient knowledge of one's location and orientation with respect to important objects in the immediate environment. In the situation above for example, Sarah did not need to recall the location of her campground in order to track and admire a nearby bird. Tasks such as speed, distance, and turn estimation, as well as online spatial updating (i.e., the ability to track one's location and orientation with respect to salient objects in the environment; see [62]) may involve a dynamic form of environmental knowledge that consists of little more than sensitivity to ongoing changes in immediately available information. Here, internal representations of the environment may be minimal—fragmented, sketchy, and schematic—or even nonexistent, because the environment itself provides easy access to necessary information. Such situations seem particularly well addressed by concepts that have developed out of the ecological approach to perception, such as information pickup, direct perception, and perception/action complementarity (see, for example, [31, 72, 107]).

On the other hand, people often rely on knowledge of remote locations to set future goals, to orient toward unseen locations, and to imagine themselves (and other people) in different places. Such knowledge consists of remembered information that is not perceptually available for immediate inspection. Although ecological approaches may be able to address and describe this type of knowledge (see for example, [41, 42]), it is generally more common in the literature to consider spatial memory as involving internal representations of environmental structure [29]. By this view, enduring spatial knowledge is the result of mental processing and storage of environmental and internal information. Most researchers accept a three-tiered description of enduring spatial knowledge that includes: (a) relatively rudimentary (and arguably nonspatial) memory for landmarks and features, (b) route knowledge of how places are interconnected, typically without regard to the metric distances and directions between them, and (c) configuration (or survey) knowledge of global metric relations among places in an environment. Of course, sensory contributions to any of these types of knowledge may be qualitatively different, which can complicate our discussion considerably.

Regardless of whether spatial knowledge enables action in the moment or is stored for subsequent use, it is clear that in order to be useful, spatial knowledge must coordinate the disposition of the knower with characteristics of the environment. Thus in laying-out the sensory bases for environmental cognition, we make a primary distinction between sensory information that provides either: (a) external information about the nature of one's environment, or (b) internal (or idiothetic) information about the status of one's body or effectors. After discussing these sources of sensory information, we turn briefly to a consideration of non-sensory information that is internally generated, such as attention allocation and other cognitive factors.

1.1 External Sensory Information

Vision provides a direct, rich, and precise source of spatial information, and it is undoubtedly the most researched sensory modality with respect to environmental knowledge and navigation. Like many other external senses, the visual system can provide detailed and useful information about the spatial layout of the immediate environment without significant bodily movement (see [24, 80] for a review of visual cues that enable apprehension of layout). The layout of an environment, including relative directions, distances, and scale can be accurately perceived and remembered from a stationary viewpoint (e.g., [92]), from brief glimpses of images of a spatial layout [118], and even from symbolic media such as maps [105] and gestures [3]. Frequently however, visual information about an environment comes from movement through it—from the optic flow generated from changes in one's position and heading. As one changes position (e.g., moving forward or backward) optic flow patterns radiate from a focus of expansion or contraction. As one changes orientation (e.g., rotating to one's left or right) flow patterns are laminar, moving across the retina in the direction opposite to the motion. Combinations of radial and laminar optic flow can signify relations among one's: (a) direction of travel (i.e., one's course), (b) head direction, and (c) gaze direction. In so doing, the online interpretation of optic flow provides information about the spatial relations in the environment that both are and are not dependent on one's viewpoint [31].

A substantial body of research (as well as informal observation) indicates that despite its availability and precision, visual information is often not necessary for performing many rudimentary spatial tasks or even for developing detailed knowledge of one's environment. The navigational abilities of congenitally blind people [34] attest to this idea. The experimental literature too is rife with demonstrations that visual information can be surprisingly unnecessary for accurate perception and memory of space, and researchers frequently conclude that internal senses (discussed below) are relatively more influential than vision for the acquisition of online environmental knowledge (e.g., [15]). On the other hand, it has been known for decades that in theory, visual information can be sufficient for performing a variety of spatial tasks [30, 31]. For online perceptual tasks such as speed [59], distance [28, 64], and heading [58] estimation, optic flow alone appears to be sufficient for enabling accurate spatial knowledge. However, when tasks require online tracking of one's rotational changes, visual information by itself tends to be much less sufficient. Klatzky et al. [55], for example, demonstrated that optic flow alone does not generally enable people to keep track of their orientation in space and that additional body-based information is required (see also [16, 76]). There is some evidence, however, that visual information can enable people to account for rotational changes, especially if it provides information about landmarks, and is available in a sufficiently large field of view [83, 85]. Finally, for memory-based tasks, even very briefly presented visual information can be sufficient for the acquisition of knowledge of spatial layout, including relative directions and distances between objects in a single scene [118]. Indeed, static visual information from photographs [4], as well as dynamic visual information from video [33] or desktop virtual environments [86, 89] can be sufficient for acquiring survey knowledge of relatively complex environments, although such knowledge may be slow to develop [111] and show especially large differences among users [87].

Other external senses such as audition, olfaction, and the somatosenses (i.e., pain, pressure, vibration, and heat) can provide spatial information for both online and offline purposes; however, the information from these sources is typically not as rich as that provided by vision. Among these non-visual external senses, two are especially useful in generating environmental knowledge. First, audition can be used to localize noise-producing objects [117] and to sense the scale of a local environment [90]. Indeed, people can learn to echolocate well enough to aid nonvisual navigation [91, 98]. People can also gain accurate knowledge of an environment based solely on auditory information such as verbal directions [21], spatial language [63] or environmental descriptions [36]. Second, a very common—though rarely investigated—source of information about the external environment comes from somatosensory pressure receptors that provide information about: (a) acceleration, through one's points of contact with a surface of support, and (b) physical contact with objects in one's environment. When acquired actively, and in conjunction with kinesthetic information (discussed below) this latter source of somatosensory information is known as *haptics*, a topic that we discuss briefly below.

The other external senses likely have a negligible impact on the acquisition of environmental knowledge. In theory, one can pickup somatosensory information from external sources and use variations in the strength of these sources to navigate or to inform one about the environment. For example, direct sunlight may heat one side of a person more than another, allowing her to determine the direction of the sun and, by extension, her orientation. Similarly, the strength of vibrations under one's feet may be used to estimate one's proximity to a large machine on a factory floor. Indeed, the fact that vibratory devices have been constructed to signal directions for navigation [108] indicates that such information sources may be useful in some circumstances. However in most everyday situations, the role of these and other sensory systems (such as olfaction and taste) in environmental cognition is probably minimal and is certainly not well researched.

1.2 Internal Sensory Information

Idiothetic information about space derives from three principal sensory systems: (a) the vestibular system, which senses angular and linear acceleration of the head; (b) the kinesthetic/proprioceptive system, which provides information about the position, orientation, and movement of the musculature; and (c) the somatogravity system [66] that informs people about the direction of gravity. A fourth source of information, efference copy, is sometimes considered a source of idiothetic information [67]; however, we will defer discussion of efference copy to our discussion below of efferent information more generally. The vestibular and kinesthetic/proprioceptive systems have direct and important ties to spatial cognition, and we discuss them briefly below; however, we do not discuss further the somatogravity system, as it is little researched in humans, and its contributions to environmental knowledge are largely unknown.

The anatomical basis of the vestibular system is a set of structures in the inner ear—otoliths and semicircular canals—that sense linear and angular acceleration, respectively. In addition to supporting several postural and oculomotor reflexes, vestibular information provides information that can be doubly-integrated in order to determine linear or angular displacement. In this way, it is thought to provide a critical input to people's ability to dead reckon and to update (see [77]). In the experimental literature, vestibular information is typically isolated by passively transporting blindfolded participants; however, we note that such procedures do not completely isolate the vestibular sense, as they do not remove somatosensory information about inertial forces (e.g., the sensation of the wheelchair pressing against one's back). These experiments have demonstrated that vestibular information enables better than chance accuracy on online spatial tasks such as turn and distance estimation [7, 38,

116], as well as for more complex online tasks such as spatial updating. However, the accuracy and precision of vestibularly-acquired spatial information is typically rather low, and degrades rapidly as the knowledge derived from it becomes more complex. For example, Sholl [94] pushed blindfolded participants in a wheelchair over paths of varying complexity. Most participants were able to track simple two-segment paths with well-better than chance accuracy (see also [38]); however, participants were generally unable to keep track of complex trajectories (i.e., those involving more than three turns). With respect to the creation of relatively complex and enduring spatial knowledge, the sufficiency of vestibular information is largely unexplored in the literature, although Sholl's results would suggest that it is difficult to acquire complex route or survey knowledge on the basis of vestibular information alone. Among healthy adults, vestibular information may also be necessary for successful interaction with the environment, as sudden loss or disruption of it through pathology [20, 32, 75, 76] or experimental manipulation [101] can significantly impair performance on basic spatial tasks such as turn and distance estimation. However, people can generally adapt to a gradual progressive degradation of the vestibular system [57], and thus its full functioning is not always necessary for acquiring spatial knowledge. The role of the vestibular system as well as its relation to other idiothetic senses is reviewed in [57].

Although the terms proprioceptive and kinesthetic are often used interchangeably in the literature, we use the former to refer to information about a relatively static position or attitude of the musculature; whereas the latter refers to information about the movement of one's limbs or effectors. By this distinction, knowledge of the location of one's unseen hand, for example, may come from a proprioceptive sense; whereas the ability to brush one's teeth without a mirror would require kinesthetic information. A relatively small research literature has found that kinesthetic information is generally sufficient for acquiring knowledge of distances [67, 102] and orientation changes [8, 13, 52]. Indeed, compared to the vestibular and somatosensory components of idiothetic information, the proprioceptive component may enable relatively more accurate performance in a variety of online tasks, such as heading estimation [73, 103], turn estimation [53], distance estimation [67], and spatial updating [94]. As with vestibular information, the sufficiency of kinesthetic information for acquiring relatively sophisticated knowledge about spatial layout is under-researched. But the ability to acquire accurate spatial information about traveled distances [28], turns [83], routes [114], and spatial layout [86] from purely visual sources demonstrate that kinesthetic (and other sources of) information is not strictly necessary for the acquisition of spatial knowledge.

1.3 Efferent Sources of Information

Finally, we consider the contributions to spatial knowledge of four other internal sources of spatial information—efference copy, attention allocation, cognitive decision making, and mental transformations. Of course it is incorrect to consider these information sources as "sensory" inasmuch as they are not considered to carry afferent information from the peripheral to the central nervous system. Nonetheless, these information sources form a critical component of active engagement with one's environment and may have a strong influence on the nature and quality of spatial information that is available for acquiring environmental knowledge. It is worth noting that these sources of information are also tightly constrained and dictated by one's goals. Our brief discussion of these information sources draws heavily on the conceptual distinctions made by Chrastil and Warren [18] who used these concepts in their recent review of the extensive and complicated literature on active and passive contributions to spatial knowledge.

Efference copy [47] is a simultaneous record of the motor commands from the central nervous system to the musculature that enables organisms to account for the difference between external stimulation and the stimulation that arises as a consequence of their own actions. For example, as we discussed above, when one turns one's head clockwise, the visual system has access to laminar optic flow in a counterclockwise direction. Logically, such a pattern of optic flow could signal a counterclockwise rotation of the visible environment around a stationary viewpoint, rather than a rotation of a viewpoint in a stationary environment. However, people are typically able to distinguish these possibilities by accounting for the fact that motor commands actively produced a set of expected visual consequences (but see [43] for situations when people cannot). This knowledge of one's motor commands constitutes efference copy. We consider the efferent copy of motor commands to contain information about both the implicit or explicit intentions used to move in and interact with the environment, as well as information about the strength of these intentions. In this way, efference copy can be used to generate a set of expectations about the consequences of one's actions (see [12]). Discrepancies between the expected and the perceived consequences of a set of motor commands can indicate a need for perceptual recalibration or can be used as an indicator of the precision of one's intentions.

Other efferent sources of information have origins in "higher-level" cognition and include factors such as constraints about how and where to allocate attention, decisions about how and where to navigate, and the ability to transform spatial information. Chrastil and Warren [18] point out that these internal sources of information are often confounded or conflated in studies that examine the relative effects of active versus passive navigation on the acquisition of spatial knowledge. However, because virtual environments (VEs) enable yoked playback of others' interactive experiences, computer simulations have recently provided researchers a helpful methodological tool for teasing some of these influences apart. For example, Christou and Bülthoff ([19], Experiment 3) asked participants to explore a model of a complex attic space on a desktop virtual reality system and to familiarize themselves with the environment and its contents. One group of participants navigated through the environment actively, by manipulating a trackball that controlled the position and orientation of their simulated viewpoint. A second group of participants had access to the same visual information as the first, but passively viewed playbacks of the explorations made by matched participants in the other condition. Differences between these groups on the ability to recognize novel and previously seen views were minimal, not statistically reliable, and in some cases numerically favored the passive learners. The researchers concluded that the acquisition of spatial knowledge was probably influenced more by the relative amounts of attention deployed to either the navigational interface or to the learning of the environment than by active decision making about where and how to navigate. Similar conclusions were reached by Chrastil and Warren in their review of the broader literature, who noted that differences between active and passive learning of environments are probably more heavily influenced by the allocation of attention than by the ability to choose one's own route through the environment.

1.4 Relative Influence of External and Internal Sensory Information

To this point, we have discussed different sources of spatial information in isolation. Yet we have also noted that in normal real-world situations, multiple sources of sensory and efferent information are mutually available, and that these sources typically provide overlapping and redundant information. Thus, in understanding the contributions of sensory systems in real-world environments, it is often more relevant and practical to consider the relative influences of these sources when many are present, rather than considering the degree to which isolated sources of sensory information are necessary or sufficient for the acquisition of spatial knowledge. On the other hand, users of VE systems often do not have access to one or more sensory sources of information. Thus, in understanding the contributions of sensory systems in VEs, it is especially relevant to consider the consequences to spatial knowledge acquisition when some sources of sensory information are absent, degraded, or uninformative. These two topics (i.e., overlapping sensory information in real world environments and degraded sensory sources in virtual environments) comprise the outline for the final section of this chapter. After a brief discussion of the relative contributions of different sources of sensory information in real-world environments, we close by summarizing what is known about the impact of unavailable or degraded sources of sensory information that comes with various types of VE systems.

1.4.1 Sensory Contributions in the Real World

As we have seen in many cases, separate sensory sources are frequently naturally coupled, and thus can be extremely difficult to isolate experimentally. For example, it is difficult for experimenters to dissociate kinesthetic information about neck movement from vestibular information about the head's rotation. As a result of these naturally conflated sensory sources, researchers commonly combine various sources of sensory or efferent information into broader categories. For example, many researchers (ourselves included) refer to *body-based* information to refer to the amalgam of vestibular, kinesthetic, and efferent information that is primarily associated with self-movement. Others may refer to the amalgamated concept *inertial information* to refer to the combination of vestibular, somatosensory, and sometimes visual information that indicates acceleration through space. Additionally, the concept of *podokinesthetic* (or *podokinetic*) information is sometimes used in the literature to refer to the grouping of somatosensory, kinesthetic, and efferent information available about foot contact with the ground during locomotion. Finally, *haptic* information [60] derives from the combination of efferent, kinesthetic and somatosensory information that arises during the active manipulation of objects. In these ways, groups of commonly occurring sensory sources can be isolated and examined for their combined contribution to spatial knowledge. It is worth noting that the advent of virtual reality technology has made much of this research relatively easy to do, although visual-idiothetic dissociations using mirrors [70] or prism glasses [39, 99] have also historically been able to address many of these questions.

A growing body of research has examined the degree to which the presence of these amalgamated sources of information facilitates the acquisition or enhances the quality of spatial knowledge when it and others are available. Probably most of this work has focused on the relative influences of body-based information when vision is available [54, 55, 73, 94, 102, 103]. With respect to online tasks that require transient knowledge, the literature generally shows that body-based senses do facilitate the acquisition of accurate spatial knowledge of one's environment [15, 53, 116]. Indeed, as mentioned above, there is some evidence that especially with respect to acquiring accurate information about turns and orientation, body-based senses may be necessary [7, 55, 88].

The relative contributions of visual and body-based senses in tasks and environments that require sophisticated enduring knowledge of space, however, is less clear, and only a handful of studies have addressed this issue. The few that have tended to conclude that body-based senses contribute minimally to environmental learning. Waller et al. [113] examined the degree to which inertial information facilitates the visual system in acquiring configural knowledge of a large environment by manipulating the amount of valid inertial information available from a 1.6 km car trip through a real-world neighborhood. Three matched groups of participants received visual information by viewing the trip through a head-mounted display (HMD). These three groups differed on the presence and quality of inertial (i.e., vestibular and somatosensory) information: one group had access to valid inertial cues, one group had access to invalid inertial cues, and a final group had access to no inertial cues. The results clearly showed no differences among these groups in their (fairly accurate) memory of the layout of the environment. Interestingly however, in a closely-related follow-up study, Waller et al. ([112]; see also [110]) showed that the additional idiothetic cues provided by proprioception and efference copy did make a small but significant improvement in the accuracy of spatial knowledge acquired from vision. The primary conclusions from all of these studies, however, has been that the role of idiothetic information in forming an enduring representation of large environments is quite small—generally much less than the facilitative effect of idiothetic on transient knowledge of local environments.

Finally, it is worth noting that questions about the relative influence of various sensory systems (and combinations of sensory information) on spatial knowledge are closely related to the question of how and whether spatial information from different sources is integrated or combined into a coherent and unitary percept of space. Recent research has provided empirical support for the combination of visual information with auditory (e.g., [1]), vestibular (e.g., [14]), and haptic information [26]. Most of this work is consistent with a Bayesian model of sensory integration in which the weights assigned to various sources of information are determined by estimates of the source's relative precision [10, 17, 68].

Having discussed the influence of various sensory systems to the acquisition of knowledge of real world environments, we now turn our attention to the spatial information available to users of VEs. As suggested earlier, a primary difference between the sensory contributions of real and virtual environments is that it is typical for the latter to involve the degradation or elimination of various sources of sensory information that are commonly available in the real world.

1.4.2 Sensory Contributions in Virtual Environments

As VE systems have proliferated over the last several decades, a variety of interfaces have emerged, with similarly diverse combinations of spatial information available to users. For example, users of the 1950s-era Sensorama [40]—a widely recognized precursor to modern VE systems—sat in front of a stationary display that presented moving stereoscopic images, but lacked any accompanying head or neck motion. The Sensorama's seat was capable of moving and thus provided a bit of sensorimotor information, but the users' feet and legs remained relatively stationary, lacking proprioceptive and kinesthetic input. Surround audio provided pre-programmed spatial auditory information, and fans created the illusion of wind blowing from a certain direction across the users' skin, which could be used to indicate the direction of one's movement. Even synthetic smells could inform the user of being near one location versus another. Of course the Sensorama was not a true VE system inasmuch as users could not actively navigate or interact with their environment, but rather passively took in a pre-programmed experience. Nonetheless, this early system captures the types of sensory limitations and tradeoffs that are inherent in all VE designs. In this final section of our chapter, we discuss three fundamentally different types of VE systems: (a) desktop; (b) CAVE (Cave Automatic Virtual Environment); and (c) HMD-based systems. Examples of these types of systems are depicted in Fig. 1.1. For each general type of system, we summarize the types of sensory information that are available as well as which sources are absent, degraded, or limited, and what is known about performance differences as a result of these limitations on sensory information. Finally, general use cases under which each of these systems excels or suffers are considered.

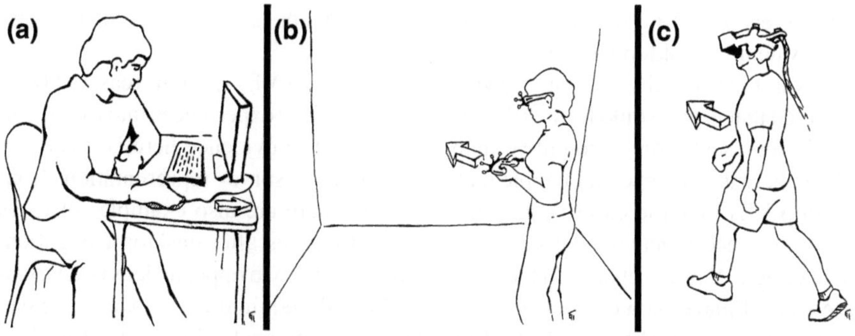

Fig. 1.1 **a** A user of a typical desktop VE system, which offers comparatively impoverished sensory information to users. The primary navigation interface in this case is the mouse and keyboard. Somatosensory, ideothetic, and inertial sensory information about virtual movements are not available, as the user's body is stationary. Visual field-of-view is also typically limited in desktop displays. **b** A user in a typical CAVE system, which visually surrounds her. The user is free to turn and look about and she can move freely over a small area. The primary navigation interface in this case is a wand-type device. CAVEs offer relatively wide and physically-accurate visual field-of-view and typically feature surround-sound audio, but sacrifice veridical somatosensory, idiothetic, and inertial sensory feedback over large ranges. **c** A user in a typical HMD system. The primary navigation interface is the user, who can walk and turn naturally. Range of navigation is limited in some systems by the cable length or, if the rendering computer is worn, the size of the tracking area. Visual field-of-view is typically limited in HMDs

1.4.2.1 Desktop Virtual Environments

Classic Desktop Systems

In a traditional desktop VE (Fig. 1.1a), a stationary user is seated in front of a display and uses a set of arbitrary controls to navigate through a virtual space (e.g., press "w" to move forward). This can include simulations that are presented on desktop computers, laptops, or on a television using a gaming console. While such simulations can contain accurate positional audio and optic flow that inform users about the simulated movement of the depicted viewpoint, the simulation presents nearly all other spatial senses with information that does not match. The user's idiothetic senses will report—quite accurately—that she is sitting still even though the images on the screen simulate her movement. Likewise, traditional efferent information about the user's intended movements (e.g., "walk 2 m forward") are replaced with information about intended button presses and joystick movements that map to unknown units of translation and rotation. This leaves the user not only in a state of sensory conflict, but also removes sensory-motor feedback that can be useful for accurate path integration [16, 76]. Above, we noted that idiothetic information seems to be particularly important for tracking rotational changes (e.g., [55]), and as such, it can logically be expected that users of desktop VE will have relative difficulty with updating their current heading and integrating multiple legs of a journey to keep track of their orientation within a virtual world. This may lead to users becoming lost and disoriented more often, which is perhaps the reason that mini-maps and virtual compasses are so prevalent in desktop game environments.

Given the difficulty of spatial updating in a desktop VE, it might reasonably be expected that users would have difficulty integrating their experiences into a coherent survey representation or cognitive map of the VE. The evidence here is less clear. While there is evidence that people may be able to form survey representations from desktop VEs in some cases (e.g., [86, 89, 111]), there are also examples of users failing to piece together relatively simple spatial layouts after desktop navigation. For example, Stanton et al. [97] had participants explore two opposite legs of a virtual kite-shaped maze on a desktop VE system. The task required participants to travel from corner 1–2 and back, and later from corner 3–4 and back without ever traveling between these two routes. Despite repeated practice locating each of the four corners and the presence of distinct visual landmarks with which to locate these places and paths relative to each other, participants were unable to take novel shortcuts at above-chance levels in subsequent testing. It is unclear whether this same result would have been found in VEs that offer wide fields of view (such as CAVES, discussed below) or that incorporate body-based senses (such as HMD-based systems, discussed below).

Desktop + Motion Controller

Many modern gaming systems and some desktop simulation systems allow the user to interact with simulations using naturalistic motions. The best known of these systems are perhaps the Nintendo Wii and Microsoft Kinect, both of which can be used either on their respective game consoles or on a computer. Such systems leverage the idea of including a user's body into the desktop simulation experience. Indeed, Microsoft marketed the Kinect system with the slogan, "You are the controller." These interfaces are still in their infancy, making it hard to draw firm conclusions about the impact that they have on the user's ability to form accurate spatial knowledge. There is also substantial variation in the way that users interact with various devices both between platforms and between different simulations on the same platform. For example, different simulations of the Wii might use gestures from one of the user's hands, posture information from a balance board, traditional button-presses, or some combination of the above. Similarly, the Kinect performs full-body tracking of the user that could be directly mapped onto the movements of one's avatar, processed to extract pre-defined control gestures, or merely recorded and post-processed. Other accessories, like the Wii balance board, Dance Revolution dance pad, or even the early Nintendo Power Pad could be leveraged to implement a walking-in-place navigation interface.

While it is difficult to make specific claims about the impact of motion controls on spatial knowledge, there are some generalities that can be drawn from these types of interaction. First, including one's body into the user interface of a desktop simulation should improve spatial sensing and navigation insomuch as the movements pertain to the user's navigation. For example, leaning left to steer leftward or to initiate a turn provides more accurate efferent information about leftward movement than, say, pushing the left arrow key. It also differentially tenses muscles in the left and right legs to indicate leftward movement and engages the posture that one might adopt when leaning into a leftward turn to counteract the impending inertia. Spatial knowledge acquisition may or may not be helped, however, by other physical actions such as waving in a defined pattern to call-up a menu or even flicking one's wrist upward to jump—an action typically performed with one's legs. Indeed, this is an area ripe for further research. It is well known that spatial congruence between controls and actions can facilitate users' behavior (e.g., [79, 93]). In the previous example, flicking one's wrist upward to jump would certainly be preferable to some other arbitrary motion—say, rotating one's wrist leftward—given that the upward motion is congruent with an upward jumping action. Whereas the research is unequivocal in favoring spatially mapped controls relative to arbitrary controls, it is unclear whether well-mapped controls are as beneficial to users as the actual motions they replace. In other words, flicking one's wrist upward to jump is certainly better than rotating it leftward, but is it as good as actually jumping? From a perceptual standpoint, it seems unlikely. Users of these systems are still largely stationary or confined to a small area with a fixed toward-the-screen orientation, so some deficits in spatial knowledge are still likely.

Uses for Desktop VEs

While navigation in a desktop VE may be the most limiting from a multi-sensory perspective, there are compelling reasons to use desktop systems, and perhaps even several uses for which it is ideal. Desktop VEs are clearly superior to CAVEs and HMDs on many non-spatial dimensions, including system cost, convenience, and availability to a wide audience. Thus, one might be willing to trade off the fidelity of spatial perception and learning in favor of other factors. Additionally, many computer simulations require little navigation or environmental knowledge. Specifically, when simulating a single virtual object or a small number of objects—a new product design, for example—it may more intuitive to present the VE as a desktop simulation. Users can sit at a desk or table and interact with the virtual object on that table in a similar manner to which they might sit and interact with a physical object on the table. If a VE does not require users to travel, or it is relatively unimportant for them to stay oriented or update their movements, then a desktop VE may be best suited to that particular task.

Conversely, for simulations in which navigation is crucial or in which accurate spatial perception and learning is desired, a desktop VE is likely to be a poor choice. For example, a training simulation designed to teach elite police forces to raid a building would be best implemented in a system that enabled users to physically move, crouch, look around, and remain keenly aware of their position and heading.

1.4.2.2 CAVE Systems

Classic CAVE Systems

A *CAVE* [23] consists of multiple large projected surfaces—traditionally at least three walls that surround a user, and sometimes a floor, ceiling, or fourth wall—that form a booth in which one or more users stand while interacting with the VE (Fig. 1.1b). CAVE systems differ from traditional desktop VEs in that users do not merely sit and view a simulation on a small screen, but rather step physically into the simulated space because the projected surfaces surround them. Such systems thus are able to engage kinesthetic, vestibular, somatosensory, and efferent systems. For example, instead of pressing a button to rotate the virtual viewpoint to one's left, a user need only turn his head or shift his feet. Perhaps most strikingly, however, CAVE displays are generally much more visually immersive than desktop VEs and can present more complete optic flow to inform users about their movements. Access to a wide field of view, and particularly peripheral vision, has been shown to be beneficial for gauging the speed [106] and accuracy of navigation [2, 95]. While typical desktop displays present a simulated view frustum[1] with a horizontal field of view (HFOV) of around 60°, a CAVE display typically displays more than 180° HFOV with three walls or a full 360° HFOV with four walls. Likewise, systems with a floor and/or ceiling will offer significantly improved vertical field of view (FOV). Indeed, a six-walled CAVE is able to simulate a full field of optical flow perfectly regardless of the user's facing direction, although a user's actual viewing angle may be artificially limited if the system requires users to wear shutter glasses. These wide FOVs not only allow peripheral visual information to be displayed, but enable it to be displayed to a physically appropriate location on the user's retina. It is well known that reducing the field of view, for example in an HMD [2] or a desktop VE with a physically matched viewing angle [5], can reduce performance in spatial tasks. Indeed, when normally-sighted users wear a reduced-FOV HMD, they effectively reproduce the inaccuracies in knowledge of spatial layout that are typical of patients who have peripheral field loss [27].

Valid and realistic body-based sensory information in CAVEs, however, is limited to head, neck, and eye movements. Users cannot typically walk or travel more than a few steps in any direction (but see below), placing real limits on the amount of podokinesthetic sensory information that can be acquired during navigation. As mentioned above, the lack of incorporation of body-based senses—particularly for navigational tasks requiring knowledge of orientation changes—can impede learning and produce less accurate spatial knowledge ([88]; see also [16]).

---

[1] Desktop VEs are rarely constructed to match the physical viewing angle (the visual extent taken up by the monitor) and the virtual field of view (the amount of the VE that is visible). A user's physical viewing angle can vary greatly depending on her distance from the display, and it has been shown both that users do not notice this discrepancy [56] and that a wider field of view is advantageous in many tasks (e.g., [5]).

CAVE + Locomotion

In some CAVE systems, users are able to navigate through a computer simulated environment by making walking motions rather than by using a joystick or other controls. This may be done either by tracking the motion of a user's legs while stepping in place or while implementing some forms of redirected walking (e.g., [82]) or by having users walk on a traditional treadmill (e.g., [69]) or omnidirectional treadmill [22, 25, 96] or even while making stepping or leaning motions on a Wii balance board (e.g., [44]). But as we discussed above with current motion controllers for desktop systems, this type of interface increases the *involvement* of idiothetic and efferent systems, but does not necessarily provide *accurate* idiothetic and efferent information. Of course vestibular information in such systems also tends to be quite limited. The sensation of stepping in place is not the same as walking forward, for example. As discussed below, this limitation is not unique to CAVEs, but also applies to any VE system with limited movement range. Tethered HMDs, single-screen projections, and motion-controlled desktop or gaming platforms (e.g., Wii, Kinect) can all be configured to permit the user some small area in which to move, but inherently prevent users from having true, natural mobility through the VE. Users are aware at some level that large movements are not possible within the physically available space, and this knowledge can interfere with even well-simulated motion. When studying illusory motion (vection) in the laboratory, for example, informing participants that their chair is incapable of moving can delay the onset of vection [61]. Alternatively, seating participants on a rotating stool that *might* rotate, raising their feet from the floor, and providing subtle vibrations to the stool to haptically simulate rotation can facilitate even weak vection effects such as auditory vection [84].

Uses for CAVEs

While CAVEs offer only limited range of motion, they are by far the most visually immersive type of VE system currently available. These attributes make a CAVE ideal for applications in which users do not need to move more than a few steps or in which peripheral vision is crucial. A CAVE is ideal for any type of vehicle simulator, in which the user remains seated in a cockpit and uses realistic controls to operate the virtual vehicle. Although even in this case, vestibular and somatosensory inputs would not be accurately simulated without some type of motion platform to generate acceleration forces. A CAVE would also be ideal to train users who will subsequently be operating in a confined space, such as medics gathered around a patient, astronauts working inside a space station capsule, an aircraft technician learning to repair a virtual jet engine, or a machine operator learning to run a dangerous press on a factory floor. Like desktop VEs, CAVEs are also well suited to visualizing small spaces such as a prototype shelf layout in a retail store, or single virtual objects such as a sculpture or an engine design. In these cases, extended foot navigation is unnecessary. Indeed, it has become increasingly common for large manufacturers to use CAVEs in the design process of creating automobiles, farm equipment, airplanes, jet engines, and other devices.

On the other hand, CAVEs are not particularly ideal when a large amount of ambulatory navigation is necessary. Navigation in these cases is conceptually similar to operating a glass elevator floating in the VE—a user can walk a few steps back and forth within the elevator, but ultimately must use some other controls to move the elevator through space. CAVEs that incorporate a natural walking interface, such as an omni-directional treadmill, are more suited to large-scale navigation tasks but are relatively rare. Stepping in place in a traditional CAVE can also approximate natural walking to a degree, but is a suboptimal solution relative to using a treadmill system or walking in an HMD.

1.4.2.3 HMD-Based Systems

Classic HMD-Based System

HMD-based VEs allow users to wear the virtual display and carry it with them during movement (Fig. 1.1c). Unlike CAVEs or desktop VEs, which are in a fixed location and require the user to remain in a relatively fixed space, HMDs can allow users to roam freely around a larger area. From the perspective of providing or simulating sensory information about space, this is an excellent way to simulate spatial sensory information accurately. Naturalistic navigation through virtual spaces is accomplished by physically walking, turning one's body, or looking around. All of the spatial senses including vision, audition, proprioception, vestibular, and efference can access veridical information about one's movements and current state in the simulated environment, with the exception that peripheral visual information is often lacking or absent due to the decreased FOV. This is particularly true relative to CAVEs, which can offer a FOV equal to that of natural viewing, minus any obstruction of shutter glasses. However with HMDs, the loss of visual information in the periphery can vary widely depending on the specific device being used (e.g., FOVs ranging from below 30° to 150° or more). Despite offering a view frustum similar to that of a desktop VE display, HMDs can create a stronger sense of visual immersion by allowing users to turn their heads and look around, giving the impression that the virtual world truly surrounds them, and also by using blinders that occlude a user's view of the surrounding physical environment. This prevents the visual experience for users of desktop VEs in which perceived movement in the optic flow of the VE conflicts with perceived stability of the world that surrounds the monitor.

Despite the potential of HMDs to provide natural walking and complete sensory feedback for navigation, this benefit has been limited historically by a lack of space available for navigation. While it is relatively easy to create a very large VE, the HMD systems that portray them have typically been confined to a small physical area due to (a) the size of one's VE facility, such as a small laboratory room, (b) the range and capabilities of available motion tracking equipment, (c) the fixed-length cable that tethers the HMD to a rendering computer, or often (d) a combination of these factors. The result, of course, is that the utility of full body-based sensory input during navigation becomes much like users of a CAVE—full body-based sensory input is available, as long as the user does not take more than a few steps. These space constrains have been addressed in much the same manner as CAVEs. While users may control their orientation by turning their head or body, linear movements are relegated to a joystick control or accomplished by walking in place (e.g., [35, 104]).

Altered HMD-Systems

A variety of methods have been employed to allow users to navigate through HMD-based VEs in a way that provides sensory information from all modalities. Perhaps the simplest of these methods is to increase the available physical space and untether the user. If sufficiently large physical space is available (e.g., a gymnasium or airplane hangar), recent advances in motion tracking technology have made it is possible to track user motion over much larger volumes with sufficient resolution and accuracy. Optical tracking systems, for example, are now available with high resolutions that can differentiate between distal objects, fast update rates, and on-board graphics processing capabilities. Software advances have also made it possible to chain a large number of cameras together. Likewise, it is possible to untether the HMD by either transmitting rendered images to the user wirelessly (e.g., [65]) or by having the user wear a high-powered but portable rendering computer (e.g., [109]). Our own large-scale HMD facility [109] for example, uses these approaches and provides users more than 1,100 m^2 of tracked space in which to walk. It is even possible to situate users in large, open, outdoor spaces while wearing rendering and tracking equipment in order to simulate very large VEs without sacrificing naturalistic navigation and full idiothetic sensory feedback by employing a combination of inertial and GPS position tracking (e.g., [6]).

An alternative, but much more common approach has been to employ specialized hardware navigation interfaces. Indeed, a wide range of devices has been created to permit navigation in a tethered HMD, including omnidirectional treadmills [25], roller skates [49], unicycles [78], stepping platforms [51], robotic floor tiles [50], or discs of ball bearings [48]. Many of these navigation interfaces can be used interchangeably with a CAVE or HMD display, with the important consideration that an HMD will occlude users' view of the navigation device. Hollerbach [46] has written a review of many such devices, along with their advantages and shortcomings for virtual navigation, and the implications of permitting (as in a CAVE), or not permitting (as in an HMD), immersed users to view the navigation interface device. In such a system, natural gaits are possible, providing accurate proprioceptive and efference information to the user. Inertial information, however, will be in conflict in such cases as the user remains relatively stationary in the treadmill's center. Consider the recently developed Virtusphere [65], for example. Untethered users walk inside of a hollow sphere that sits on a base of rollers. Because the sphere has its own mass, it will not stop, start, or change directions with a high degree of responsiveness, and users must essentially recalibrate their movements to account for the movement of the surface under their feet.

As an alternative to these hardware-based solutions, several researchers have undertaken the development of software-based systems to allow physical navigation through VEs that are larger than the available physical space. Such techniques include combinations of redirected walking [81], motion compression (e.g., [71]), and reorienting users who approach the space's limits (e.g., [74]). These techniques, which we will refer to in aggregate as redirected walking, work by subtly altering the virtual display in an effort to induce a change in the user's movements. For example, by slowly rotating the VE leftward about a user's viewpoint while she is attempting to walk forward, she will—without realizing it—adjust her course to the left. When done correctly, a user can be induced essentially to walk in circles and thus to use a relatively small physical space in navigating a much larger virtual one. Again, these techniques introduce non-veridical idiothetic sensory input that conflicts with the visual and auditory sensory input that users perceive regarding their movements. However, these conflicts are typically designed to be below consciously detectable thresholds and seem not to impact spatial learning [45] or increase motion sickness [100].

Uses for HMD Systems

Like other VE systems, there are situations for which an HMD-based system may be more or less appropriate. If it is important for users to be able to navigate by means of natural ambulation—for example, when familiarizing users with a large-scale environment—then an un-tethered HMD would be particularly appropriate. Additionally, HMDs are ideal for occluding users' view of the surrounding environment, thereby removing visual distractions and increasing the sense of immersion. For example, if a user is seated on a motion platform (e.g., [9]), it may be ideal to remove it from view so that the resulting motion appears to be coming from a virtual source and not a visible physical source.

Conversely, HMDs are not ideal for situations in which it would be advantageous for users to see their own hands, feet, or body within the VE—an inherent feature of CAVEs. Similarly, multiple users can gather around a desktop VE or step into the same CAVE to share a virtual space. Because a user wearing an HMD is essentially blindfolded to the physical world, any additional users and the user's own body must be simulated within the VE, which can require substantial additional resources in terms of motion tracking, communication, and computational load. Finally, HMDs are adequate but perhaps not ideal at rendering close booth-sized environments that require little or no navigation, such as a cockpit or driving simulator. As mentioned above, the visual expanse of a CAVE makes it very well suited to these situations. Because an HMD typically restricts the user's FOV, the same scene will require substantially more head movement to apprehend in an HMD relative to a CAVE, which probably adversely affects the quality of acquired spatial knowledge [27].

1.5 Conclusion

In this chapter, we have enumerated and briefly described the contributions of three external (visual, auditory, and somatosensory), two internal (vestibular and kinesthetic), and two efferent (efference copy, and attention allocation) sources of information that collectively and routinely inform humans about their place in the environment. With few exceptions, each of these information sources in isolation can be sufficient for acquiring at least rudimentary knowledge of the environment and one's movement through it. At the same time, because of our flexibility in handling various combinations of spatial information, none appears to be truly necessary, even for generating relatively abstract and sophisticated knowledge about one's environment. Thus it is a complex endeavor for researchers to determine the relative contributions of different sources of sensory information when—as is typical in real world situations—most or all are differentially available. Nonetheless, a great deal of exciting contemporary research has begun to describe and understand how different sources of information are acquired, weighted, and combined in the creation and expression of our environmental knowledge.

These issues have very real consequences for users of virtual environments. Currently all VE systems must make some sacrifices on the information that they provide to users. Desktop systems and CAVES eliminate or degrade many body-based sources of spatial information, while HMD systems limit the quality of visual information and (typically) the range of full-body movement. Research on the relative importance of these different sources of sensory information is critical for making intelligent decisions about how to use and choose among the many different options for VE systems. The issue of conflicting or degraded sensory information in VEs has additional practical importance when one notes that discrepancies among sensory information are widely believed to underlie the onset of simulator sickness (e.g., [37, 115]) and are also thought to impact user's sense of "presence" and "immersion" in the VE [11].

Perhaps equally important for the application of sensory psychology to VE technology is an appreciation of the role of efferent information—particularly attention allocation—to the acquisition of knowledge of one's environment. The literature is generally consistent with the notion that attention allocation is perhaps the strongest contributor to spatial knowledge among efferent sources [18]. Navigation interfaces for VEs that demand a user's limited attentional resources are thus especially likely to impede one's ability to learn about their environment. More generally, because no VE is yet able to present a perfectly full and veridical sensory experience to its user, it is important for researchers and VE developers to understand the circumstances, tasks, and goals for which these numerous information sources are most critical.

References

1. Alais D, Burr D (2004) No direction-specific bimodal facilitation for audiovisual motion detection. Cogn Brain Res 19:185–194
2. Alfano PL, Michel GF (1990) Restricting the field of view: perceptual and performance effects. Percept Motor Skills 70:25–45
3. Alibali MW (2005) Gesture in spatial cognition: expressing, communicating and thinking about spatial information. Spatial Cogn Comput 5:307–331
4. Allen GL, Siegel AW, Rosinski R (1978) The role of perceptual context in structuring spatial knowledge. J Exp Psychol Hum Learn Mem 4:617–630
5. Arsenault R, Ware C (2002) Frustum view angle, observer view angle and VE navigation. In: Vidal CA, Kimer C (eds) Proceedings of the V Simposio de Realidade Virtual, Brazilian Computer Society, pp 15–25
6. Bachmann E, Calusdian J, Hodgson E, Yun X, Zmuda M (2012) Going anywhere anywhere—creating a low cost portable immersive VE system. Manuscript submitted for publication
7. Bakker NH, Werkhoven PJ, Passenier PO (1999) The effects of proprioceptive and visual feedback on geographical orientation in virtual environments. Presence Teleoperators Virtual Environ 8:36–53
8. Becker W, Nasios G, Raab S, Jürgens R (2002) Fusion of vestibular and podokinesthetic information during self-turning towards instructed targets. Exp Brain Res 144:458–474
9. Berger DR, Schulte-Pelkum J, Bülthoff HH (2010) Simulating believable forward accelerations on a Steward motion platform. ACM Trans Appl Percept 7(1):1–27 (Article 5)
10. Besson P, Richiardi J, Bourdin C, Bringoux L, Mestre DR, Vercher J (2010) Bayesian networks and information theory for audio-visual perception modeling. Biol Cybern 103:213–226
11. Biocca F, Kim J, Choi Y (2000) Visual touch in virtual environments: an exploratory study of presence, multimodal interfaces, and cross-modal sensory illusions. Presence Teleoperators Virtual Environ 10:247–265
12. Blakemore SJ (2003) Deluding the motor system. Conscious Cogn 12:647–655
13. Bruggeman H, Piuneu VS, Rieser JJ (2009) Biomechanical versus inertial information: stable individual differences in perception of self-rotation. J Exp Psychol Hum Percept Perform 35:1472–1480
14. Butler JS, Smith ST, Campos JL, Bülthoff HH (2010) Bayesian integration of visual and vestibular signals for heading. J Vis 10:1–13
15. Campos JL, Byrne P, Sun HJ (2010) The brain weights body-based cues higher than vision when estimating walked distances. Eur J Neurosci 31:1889–1898
16. Chance SS, Gaunet F, Beall AC, Loomis JM (1998) Locomotion mode affects the updating of objects encountered during travel: the contribution of vestibular and proprioceptive inputs to path integration. Presence Teleoperators Virtual Environ 7:168–178
17. Cheng K, Shettleworth SJ, Huttenlocher J, Rieser JJ (2007) Bayesian integration of spatial information. Psychol Bull 133:625–637
18. Chrastil ER, Warren WH (2012) Active and passive contributions to spatial learning. Psychon Bull Rev 19:1–23
19. Christou CG, Bülthoff HH (1999) View dependence in scene recognition after active learning. Mem Cogn 27:996–1007
20. Cohen HS (2000) Vestibular disorders and impaired path integration along a linear trajectory. J Vestib Res 10:7–15
21. Couclelis H (1996) Verbal directions for way-finding: space, cognition, and language. In: Portugali J (ed) The construction of cognitive maps. Kluwer, Dordrecht, The Netherlands, pp 133–153
22. Cruz-Neira C, Reiners D, Springer JP (2010) An affordable surround-screen virtual reality display. Soc Inf Disp 18:836–843
23. Cruz-Neira C, Sandin DJ, DeFanti TA, Kenyon RV, Hart JC (1992) The cave: audio visual experience automatic virtual environment. Commun ACM 35:64–72

24. Cutting JE, Vishton PM (1995) Perceiving layout: the integration, relative dominance, and contextual use of different information about depth. In: Epstein W, Rogers S (eds) Handbook of perception and cognition: Vol. 5: perception of space and motion. Academic Press, New York
25. Darken RP, Cockayne WR, Carmein D (1997) The omni-directional treadmill: a locomotion device for virtual worlds. In: Proceedings of ACM symposium on user interface software and technology, pp 213–221
26. Ernst MO, Banks MS (2002) Humans integrate visual and haptic information in a statistically optimal fashion. Nature 415:429–433
27. Fortenbaugh FC, Hicks JC, Turano KA (2008) The effect of peripheral visual field loss on representations of space: evidence for distortion and adaptation. Investig Ophthalmol Visual Sci 49:2765–2772
28. Frenz H, Bremmer F, Lappe M (2003) Discrimination of travel distance from 'situated' optic flow. Vis Res 43:2173–2183
29. Gallistel CR (1990) The organization of learning. MIT, Cambridge
30. Gibson JJ (1958) Visually controlled locomotion and visual orientation in animals. Br J Psychol 49:182–194
31. Gibson JJ (1979) The ecological approach to visual perception. Houghton Mifflin, Boston
32. Glasauer S, Amorim MA, Vitte E, Berthoz A (1994) Goal directed linear locomotion in normal and labyrinthine-defective subjects. Exp Brain Res 98:323–335
33. Goldin SE, Thorndyke PW (1982) Simulating navigation for spatial knowledge acquisition. Hum Factors 24:457–471
34. Golledge R, Klatzky R, Loomis J (1996) Cognitive mapping and wayfinding by adults without vision. In: Portugali J (ed) The construction of cognitive maps. Kluwer, The Netherlands, pp 215–246
35. Grant SC, Magee LE (1998) Contributions of proprioception to navigation in virtual environments. Hum Factors 40:489–497
36. Guidice NA, Bakdash JZ, Legge GE (2007) Wayfinding with words: spatial learning and navigation using dynamically updated verbal descriptions. Psychol Res 71:347–358
37. Harm DL (2002) Motion sickness neurophysiology, physiological correlates, and treatment. In: Stanney KM (ed) Handbook of virtual environments: design, implementation, and applications. Erlbaum, Mahwah, pp 637–661
38. Harris LR, Jenkin M, Zikovitz DC (2000) Visual and non-visual cues in the perception of linear self-motion. Exp Brain Res 135:12–21
39. Hay JC, Pick HL Jr (1965) Visual capture produced by prism spectacles. Psychon Sci 2:215–216
40. Heilig M (1992) El cine de futro: the cinema of the future. Presence Teleoperators Virtual Environ 1:279–294 (Original work published 1955)
41. Heft H (1996) The ecological approach to navigation: A Gibsonian perspective. In: Portugali J (ed) The construction of cognitive maps. Kluwer Academic Publishers, Dordrect, pp 105–132
42. Heft H (1997) The relevance of Gibson's ecological approach for environment-behavior studies. In: Moore GT, Marans RW (eds) Advances in environment, behavior, and design, vol 4. Plenum, New York, pp 71–108
43. Hettinger LJ (2002) Illusory self-motion in virtual environments. In: Stanney KM (ed) Handbook of Virtual Environments. Erlbaum, New Jersey, pp 471–492
44. Hilsendeger A, Brandauer S, Tolksdorf J, Fröhlich C (2009) Navigation in virtual reality with the wii balanceboard^{TM}. In: GI-Workshop Virtuelle und Erweiterte Realität, ACM
45. Hodgson E, Bachmann E, Waller D (2011) Redirected walking to explore virtual environments: assessing the potential for spatial interference. ACM Trans Appl Percept 8, Article 22
46. Hollerbach JM (2002) Locomotion interfaces. In: Stanney KM (ed) Handbook of virtual environments: design, implementation, and applications. Erlbaum, New Jersey, pp 239–254
47. von Holst E, Mittlestaedt H (1950) Das reafferenz princip: (Wedlselwirkungen zwischen Zentrainervensystem und Peripherie.) Die Naturwissenschften, 37, 464–476. Translated in:

Dodwell PC, ed (1971) Perceptual processing: stimulus equivalence and pattern recognition (pp. 41–72). Appleton-Century-Crofts, New York
48. Huang J, Chiu W, Lin Y, Tsai M, Bai H, Tai C, Gau C, Lee H (2000) The gait sensing disc: a compact locomotion device for the virtual environment. In: Proceedings of the international conference in central Europe on computer graphics, visualization and interactive digital media. Pilsen, Czech Republic
49. Iwata H (2000) Locomotion interface for virtual environments. In: Hollerbach J, Koditshek D (eds) Robotics research: the ninth international symposium. Springer-Verlag, London, pp 275–282
50. Iwata H, Yano H, Fukushima H, Noma H (2005) CirculaFloor. IEEE Comput Graph Appl 25:64–67
51. Iwata I, Yano H, Nakaizumi F (2001) Gait master: a versatile locomotion interface for uneven virtual terrain. In: Proceedings of IEEE virtual reality conference, pp. 131–137
52. Jürgens R, Becker W (2006) Perception of angular displacement without landmarks: evidence for Bayesian fusion of vestibular, optokinetic, podokinesthetic, and cognitive information. Exp Brain Res 174:528–543
53. Jürgens R, Boß T, Becker W (1999) Estimation of self-turning during active and passive rotation in the dark. Exp Brain Res 128:491–504
54. Kearns MJ, Warren WH, Duchon AP, Tarr MJ (2002) Path integration from optic flow and body senses in a homing task. Perception 31:349–374
55. Klatzky RL, Loomis JM, Beall AC, Chance SS, Golledge RG (1998) Updating an egocentric spatial representation during real, imagined, and virtual locomotion. Psychol Sci 9:293–298
56. Kubovy M (1986) The psychology of linear perspective and renaissance art. Cambridge University Press, Cambridge
57. Lackner JP, DiZio P (2005) Vestibular, proprioceptive, and haptic contributions to spatial orientation. Ann Rev Psychol 56:115–147
58. Lappe M, Bremmer F, van den Berg AV (1999) Perception of self-motion from visual flow. Trends Cogn Sci 3:329–336
59. Larish JF, Flach JM (1990) Sources of optical information useful for perception of speed of rectilinear self-motion. J Exp Psychol Hum Percept Perform 16:295–302
60. Lederman SJ, Klatzky RL (2009) Haptic perception: a tutorial. Atten Percept Psychophys 71:1439–1459
61. Lepecq J, Giannopulu I, Baudonnière P (1995) Cognitive effects on visually induced body motion in children. Perception 24:435–449
62. Loomis JM, Klatzky RL, Golledge RG, Philbeck JW (1999) Human navigation by path integration. In: Golledge RG (ed) Wayfinding: cognitive mapping and other spatial processes. Johns Hopkins, Baltimore, pp 125–151
63. Loomis JM, Klatzky RL, Avraamides M, Lippa Y, Golledge RG (2007) Functional equivalence of spatial images produced by perception and spatial language. In: Mast F, Jäncke L (eds) Spatial processing in navigation, imagery, and perception. Springer, New York, pp 29–48
64. May M, Klatzky RL (2000) Path integration while ignoring irrelevant movement. J Exp Psychol Learn Mem Cogn 26:169–186
65. Medina E, Fruland R, Weghorst S (2008) Virtusphere: walking in a human size VR "hamster ball". In: Proceedings of the human factors and ergonomics society, pp 2102–2106
66. Mittelstaedt H (1996) Somatic graviception. Biol Psychol 42:53–74
67. Mittelstaedt ML, Mittelstaedt H (2001) Idiothetic navigation in humans: estimation of path length. Exp Brain Res 139:318–332
68. Nardini M, Jones P, Bedford R, Braddick O (2008) Development of cue integration in human navigation. Curr Biol 18:689–693
69. Neider MB, Gaspar JG, McCarley JS, Crowell J, Kaczmarski H, Kramer AF (2011) Walking and talking: dual-task effects on street crossing behavior in older adults. Psychology and Aging (Advance online publication)
70. Nielsen TI (1963) Volition: a new experimental approach. Scand J Psychol 4:225–230

71. Nitzshe N, Hanebeck UD, Schmidt G (2004) Motion compression for telepresent walking in large target environments. Presence Teleoperators Virtual Environ 13:44–60
72. Noë A (2004) Action in perception. MIT Press, Boston
73. Ohmi M (1996) Egocentric perception through interaction among many sensory systems. Cogn Brain Res 5:87–96
74. Peck TC, Whitton MC, Fuchs H (2009) Evaluation of reorientation techniques for walking in large virtual environments. IEEE Trans Vis Comput Graph 15:121–127
75. Péruch P, Borel L, Magnan J, Lacour M (2005) Direction and distance deficits in path integration after unilateral vestibular loss depends on task complexity. Cogn Brain Res 25:862–872
76. Péruch P, May M, Wartenberg F (1997) Homing in virtual environments: effects of field of view and path layout. Perception 26:301–311
77. Potegal M (1982) Vestibular and neostriatal contributions to spatial orientation. In: Potegal M (ed) Spatial abilities development and physiological foundations. Academic Press, New York, pp 361–387
78. Pratt DR, Barham PT, Locke J, Zyda MJ, Eastman B, Moore T et al (1994) Insertion of an articulated human into a networked virtual environment. In: Proceedings of the 1994 AI, simulation, and planning in high autonomy systems conference. University of Florida, Gainesville
79. Prinz W (1997) Perception and action planning. Eur J Cogn Psychol 9:129–154
80. Proffitt DR (2006) Distance perception. Curr Dir Psychol Sci 15:131–135
81. Razzaque S (2005) Redirected walking. Doctoral dissertation, University of North Carolina, Chapel Hill
82. Razzaque S, Swapp D, Slater M, Whitton MC, Steed A (2002) Redirected walking in place. In: Muller S, Stuzlinger W (eds) Proceedings of the eurographics workshop on virtual environments. Eurographics Association, pp 123–130
83. Riecke BE, Cunningham DW, Bülthoff HH (2007) Spatial updating in virtual reality: the sufficiency of visual information. Psychol Res 71:298–313
84. Riecke B, Feuereissen D, Rieser JJ (2009) Auditory self-motion simulation is facilitated by haptic and vibrational cues suggesting the possibility of actual motion. ACM Trans Appl Percept 6, Article 20
85. Riecke BE, van Veen HAHC, Bülthoff HH (2002) Visual homing is possible without landmarks: a path integration study in virtual reality. Presence Teleoperators Virtual Environ 11:443–473
86. Rossano MJ, West SO, Robertson TJ, Wayne MC, Chase RB (1999) The acquisition of route and survey knowledge from computer models. J Environ Psychol 19:101–115
87. Ruddle RA (2001) Navigation: am I really lost or virtually there? In: Harris D (ed) Engineering psychology and cognitive ergonomics, vol 6. Ashgate, Burlington, pp 135–142
88. Ruddle RA, Lessels S (2009) The benefits of using a walking interface to navigate virtual environments. ACM Trans Comput Hum Interact 16:1–18
89. Ruddle RA, Payne SJ, Jones DM (1997) Navigating buildings in "desk-top" virtual environments: experimental investigations using extended navigational experience. J Exp Psychol Appl 3:143–159
90. Sandvad J (1999) Auditory perception of reverberant surroundings. J Acoust Soc Am 105:1193
91. Schenkman BN, Nilsson ME (2010) Human echolocation: blind and sighted persons' ability to detect sounds recorded in the presence of a reflected object. Perception 39:483–501
92. Shelton AL, McNamara TP (1997) Multiple views of spatial memory. Psychon Bull Rev 4:102–106
93. Shin YK, Proctor RW, Capaldi EJ (2010) A review of contemporary ideomotor theory. Psychol Bull 136:943–974
94. Sholl MJ (1989) The relation between horizontality and rod-and-frame and vestibular navigational performance. J Exp Psychol Learn Mem Cogn 15:110–125
95. Sholl MJ (1996) From visual information to cognitive maps. In: Portugali J (ed) The construction of cognitive maps. Kluwer, Dordrecht, pp 157–186

96. Souman JL, Giordano PR, Schwaiger M, Frissen I, Thümmel T, Ulbrich H, De Luca A, Bülthoff HH, Ernst MO (2011) CyberWalk: enabling unconstrained omnidirectional walking through virtual environments. ACM Trans Appl Percept 8, Article 25
97. Stanton DEB, Wilson PN, Foreman N (2003) Human shortcut performance in a computer-simulated maze: a comparative study. Spatial Cogn Comput 3:315–329
98. Stoffregen TA, Pittenger JB (1995) Human echolocation as a basic form of perception and action. Ecol Psychol 7:181–216
99. Stratton G (1896) Some preliminary experiments on vision without inversion of the retinal image. Psychol Rev 3:341–360
100. Steinicke F, Bruder G, Jerald J, Frenz H, Lappe M (2009) Estimation of detection thresholds for redirected walking techniques. IEEE Trans Vis Comput Graph 16(1):17–27
101. St. George RJ, Fitzpatrick RC (2011) The sense of self-motion, orientation and balance explored by vestibular stimulation. J Physiol 589:807–813
102. Sun HJ, Campos JL, Chan GSW (2004) Multisensory integration in the estimation of relative path length. Exp Brain Res 154:246–254
103. Telford L, Howard IP, Ohmi M (1995) Heading judgments during active and passive self-motion. Exp Brain Res 104:502–510
104. Templeman JN, Denbrook PS, Sibert LE (1999) Virtual locomotion: walking in place through virtual environments. Presence Teleoperators Virtual Environ 8:598–619
105. Thorndyke PW, Hayes-Roth B (1982) Differences in spatial knowledge acquired from maps and navigation. Cogn Psychol 14:560–589
106. Toet A, Jansen SEM, Delleman NJ (2007) Effects of field-of-view restrictions on speed and accuracy of manoeuvring. Percept Motor Skills 105:1245–1256
107. Turvey MT, Carello C (1986) The ecological approach to perceiving-acting: a pictorial essay. Acta Psychol 63:133–155
108. Van Erp JBF, Van Veen HAHC, Jansen C, Dobbins T (2005) Waypoint navigation with a vibrotactile waist belt. ACM Trans Appl Percept 2:106–117
109. Waller D, Bachmann E, Hodgson E, Beall AC (2007) The HIVE: A Huge Immersive Virtual Environment for research in spatial cognition. Behav Res Methods 39:835–843
110. Waller D, Greenauer N (2007) The role of body-based sensory information in the acquisition of enduring spatial representations. Psychol Res 71:322–332
111. Waller D, Hunt E, Knapp D (1998) The transfer of spatial knowledge in virtual environment training. Presence Teleoperators Virtual Environ 7:129–143
112. Waller D, Loomis JM, Haun DBM (2004) Body-based senses enhance knowledge of directions in large-scale environments. Psychon Bull Rev 11:157–163
113. Waller D, Loomis JM, Steck S (2003) Inertial cues do not enhance knowledge of environmental layout. Psychon Bull Rev 10:987–993
114. Witmer BG, Bailey JH, Knerr BW, Parsons KC (1996) Virtual spaces and real world places: transfer of route knowledge. Int J Hum Comput Stud 45:413–428
115. Yardley L (1992) Motion sickness and perception: a reappraisal of the sensory conflict approach. Br J Psychol 83:449–471
116. Yardley L, Higgins M (1998) Spatial updating during rotation: the role of vestibular information and mental activity. J Vestib Res 8:435–442
117. Yost WA (2001) Auditory localization and scene perception. In: Goldstein EB (ed) Blackwell handbook of perception. Blackwell Publishers, Malden, pp 437–468
118. Zhang H, Mou W, McNamara TP (2011) Spatial updating according to the intrinsic reference direction of a briefly viewed layout. Cognition 119:419–429

Chapter 2
Perceptual and Cognitive Factors for Self-Motion Simulation in Virtual Environments: How Can Self-Motion Illusions ("Vection") Be Utilized?

Bernhard E. Riecke and Jörg Schulte-Pelkum

Abstract How can we convincingly simulate observer locomotion through virtual environments without having to allow for full physical observer movement? That is, how can we best utilize multi-modal stimulation to provide the compelling illusion of moving through simulated worlds while reducing the overall simulation effort? This chapter provides a review on the contribution and interaction of visual, auditory, vibrational, and biomechanical cues (e.g., walking) for self-motion perception and simulation in VR. We propose an integrative framework and discuss potential synergistic effects of perceptual and cognitive influences on self-motion perception in VEs. Based on this perspective, we envision a lean-and-elegant approach that utilizes multi-modal self-motion illusions and perceptual-cognitive factors in a synergistic manner to improve perceptual and behavioral effectiveness and reduce the demand for physical (loco-)motion interfaces to a more affordable level.

2.1 Introduction: The Challenge of Walking in VR

Walking is probably the oldest and still most common mode of transportation for humans. Walking allows for easy and intuitive locomotion, and even with eyes closed enables us to remain oriented in our immediate environment with little cognitive effort [80, 97]. This phenomenon is typically ascribed to an (at least partially) automated mental process that spatially updates our egocentric mental spatial representation such as to stay aligned with where we are with respect to our immediate

B. E. Riecke (✉)
Simon Fraser University, Surrey, BC, Canada
e-mail: ber1@sfu.ca
web: iSpaceLab.com/Riecke

J. Schulte-Pelkum
Vechta University, Vechta, Germany
e-mail: joerg.schulte-pelkum@uni-vechta.de

F. Steinicke et al. (eds.), *Human Walking in Virtual Environments*,
DOI: 10.1007/978-1-4419-8432-6_2, © Springer Science+Business Media New York 2013

surroundings. Thus, it seems to make sense that we should be able to walk through virtual environments in a similar manner, in the hope that walking will enable us to more easily remain oriented and reach our destination with little effort or cognitive load, just like in the real world. As several chapters in this book discuss in detail, however, enabling humans to use this most intuitive mode of transportation in VR bears many challenges, both from technical and perceptual points of view (see also [37] for a review). Allowing VR users to walk naturally requires them to carry the visual display with them, typically using position-tracked head-mounted displays (HMDs). Although technology is advancing, there are still major technical limitations (e.g., pixel resolution, limited (FOV) of view, and tracking/display latencies) as well as perceptual challenges including spatial misperception such as underestimation of distance [59] or motion sickness [31, 66]. Moreover, allowing for actual and unencumbered walking requires huge tracked free-space walking areas, especially if virtual environments larger than room-sized are intended.

A variety of techniques have been proposed to address these fundamental issues, including virtual walking interfaces, walking-in-place metaphors, or redirected walking. While many of these approaches are promising and discussed in detail in other chapters of this book, they include non-trivial technical challenges, and often either restrict the walking motions or possible trajectories as in the case for re-directed walking (e.g., [111], and Chap. 10 of this book), change the biomechanics of walking fundamentally (as in the case for walking-in-place interfaces, see Chap. 11 of this book) and/or require considerable technical, financial, and safety efforts to implement (as in the case for larger or omni-directional treadmills, where additional safety measures like harnesses are needed). Many of these issues are actively researched, and we are hopeful that most of these issues might be solved eventually.

Treadmills are probably the most promising and most widely used and researched approaches to allow for walking in VEs, as they are commercially available for relatively affordable prices and allow for fairly natural biomechanical cues from walking, especially when augmented with a force-feedback harnesses for linear or omni-directional locomotion ([37], and Chap. 6 of this book). Somewhat counter-intuitively, though, despite allowing for fairly natural walking motions, even the most advanced treadmills do not seem to provide the user with an actual compelling sensation of self-motion unless accompanied with wide-FOV visual motion cues. That is, while actual walking is naturally accompanied with an embodied sensation of self-motion through the environment, even in the absence of visual or auditory cues, walking on a linear treadmill is typically not. Walking can, however, sometimes affect our visual perception: for example, Yabe and Taga [131] showed that walking on a linear treadmill can affect the perception of ambiguous visual motion, similar to motion or action capture phenomena. This "treadmill capture" effect seems to disappear, however, for extended experience of treadmill locomotion in regular treadmill runners [132].

There is little published research on the perception or illusion of self-motion ("vection") on linear treadmills. Durgin et al. [26] observed, for example, that "during treadmill locomotion, there is rarely any illusion that one is actually moving forward" (p. 401) and continues to state that "people do not have the illusion that they are moving when running on a treadmill, nor do their inertial systems experience any net acceleration" (p. 415). Informal observation, discussions with colleagues, and pilot studies by the authors corroborate the notion that biomechanical cues from walking on linear treadmills hardly ever lead to compelling and reliable sensations of self-motion that matches the walker's biomechanical motion, even for the most advanced linear treadmills that include force-feedback harnesses.

This might, of course, be related to the lack of any net acceleration cues as Durgin et al. pointed out [26]. Most treadmills simply do not seem to be long enough to allow for sufficient motion cueing and physical translations that would allow for sustained biomechanically-induced linear vection that would approach the intensity and compellingness of self-motion illusions induced by moving visual stimuli (for recent reviews in the context of VR, see [34, 86, 100].

Hence, for the current chapter we will pursue an alternate approach, by focusing not on how to enable realistic walking in VR (which is covered in depth by other chapters in this book), but on how to provide a compelling and embodied sensation of self-motion through computer-mediated environments with minimal or no physical motion of the observer, with or without walking. In particular, we will review and discuss how we can utilize and maximize illusory self-motions ("vection") that can be induced by visual, auditory, and sometimes biomechanical/somatosensory cues, and how these different cues contribute and interact, often in a synergistic manner. Especially for visually-induced vection, there is a large body of literature that will provide essential guidelines, and dates back to more than a century ago [33, 60]. Here, we will start with a brief review on visually-induced self-motion illusions, as they have received by far the most attention in research and are known to induce quite compelling vection (Sect. 2.2). After this general introduction to vection, we will review potential relations between walking and perceived self-motion and self-motion illusions (Sect. 2.3). In particular, we will discuss how walking interacts with other sensory information such as visual or auditory motion cues (see Sect. 2.4) and briefly cover further cross-modal effects (Sect. 2.5) and potential relations between vection and simulator sickness in VR (Sect. 2.6). We will discuss both perceptual factors and cognitive contributions (such as participants' perception/knowledge of whether or not actual self-motion might be possible), and how to best utilize such factors and interactions in VR to provide a compelling and embodied sensation of self-motion through computer-simulated environments while trying to minimize overall costs and efforts (Sect. 2.7). We will continue by discussing how self-motion illusions might facilitate spatial orientation in VR (Sect. 2.8), and conclude by proposing a conceptual framework that integrates perceptual and cognitive factors and is centered on perceptual as well as behavioral effectiveness of VR simulations (Sects. 2.9–2.10).

2.2 Visually Induced Self-Motion Illusions

In this section, we will provide a brief review of the literature on self-motion illusions that is relevant for the current context. More comprehensive reviews on visually induced vection are provided by, e.g., [2, 23, 38, 39, 61, 123]. Vection with a specific focus on VR, motion simulation, and undesirable side-effects has more recently been reviewed in [34, 86, 100].

When stationary observers view a moving visual stimulus that covers a large part of the field of view (FOV), they can experience a very compelling and embodied illusion of self-motion in the direction opposite to the visual motion. Many of us have experienced this illusion in real life: For example, when we are sitting in a stationary train and watch a train pulling out from the neighboring track, we will often (erroneously) perceive that the train we are sitting in is starting to move instead of the train on the adjacent track [33]. This phenomenon of illusory self-motion has been termed "vection" and has been investigated for well over a century [33, 60, 114, 122, 127]. Vection has been shown to occur for all motion directions and along all motion axes: Linear vection can occur for forward-backward, up-down, or sideways motion [38]. Circular vection can be induced for upright rotations around the vertical (yaw) axis, and similarly for the roll axis (frontal axis along the line of sight, like in a "tumbling room"), and also around the pitch axis (an imagined line passing through the body from left to right). The latter two forms of circular vection are especially nauseating, since they include a strong conflict between visual and gravitational cues and in particular affect the perceived vertical [11].

2.2.1 Circular Vection

In a typical classic circular vection experiment, participants are seated inside an upright rotating drum that is painted with black and white vertical stripes (see illustration in Fig. 2.1a), a device called optokinetic drum [16, 23]. After the optokinetic drum starts to rotate around the earth-vertical axis, the onset latency until the participant reports perceiving self-motion is measured, which ranges from about 2–20 s typically, depending on various stimulus and procedural parameters as discussed below.

Note that vection typically does *not* occur instantly with the stimulus motion, and takes some time to saturate, as sketched in Fig. 2.2. The strength of the illusion can be measured by a variety of introspective measures including the onset latency and duration of the illusion, or by some indication of perceived speed, intensity, or compellingness of self-rotation, e.g., by magnitude estimation or by letting the participant press a button every time they think they have turned 90° [8]. As Riecke et al. point out, one of the challenges for utilizing self-motion illusions in VR is to reduce the vection onset latency and increase the intensity and compellingness of the illusion [94].

The most frequently investigated type of vection is circular vection around the earth-vertical axis (see illustrations in Fig. 2.1). In this special situation where the observer perceives self-rotation around the earth-vertical axis, there is no interfering effect of gravity, since the body orientation always remains aligned with gravity during illusory self-rotation. Roll and pitch vection are consequently harder to induce and

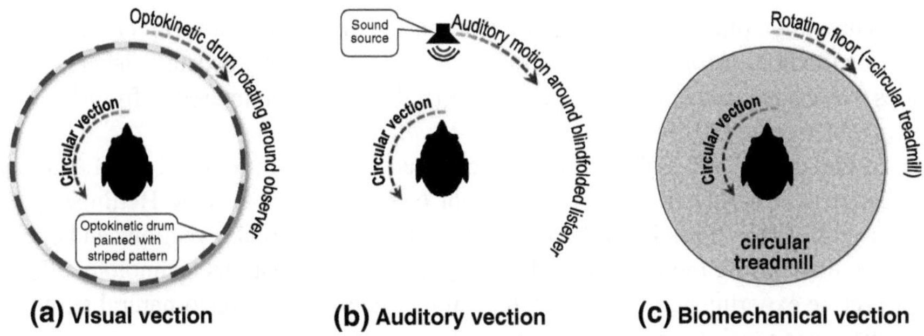

Fig. 2.1 Top-down sketch of different circular vection conditions. **a** Visual vection induced by an optokinetic drum rotating around the stationary observer. **b** Auditory vection induced by sound sources rotating around blindfolded listeners. **c** Biomechanical or "apparent stepping around" vection induced by blindfolded participants stepping along a rotating floor ("circular treadmill")

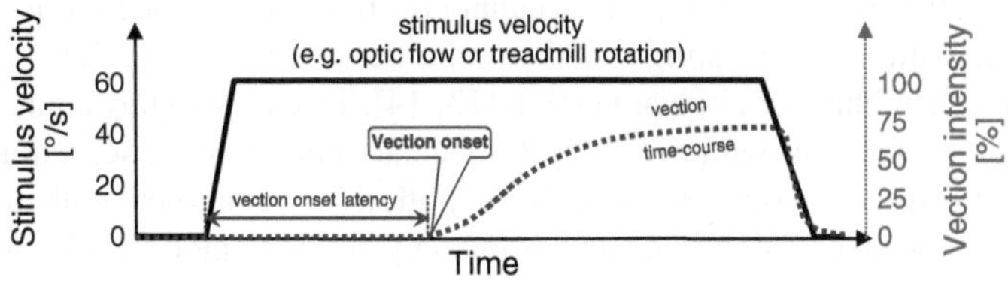

Fig. 2.2 Schematic depiction of typical stimulus motion and resulting vection time course

can lead to paradoxical sensations of continuous illusory self-rotation despite perceiving only limited overall body tilt of generally no more than 20° [1, 32, 133]. Complete head-over-heals tumbling sensations can, however, be induced when the conflict between rotating visual cues and gravitational cues (from otoliths and somatosensory system) is reduced, e.g., in bilateral vestibular loss patients [22] or in micro-gravity conditions [21, 134]. Alternatively, even under normal gravitational conditions, 360° head-over-heals tumbling sensations can be induced in most observers when a fully furnished naturalistic room is rotated around a stationary observer [1, 40, 43, 71].

2.2.2 Linear Vection

In a similar manner, linear vection can be induced by presenting optic flow patterns that simulate translational motion. The traditional method used to induce linear vection in the laboratory is to use two monitors or screens facing each other, with the participant's head centered between the two monitors and aligned parallel to the screens, such that they cover a large part of the peripheral visual field [10, 47, 58]. Optic flow presented in this peripheral field induces strong linear vection. For example, Johansson showed that observers perceive an "elevator illusion", i.e., upward linear vection, when downward optic flow is shown [47]. More recent studies often use a monitor or projection screen in the front of the participant to show expanding or contracting optic flow fields [3, 70]. Comparing different motion directions shows vection facilitation for up-down (elevator) vection, presumably because visual motion does not suggest a change in the gravito-inertial vector as compared to front-back or left-right motion [30, 112].

In recent times, VR technology has been successfully introduced to perceptual research as a highly flexible research tool (see recent reviews by Hettinger [34] and Riecke [86]). It has been shown that both linear and circular vection can be reliably induced using modern VR technology, and the fact that this technology allows for precise experimental stimulus control under natural or close-to-natural stimulus conditions is much appreciated by researchers.

2.3 Self-Motion Sensation from Walking

Although walking on a linear treadmill cannot itself reliably induce vection, walking in a circular pattern on a rotating disc ("circular treadmill", see Fig. 2.1c) can induce compelling curvilinear or circular vection [13, 14]. That is, stepping along a circular treadmill in darkness or with eyes blindfolded can induce strong sensations of self-rotation in the direction opposite to the floor motion (i.e., congruent with the walking motion), irrespective of step size and without any net body motion [13, 14]. Several names have been used to refer to this phenomenon, including "***apparent stepping around***" by Bles and colleagues [13, 14], "***podokinetic vection***" by Becker et al. [8], or "***biomechanical vection***" by Bruggeman et al. [18] and Riecke et al. [92]. Note that the mere act of moving one's leg as if walking but without floor contact does not induce any vection. While the above-mentioned studies reported reliable and consistent biomechanical vection for circular treadmill walking without net body motion, Becker and colleagues observed biomechanical vection only in rare cases: only 25 % of participants occasionally reported biomechanical vection, suggesting that their procedure did not reliably induce vection [7]. As suggested by Becker et al. [18], this unusually low rate of biomechanical circular vection occurrences might be related to the specific instructions used by Becker et al., in that they asked participants to "track angular self-displacement relative to the platform" (p. 461), not relative the surrounding stationary room.

In addition to biomechanically-induced self-motion illusions, Bles and colleagues also reported nystagmus and Coriolis-like effects when participants performed active head tilts, corroborating the strength of vection that can be induced by biomechanical cues [13, 14]. Biomechanical vection from stepping-around occurs similarly in labyrinth-defective patients, although their somatosensory nystagmus was stronger [12]. While actual rotation results in self-rotation illusion after-effects in the direction *opposite* to the prior motion, circular vection induced by blindfolded stepping along a rotating disc results in illusory self-rotation after-effects in the *same* direction as the prior perceived self-motion [44].

Apart from walking on a circular treadmill, passive arm or foot movement can induce similar circular vection [15]: Participants sat stationary in complete darkness inside a slowly rotating optokinetic drum (10°/s). When they touched the rotating surrounding wall with their extended hand such that it was passively rotated around their shoulder joint, compelling ***arthrokinetic circular vection*** in the direction opposite to the arm movement occurred. Illusory self-rotation occurred within 1–3 s and was indistinguishable from actual self-motion. Arthrokinetic vection was accompanied by arthrokinetic nystagmus and resulted in considerable after-effects [24]. remarked that "actively pedaling the free wheeling floor while seated or turning the railing with a hand-over-hand motion makes the experience very powerful" (p. 766). We are currently investigating the feasibility of such a circular walking paradigm for rotational self-motion simulation in VR (http://iSpaceLab.com/iSpaceMecha).

2.4 Interaction of Walking and Other Modalities for Vection

2.4.1 Walking and Auditory Cues

While both biomechanical and visual cues can induce compelling vection, moving auditory cues can elicit self-motion illusions only in 1/4–3/4 of participants, and such auditory vection is much weaker, less compelling, and only occurs when participants are blindfolded (for reviews, see [95, 118]). Despite their low vection-inducing potential, however, moving auditory cues have recently been shown to significantly enhance visually induced vection [88, 118] as well as biomechanically induced circular vection [86]. In the latter study, participants were blindfolded and seated stationary above the center of a circular treadmill. Auditory circular vection was induced by binaural recordings of rotating sound fields presented via headphones (Fig. 2.1b), and biomechanical circular vection was induced by stepping along the floor disc that rotated at the same velocity (60 °/s) as the auditory stimulus (Fig. 2.1c). Although auditory vection by itself was weak and occurred in less than half of the trials, adding rotating sound fields significantly enhanced biomechanically-induced vection. Moreover, there were synergistic, super-additive effects when combining auditory and biomechanical vection-inducing stimuli, in that bi-modal stimulation resulted in vection intensities and perceived rotation realism that was higher than the sum of the uni-modal vection ratings. This corroborates the importance of consistent multi-modal simulation and suggests that even a fairly weak stimulus can sometimes make a significant contribution. This is also promising from an applied perspective of improving VR simulations, as sound spatialization can be of high fidelity while still being affordable and technically feasible.

2.4.2 Walking and Visual Cues

2.4.2.1 Circular Vection

Lackner and DiZio [55] used a circular treadmill inside an optokinetic drum to demonstrate that visual cues that did not match treadmill (i.e., walking) speed systematically affected not only perceived self-motion, but also the perceived stride length and frequency and even the perceived stepping direction. Of particular interest in our context is condition 3 in their experiment, in which participants were stationary and stepped along with the rotating floor disc while the optokinetic drum did not move. Whereas half of the participants 'correctly' perceived to be stationary while stepping along a rotating disc, the other half experienced illusory self-motion in the sense that they (erroneously) reported walking forward on a stationary disc while the optokinetic drum was moving along with them. This suggests that biomechanical cues from walking can (at least for some participants) induce self-motion illusions even in the presence of conflicting visual cues, illustrating that visual cues do not necessarily dominate in cross-modal cue conflict situations. This further corroborates the different vection-inducing potential of walking in circular patterns (where biomechanical vection is strong and can even overpower conflicting visual cues) as compared to linear walking, where biomechanical vection does not reliably occur at all. DiZio and Lackner [24] reported that combining biomechanical and visual vection by rotating the disc of a circular treadmill together with the optokinetic drum could even yield immediate vection onset.

Although Jürgens and Becke [49] demonstrated that a Bayesian sensor fusion could be successfully applied to model the rotation perception based on vestibular, biomechanical, visual, and cognitive information, further research is needed to fully explain and predict cross-modal and higher-level effects and contributions. The current data predicts substantial vection benefits for consistent multi-modal stimulation, at least for the case of self-rotation perception. Surprisingly, however, cue combination benefits are much more ambiguous for translational vection, as we will discuss below.

2.4.2.2 Linear Vection

Whereas walking on a linear treadmill apparently cannot by itself induce a compelling sensation of self-motion (linear vection), it can modulate the occurrence and strength of visually-induced linear vection: Although one would normally assume that perceived self-motion during visual motion simulation in VR should benefit from additional walking cues, a recent study by Kitazaki et al. [52] suggests that providing biomechanical cues from walking on a linear treadmill might, in fact, impair visually-induced vection (see also [51]). Participants watched expanding or contracting optic flow patterns on a 2.4×1.8 m projection screen while either standing still or walking forward on a linear treadmill with the same 4 km/h velocity as the visually simulated self-motion. When the visual cues simulated a forward motion, vection occurred later when participants also walked forwards as compared to standing still. An additional study extended these findings by including backwards walking on the linear treadmill [69]. Vection onset was delayed when the visually simulated self-motion matched participants walking direction, that is, in the condition that most closely matches real-world walking.

The authors suggest that this surprising finding might be caused by a decrease of the relative weight of the visual cues when observers are walking as compared to standing still. We propose that this effect might also be related to Wertheim and Reymond's explanation of the freezing illusion (where an optic flow pattern suddenly appears to freeze when vestibular stimulation is added) and the Pavard and Berthoz effect, in that the perceived relative velocity of the visual motion might be reduced by the biomechanical motion [124]. Additional factors might also have contributed: Apart from affecting the occurrence and amount of vection, differences in the velocity of treadmill walking and visually presented motion can also induce changes in perceived self-motion and stepping movements [25, 55] as well as adaptation and re-calibration (e.g., [25, 98]).

While Kitazaki and colleagues observed an inhibition of vection when locomotor cues matched the direction of visual motion, Seno et al. recently reported the opposite effect [104]: Using visual motions that were 30 times faster than the treadmill walking motions (58 km/h as compared to 2 km/h, respectively), they observed that visually-induced forward vection was facilitated by consistent biomechanical cues, whereas inconsistent walking cues impaired vection. In addition, they showed that locomotion cues from walking on a linear treadmill could systematically bias the strength and direction of vection perceived for up-down and left-right translational visual motion. Comparing the results from Kitazaki et al. and Seno et al. suggests that the differences between visual and walking speed might be critical, with vection facilitation occurring for larger visual motion speeds, and impairment found for matching visual speeds.

A recent study confirmed that forward walking on a linear treadmill can indeed impair visually induced vection when visual and treadmill velocities are matched [4]. Similar impairments of visually-induced linear were observed when the visual display depicted backward motion while participants walked forwards (exp. 2) or when participants simply walked on the spot while viewing forward vection displays (exp. 3). When the head motions that naturally occurred during treadmill walking were tracked and used to update the visual stimulus according to the changed viewpoint (thus mimicking real-world walking), vection strength increased [4]. However, a similar facilitation of vection was observed in passive viewing conditions when participants stood still and simulated viewpoint jitter was added to the visual display, thus confirming earlier studies (see review by Palmisano et al. [72]). Thus, even when head motions were tracked during treadmill walking, vection was still reduced compared to standing still and passively viewing the jittered display.

In conclusion, it remains puzzling how adding velocity-matched treadmill walking to a visual motion simulation can impair vection [4, 52, 69] while active head motions and simulated viewpoint jitter clearly enhance vection [72]. More research is needed to better understand under what conditions locomotion cues facilitate or impair linear vection, and what role the artificiality of treadmill walking might play. Nevertheless, the observation that self-motion perception can, at least under some circumstances, be impaired if visual and biomechanical motion cues are matched seems paradoxical (as it corresponds to natural eyes-open walking) and awaits further investigation. These results do, however, suggest that adding a walking interface to a VR simulator might potentially (at least in some cases) *decrease* instead of increase the sensation of self-motion and thus potentially decrease the overall effectiveness of the motion simulation. Thus, caution should be taken when adding walking interfaces, and each situation should be carefully tested and evaluated as one apparently cannot assume that walking will always improve the user experience and simulation effectiveness.

2.5 Further Cross-Modal Effects on Self-Motion Perception in VR

Helmholtz suggested already in 1866 that vibrations and jerks that naturally accompany self-motions play an important role for self-motion illusions, in that we expect to experience at least some vibrations or jitter [33]. Vibrations can nowadays easily be included in VR simulations and are frequently used in many applications. Adding subtle vibrations to the floor or seat in VR simulations has indeed been shown to enhance both visually-induced vection [94, 100] and auditory vection [85, 88], especially if accompanied by a matching simulated engine sound [119, 120].

Vection can also be substantially enhanced when the vection onset is accompanied by a small physical motion (such as a simple jerk of a few centimeters or degrees) in the direction of visually-simulated self-motion. This has been shown for both passive movements of the observer [9, 93, 100, 126] and for active, self-initiated motion cueing using a modified manual wheelchair [84] or a modified Gyroxus gaming chair where participants controlled the virtual locomotion by leaning into the intended motion direction [87]. For passive motions, combining vibrations and small physical movements (jerks) together was more effective in enhancing vection than either vibrations or jerks alone ([100], exp. 6).

These findings are promising for VR applications, as both vibrations and minimal motion cueing can be added to existing VR simulations with relatively little effort and cost. Moreover, these simple means of providing vibrations or jerks were shown to be effective despite being physically incorrect—while jerks normally need to be in the right direction to be effective and be synchronized with the visual motion onset, their magnitude seems to be of lesser importance. Indeed, for many applications there seems to be a surprisingly large coherence zone in which visuo-vestibular cue conflicts are either not noticed or at the least seem to have little detrimental effect [115]. Surprisingly, physical motion cues can enhance visually-induced vection even when they do not match the direction or phase of the visually-displayed motion [128]: When participants watched sinusoidal linear horizontal (left-right) oscillations on a head-mounted display, they reported more compelling vection and larger motion amplitudes when they were synchronously moved (oscillated) in the vertical (up-down) and thus orthogonal direction. Similar enhancement of perceived vection and motion amplitude was observed when both the visual and physical motions were in the vertical direction, even though visual and physical motions were always in *opposite* directions and thus out of phase by 180° (e.g., the highest visually depicted view coincided with the lowest point of their physical vertical oscillatory motion). In fact, the compellingness and amplitude of the perceived self-motion was not significantly smaller than in a previous study where visual and inertial motion was synchronized and not phase-shifted [129]. Moreover, for both horizontal and vertical visual motions, perceived motion directions were almost completely dominated by the visual, not the inertial motion. That is, while there was some sort of "visual capture" of the perceived motion direction, the extent and convincingness of the perceived self-motion was modulated by the amount of inertial acceleration.

Recently, Seno et al. [106] demonstrated that air flow provided by a fan positioned in front of observers' face significantly enhanced visually induced forward linear vection. Backward linear vection was not facilitated, however, suggesting that the air flow needs to at least qualitatively match the direction of simulated self-motion, similar to head wind.

In two recent studies, Ash et al. showed that vection is enhanced if participants' active head movements are updated in the visual self-motion display, compared to a condition where the identical previously recorded visual stimulus was replayed while observers did not make any active head-movements [5, 6]. This means that vection was improved by consistent multisensory stimulation where sensory information from own head-movements (vestibular and proprioceptive) matched visual self-motion information on the VR display [6]. In a second study with similar setup, [5] found that adding a deliberate display lag between the head and display motion modestly impaired vection. This finding is highly important since in most VR applications, end-to-end system lag is present, especially in cases of interactive, multisensory, real-time VR simulations. Despite technical advancement, it is to be expected that this limitation cannot be easily overcome in the near future.

In conclusion, there can often be substantial benefits in providing coherent self-motion cues in multiple modalities, even if they can only be matched qualitatively. Budget permitting, allowing for actual physical walking or full-scale motion or motion cueing on 6DoF motion platforms is clearly desirable and might be necessary for specific commercial applications like flight or driving simulation. When budget, space, or personnel is more limited, however, substantial improvements can already be gained by relatively moderate and affordable efforts, especially if consistent multi-modal stimulation and higher-level influences are thoughtfully integrated. Although they do not provide physically accurate simulation, simple means such as including vibrations, jerks, spatialized audio, or providing a perceptual-cognitive framework of movability (see Sect. 7.2) can go a long way. Even affordable, commercially available motion seats or gaming seats can provide considerable benefits to self-motion perception and overall simulation effectiveness [87].

As we will discuss in our conceptual framework in Sect. 2.9 in more detail, it is essential to align and tailor the simulation effort with the overarching goal: e.g., is the ultimate goal physical correctness, perceptual effectiveness, or behavioral realism? Or is there a stronger value put on user's overall enjoyment, engagement, and immersion, as in the case of many entertainment applications, which represent a considerable and increasing market share?

2.6 Simulator Sickness and Vection in VR

While a compelling sensation of self-motion in VR clearly increases the overall believability and realism of a simulation, the occurrence and strength of vection can sometimes also correlate with undesirable side-effects like motion after-effects or motion/simulator sickness [34, 35, 50, 73]. It remains unclear, however, whether and how vection might be causally related to simulator sickness, as vection is more easily observed when visuo-vestibular cue conflicts are small, whereas motion sickness tends to increase for larger cue conflicts [50, 73]. Moreover, visually-induced motion sickness can occur without either vection or optokinetic nystagmus [46], indicating that vection cannot be a necessary pre-requisite of visually-induced motion sickness.

Carefully planned research is needed to investigate and disambiguate underlying factors promoting desirable outcomes (like compelling self-motion perception with reduced simulation cost) versus undesirable side-effects (like simulator sickness, after-effects, or (re)adaptation effects) and their potential interactions. As displays become more effective in inducing vection, they might also become more powerful in inducing undesirable side-effects. Thus, applications should be carefully evaluated in terms of not only intended benefits but also potential undesirable side-effects (see also conceptual framework in Sect. 2.9).

2.7 Perceptual Versus Cognitive Contributions to Vection

While self-motion illusions have traditionally been explained by perceptual (lower-level) factors and bottom-up processes (e.g., stimulus frequency, velocity, or field of view), recent studies provide converging evidence that self-motion illusions can also be affected by cognitive (higher-level) factors and top-down processes. In the following, we will briefly review and discuss relevant findings before attempting to integrate them into a conceptual framework in the final sections of this chapter.

2.7.1 Lower-Level and Bottom-Up Contributions to Vection

Visually-induced self-motion illusions have clearly received the most attention in vection research so far, and a number of lower-level/perceptual factors and bottom-up processes have been shown to facilitate visually-induced vection, which will be briefly discussed below. More in-depth discussion of lower-level factors and bottom-up contributions for vection can be found in [2, 23, 38, 39, 61, 86, 123].

Visual field of view. Although vection can sometimes be induced using field of views as small as 7.5° [3], increasing the field of view subtended by the moving stimulus generally enhances all aspects of vection [10, 16, 23, 32]. Strongest vection is observed with full-field stimulation, up to a point where illusory self-motion cannot be distinguished from physical self-motion any more. When perceived depth is held constant, vection strength linearly increases with increasing stimulus size, independent of stimulus eccentricity [63]. This suggests that most affordable fishtank VR (desktop-monitor-based) and HMDs are unsuitable for reliably inducing compelling vection, as their field of view is typically not sufficiently large.

Eccentricity of moving stimulus. Earlier studies argued that visual motion in the periphery is more effective in inducing vection than central motion [16, 23, 47]. When display areas are equated, however, central and peripheral stimulus areas have similar vection-inducing potential [3, 41, 63, 79, 125]. However, peripheral stimuli need to be of lower spatial frequency to be maximally effective in inducing vection, as our visual acuity systematically decreases in the periphery [76]. From an applied perspective, this suggests that peripheral displays need not be of high resolution unless users frequently need to focus there [125].

Stimulus velocity. Increasing stimulus velocities generally tends to enhance both the perceived velocity and intensity of vection, at least up to an optimal stimulus velocity of, e.g., around 120 °/s for circular visual vection [1, 16, 23, 39, 101]. Note that these maximum effective velocities are larger then the maximum stimulus velocities that can easily be displayed in VR without noticeable and disturbing image artifacts (such as motion blur or seeing multiple images) due to the limited update/refresh rate of typically 60 Hz.

Density of moving contrasts. The occurrence and strength of vection in general increases with the number and density of moving objects and contrasts [17, 23]. This suggests that VR simulations that are too sparse (e.g., driving in fog, or flight simulations in clouds with low density of high-contrast objects) might not be able to reliably induce vection without artificially increasing contrast and/or the density of moving objects.

Viewpoint jitter. A common explanation why vection does not occur instantaneously is the inter-sensory conflict between those cues indicating stationarity (e.g., vestibular cues) and those suggesting self-motion (e.g., moving visual cues or circular treadmill walking). This cue conflict account is corroborated by showing that bilaterally labyrinthine defective participants perceive visual vection much earlier and more intensely [48], and can perceive unambiguous roll or pitch vection through head-over-heels orientations [22]. All the more surprisingly, however, there are situations where increasing visuo-vestibular conflicts can enhance vection, as reviewed in [72]: In a series of carefully designed experiments, Palmisano and colleagues demonstrated that forward linear vection occurred earlier, lasted longer, and was more compelling when coherent viewpoint jitter[1] was added to the expanding optic flow display [77],

[1] *Viewpoint jitter* refers to a specific optic flow pattern that simulates the visual "jittering" effects of small head movements of the observer, similar to "camera shake": For example, a constant, radially expanding optic flow pattern that simulates forward linear motion would get an additional jittering optic flow component on top if the visual effects of oscillating up-down head movements that occur during normal walking is added to the expanding optical flow field.

whereas incoherent jitter impaired vection [74]. This was found even when the display was perceived as flat and did not contain any depth cues [64]. Overall, simulated viewpoint jitter shows a larger vection-facilitating effect if it is orthogonal to the main vection direction [64, 73, 78]. In VR, such findings could be used to enhance vection by, for example, adding viewpoint oscillations induced by walking or head motions [4, 19] as is sometimes done in gaming. This should be carefully tested, however, as adding image jitter or oscillations can increase not only vection, but also motion sickness [73].

2.7.2 Cognitive and Top-Down Contributions to Vection

While earlier vection research focused predominately on perceptual and lower-level factors, there is increasing evidence that vection can also be affected by what is outside of the moving stimulus itself, by the way we move and look at a moving stimulus, our pre-conceptions, intentions, and how we perceive and interpret the stimuli, which is of particular importance in the context of VR. Vection might even be directly or indirectly affected by cognitive/top-down processes [3, 57, 61, 96]. Below we will discuss some of these examples. More comprehensive reviews are provided by [85, 86, 100].

Viewing pattern and perceived foreground-background relationship. Fixation on a stationary foreground object or simply staring at the moving visual stimulus has long been known to enhance visual vection, as compared to natural viewing or smooth pursuit [28, 60, 121, 122]. Suppressing the optokinetic reflex seems to play a central role here, and this is facilitated when a fixation object is provided [8]. Potentially related to this, stationary foreground objects facilitate vection (especially if centrally presented), whereas stationary background stimuli reduce vection, especially if presented peripherally [17, 42, 62]. Of particular importance seems to be the perceived foreground-background or figure-ground relationship, in that vection tends to be dominated by motion of the perceived background, even if the background is not physically further away than the perceived foreground [17, 45, 53, 63, 65, 67, 68]. This "object and background hypothesis for vection" has been elaborated upon and confirmed in an elegant set of experiments using perceptually bistable displays like the Rubin's vase that can be perceived either as a vase or two faces [103].

In VR simulations, these findings could be used to systematically reduce or enhance illusory self-motions depending on the overall simulation goal, e.g., by modifying the availability of real or simulated foreground objects (e.g., dashboards), changing peripheral visibility of the surrounding room (e.g., by controlling lighting conditions), or changing tasks/instructions (e.g., instructions to pay attention to instruments which are typically stationary and in the foreground).

Naturalism, presence, and interpretation of the moving stimulus. Naturalism and ecological validity of the moving stimulus has also been suggested to affect vection [84, 116], potentially due to our inherent assumption of a stable environment

[23, 81, 82]. For example, auditory vection was enhanced when the moving sounds represented "acoustic landmarks", i.e., objects that do not normally move such as church bells, as compared to typically moving objects like cars or artificial sounds like pink noise [56, 96, 118].

For visual vection Riecke et al. [84] demonstrated that vection as well as presence were impaired when the naturalistic stimulus of a city environment was systematically degraded by mosaic-like scrambling. Different aspects of presence were correlated with specific aspects of vection: Whereas spatial presence correlated most strongly with the convincingness of illusory self-motion, attention/involvement in the simulation correlated predominately with vection onset latency. In a second experiment, the visual stimulus of a natural scene was compared to an upside-down version of the same stimulus. Even though the inversion of the stimulus left the physical stimulus characteristics (i.e., the image statistics and thus perceptual/bottom-up factors) essentially unaltered, both presence and the convincingness of vection were significantly reduced. This strongly suggests a cognitive or top-down contribution to presence and the convincingness of self-motion illusions. We posit that the natural, ecologically more plausible upright stimulus might have more easily been accepted as a stable "scene", which in turn facilitated both presence and the convincingness of vection.

These findings are supported by tumbling room studies, where the tumbling sensation (roll vection) is enhanced for naturalistic environments that include a clear visual frame of reference and objects with an obvious intrinsic upright orientation [1, 40]. That is, whereas simple textured displays only tend to produce limited tilting sensations [1, 32, 133], observing a fully furnished natural room rotating around stationary participants can induce compelling 360° head-over-heals tumbling sensation in most people [1, 40, 43, 71]. Moreover, Palmisano et al. stated that "the 360° illusory self-rotations produced by rotating a furnished room around the stationary observer's roll axis were very similar to the sensations of self-rotation produced by rotating the observer inside the stationary room" (p. 4057). The importance of a naturalistic visual stimulus is corroborated by Wright et al. who demonstrated that visual motion of a photo-realistic visual scene can dominate even conflicting inertial motion cues in the perception of self-motion [128, 129].

Metaphorical cross-modal facilitation of vection. Recently, Seno et al. demonstrated that linear visual vection could even be facilitated by auditory cues that do not move by themselves, but only match the visual motion metaphor [102]. For example, sounds increasing in amplitude (as if coming closer) facilitated visually-induced forward vection, but not backwards, sideways (left-right) or vertical (up-down) vection. Sounds decreasing in amplitude did not show any clear effects on vection, though. Whereas forward motions in normal life are often accompanied by increasing sound amplitudes for sounding stationary objects in front of us, this physical correspondence to real-world situations does not seem to be necessary for sound to facilitate visually-induced vection: Sounds ascending ("going up") in frequency facilitated upwards vertical vection, but had no influence on downwards, sideways (left-right), or forward-backwards vection [102]. Correspondingly, sounds decreasing in frequency ("going down") facilitated downwards vertical vection, but had no effect on any other vection direction. Similar effects of spatial metaphor mapping have been observed for the emotional connotation of sounds, in that emotionally "positive" sounds facilitated upwards vection compared to neural sounds [99]. Together, these findings further corroborate the proposition that multi-modal consistency between different stimuli can facilitate vection [86, 94, 96, 102], even in situations where this correspondence is only metaphorical and not purely sensorial. However, as vection is an inherently subjective phenomenon, vection researchers need to carefully assess potential experimental biases such as perceived demand characteristics of the experimental situation and participants' expectations and prior knowledge.

Cognitive-perceptual framework of movability. A number of studies demonstrated that merely knowing/perceiving that actual motion is impossible versus possible can reduce visual vection, both in the real world and VR [3, 57, 130]. For example, Andersen and Braunstein [3] remark that pilot experiments had shown that in order to perceive any self-motion, participants had to believe that they could actually be moved in the direction of perceived vection. Accordingly, participants were asked to stand in a movable booth and looked out of a window to view the optic flow pattern. This procedure allowed them to elicit vection with a visual FOV as small as 7.5°. Lepecq et al. [57] demonstrated that seven year old children perceive vection earlier when they were previously shown that the chair they were seated on could physically move in the direction of simulated motion—even though this never happened during the actual experiment. Similarly, knowing that actual motion is possible in VR (by demonstrating the motion capabilities of a motion platform prior to testing) can make people believe that they actually moved, even though they never did [86, 100]. Recently, Riecke et al. [85] demonstrated that providing such a cognitive-perceptual framework of movability can also enhance auditory vection. When blindfolded participants were seated on a hammock chair while listening to binaural recordings of rotating sound fields, auditory circular vection was facilitated when participants' feet were suspended by a chair-attached footrest as compared to being positioned on solid ground. This supports the common practice of seating participants on potentially moveable platforms or chairs in order to elicit auditory vection [54, 117, 118].

Attention and cognitive load. There seems to be mixed evidence about the potential effects of attention and cognitive load on vection. Whereas Trutoiu et al. [113] observed vection facilitation when participants had to perform a cognitively demanding secondary task, vection inhibition was reported by Seno and colleagues [105]. When observers in [53] were asked to specifically attend one of two simultaneously presented upward and downward optic flow fields of different colors, the non-attended flow field was found to determine vection direction. This might, however, also be explained by attention modulating the perceived depth-ordering and foreground-background relationship, as discussed in detail in [75, 103] demonstrated that cognitive priming can also affect the time course of vection: Adult participants experienced vection earlier when they were seated on a potentially movable chair and were primed towards paying attention to self-motion sensation, compared to a condition where they were seated on a stationary chair and instructed to attend to object motion, not self-motion. Thus, while attention and cognitive load can clearly affect self-motion illusions, further research is needed to elucidate underlying factors and explain seemingly conflicting findings. A recent study suggests that vection can even be induced when participants are not consciously aware of any global display motion, which was cleverly masked by strong local moving contrasts [107].

Finally, the occurrence, onset latency, and perceived strength of vection tend to vary considerably between participants. Although there is little research investigating potential underlying factors, recent research suggests that personality traits might be a contributing factor. In a linear visual vection study, more narcissistic observers reported weaker vection, indicated by increased vection onset latencies, reduced vection duration, and decreased vection magnitude [108]. Future research is needed to investigate if differences in personality traits indeed directly affect the self-motion illusions, and/or if the observed vection reduction for increasing narcissism might also be related to a criterion shift for reporting vection.

In general, cognitive factors seem to become more relevant when stimuli are ambiguous or have only weak vection-inducing power, as in the case of auditory vection [85] or sparse or small-FOV visual stimuli [3]. It is conceivable that cognitive factors generally have an effect on vection, but that this has not been widely recognized for methodological reasons. For example, the cognitive manipulations might not have been powerful enough or free of confounds, or sensory stimulation might have been so strong that ceiling level was already reached, which is likely the case in an optokinetic drum that covers the full visible FOV.

2.8 Does Vection Improve Spatial Updating and Perspective Switches?

Spatial updating is seemingly automatic and requires little cognitive resources if participants physically move to the new position [80, 97]. For example, humans can continuously and accurately point to a previously-seen target when either walking or being passively transported, both for linear motions [20, 109] and curvilinear motions [29]. However, when participants in Frissen et al. [29] were stationary and only biomechanical cues from stepping along a circular treadmill indicated the curvilinear motion, spatial updating performance (quantified using continuous pointing) declined and showed systematic errors. The authors did not assess whether participants in some trials might have perceived biomechanical vection. In a follow-up study by Frissen et al. continuous pointing responses indicated that participants can indeed perceive a slow drift (about 7°/s) for curvilinear off-center walking-in-place on a large (3.6 m diameter) circular treadmill, but only at about 16 % of their actual walking speed of 40°/s (cf. Chap. 6 of this book). Surprisingly, although participants were always walking *forward*, pointing responses indicated *backward* self-motion in 42 % of the trials. This suggests that biomechanical cues from curvilinear forward walking were not sufficient for inducing a reliable sensation of forward self-motion.

Indeed, when averaged over trial repetitions, participants did not report any substantial net self-motion. This might have contributed to the above-mentioned decline in spatial updating performance when participant did not physically move [29].

It is, however, conceivable that a compelling illusion of self-motion (even without any actual physical motion) might be sufficient to enable spatial updating performance similar to physical motions, or at least better than in purely imagined perspective switches. Riecke et al. [90] tested this hypothesis and provide first evidence that self-motion illusions might indeed help us to update target locations in the absence of physical self-motions. After learning the layout of nine irregularly arranged objects in the lab, participants were blindfolded and asked to point to those previously-learned objects from novel imagined perspectives (e.g., "imagine facing 'mic', point to 'hat' "). As predicted by prior research [80, 97], imagined perspective switches were difficult when participants remained stationary and simply had to imagine the perspective switch. Both pointing accuracy and consistency ("configuration error") improved, however, when participants had the illusion of rotating to the to-be-imagined perspective, despite not physically moving. Circular vection in this study was induced by combining auditory vection (induced via rotating sound fields) with biomechanical vection (induced by stepping on a circular treadmill, similar to sitting stationary above a turning carousel) in order to avoid visual cues that might interfere with imagined perspective-taking.

While further studies are needed to corroborate these findings, these data suggest that providing the mere illusion of self-motion might provide similar benefits in terms of spatial orientation and perspective switches as actual self-motion. This could ultimately enable us to design effective yet affordable VR simulations, as the need for physical motion of the observer could be largely reduced, which, in turn, reduces overall costs, space and equipment needs, and required safety and simulation effort.

2.9 Conclusions and Conceptual Framework

In conclusion, the above review of the literature supports the notion that cognitive or top-down mechanisms like spatial presence, the cognitive-perceptual framework of movability, as well as the interpretation of a stimulus as stable and/or belonging to the perceptual background, do all affect self-motion illusions, a phenomenon that was traditionally believed to be mainly bottom-up driven ([85], for reviews, see [86], [100]). This adds to the small but growing body of literature that suggests cognitive or top-down contributions to vection, as discussed in Sect. 7.2. Furthermore, correlations between the amount of presence/immersion/involvement and self-motion perception [91, 92] suggests that these factors might mutually affect or support each other. While still speculative, this would be important not only for our theoretical understanding of self-motion perception, presence, and other higher-level phenomena, but also from an applied perspective of affordable yet effective self-motion simulation. In the following, we would like to broaden our perspective by trying to embed these ideas and findings into a more comprehensive tentative framework. This conceptual framework is sketched in Fig. 2.3 and will be elaborated upon in more detail below. It is meant not as a "true" theoretical model but as a tentative framework to support discussion and reasoning about these concepts and their potential interrelations.

Any application of VR, be it more research-oriented or application-oriented, is typically driven by a more or less clearly defined goal. In our framework, this is conceptualized as the ***effectiveness concerning a specific goal or application*** (Fig. 2.3, bottom box). Possible examples include the effectiveness of a specific pilot training program in VR, which includes how well knowledge obtained in the simulator transfers to corresponding real world situations, or the degree to which a given VR hardware and software can be used as an effective research tool that provides ecologically valid stimulation of the different senses.

So how can a given goal be approached and the goal/application-specific effectiveness be better understood and increased? There are typically a large number of potential contributing factors, which span the whole range from perceptual to cognitive aspects (see Fig. 2.3, top box). Potentially contributing factors include straight-forward technical factors like the FOV and update rate of a given VR setup or the availability of biomechanical cues from walking, the quality of the sensory stimulation with respect to the different individual modalities and their cross-modal consistency, and task-specific factors like the cognitive load or the users' instructions.

All of these factors might effect both our perception and our action/behavior in the VE. Here, we propose a framework where the different factors are considered in the context of both their ***perceptual effectiveness*** (e.g., how they contribute to the perceived self-motion) and their ***behavioral effectiveness*** (e.g., how they contribute by empowering the user to perform a specific behavior like robust and effortless spatial orientation and navigation in VR), as sketched in Fig. 2.3, middle box.

Perception and action are interconnected via the ***perception-action loop***, such that our actions in the environment will also change the input to our senses. State-of-the art VR and human-computer interface technology offer the possibility to provide highly realistic multi-modal stimuli in a closed perception-action loop, and the different contributing factors summarized in the top box of Fig. 2.3 could be evaluated in terms of the degree to which they support an effective perception-action loop [27].

Apart from the perceptual and behavioral effectiveness, we propose that ***psychological and physiological responses*** might also play an important role. Such responses could be emergent and higher-level phenomena like spatial presence, immersion, enjoyment, engagement, or involvement in the VE, but also other psychological responses like fear, stress, or pleasure on the one hand and physiological responses like increased heart rate or adrenalin level on the other hand. In the current framework, we propose that such psychological and physiological responses are not only affected by the individual factors summarized in the top box in Fig. 2.3, but also by our perception and our actions themselves. Slater et al. [110] demonstrated, for example, that increased body and head motions can result in an increased presence in the VE. Presence might also be affected by the strength of the perceived self-motion illusion [81, 91]. Conversely, certain psychological and physiological responses might also affect our perception and actions in the VE. By systematically

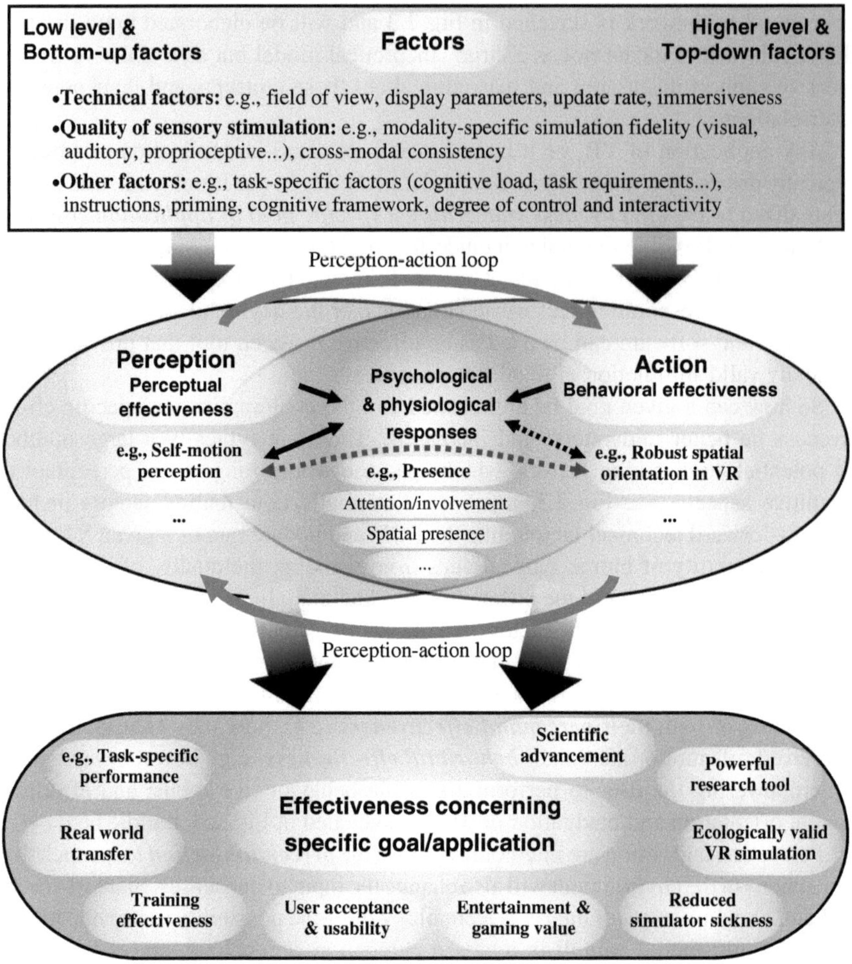

Fig. 2.3 Tentative conceptual framework that sketches how different factors that can be manipulated for a given VR/research application (*top box*) might affect the overall effectiveness with respect to a specific goal or application (*bottom box*). Critically, we posit that the factors affect the overall goal not (only) directly, but also mediated by the degree to which they support both the perceptual effectiveness and behavioral effectiveness and the resulting perception-action loop (*middle box*)

manipulating the naturalism and global scene consistency of a visually simulated scene, Riecke et al. [84] showed that the degree of presence in a simulated scene might also affect self-motion perception. Our actions and behaviors in a VE might, however, also be affected by our psychological and physiological responses. Von der Heyde and Riecke proposed, for example, that spatial presence might be a necessary prerequisite for robust and effortless spatial orientation based on automatic spatial updating or certain obligatory behaviors like fear of height or fear of narrow enclosed spaces [36, 83].

In summary, we posit that our understanding of the nature and usefulness of the cognitive factors and higher-level phenomena and constructs such as presence, immersion, or a perceptual-cognitive framework of movability might benefit if they are embedded in a larger conceptual framework, and in particular analyzed in terms of possible relations to perceptual and behavioral aspects as well as goal/application-specific effectiveness. Similar benefits are expected if other higher-level phenomena are analyzed in more detail in the context of such a framework.

2.10 Outlook

A growing body of evidence suggests that there is a continuum of factors that influence the perceptual and behavioral effectiveness of VR simulations, ranging from perceptual, bottom-up factors to cognitive, top-down influences. To illustrate this, we reviewed recent evidence suggesting that self-motion illusions can be affected by a wide range of parameters including attention, viewing patterns, the perceived depth structure of the stimulus, perceived foreground/background distinction (even if there is no physical separation), cognitive-perceptual frameworks, ecological validity, as well as spatial presence and involvement. While some of the underlying research is still preliminary, findings are overall promising, and we propose that these issues should receive more attention both in basic research and applications.

These factors might turn out to be crucial especially in the context of VR applications and self-motion simulations, as they have the potential of offering an elegant and affordable way to optimize simulations in terms of perceptual and behavioral effectiveness. Compared to other means of increasing the convincingness and effectiveness of self-motion simulations like increasing the visual field of view, using a motion platform, or building an omni-directional treadmill, cognitive factors can often be manipulated rather easily and without much cost, such that they could be an important step towards a lean and elegant approach to effective self-motion simulation [86, 94, 96]. This is nicely demonstrated by many theme park rides, where a conducive cognitive-perceptual framework and expectations are set up already while users are standing in line. Although there seems to be no published research on these priming phenomena in theme parks, they likely help to draw users more easily and effectively into the simulation and into anticipating and "believing" that they will actually be moving. Thus, we posit that an approach that is centered around the perceptual and behavioral effectiveness and not only the physical realism is important both for gaining a deeper understanding in basic research and for offering a lean and elegant way to improve a number of applications, especially in the advancing field of virtual reality simulations. This might ultimately allow us to come closer to fulfilling the promise of VR as an alternate reality, that enables us to perceive, behave, and more specifically locomote and orient as easily and effectively in virtual worlds as we do in our real environment.

Acknowledgments This work was funded by Simon Fraser University, NSERC, the European Community (IST-2001-39223), and the Max Planck Society.

References

1. Allison RS, Howard IP, Zacher JE (1999) Effect of field size, head motion, and rotational velocity on roll vection and illusory self-tilt in a tumbling room. Perception 28(3):299–306
2. Andersen GJ (1986) Perception of self-motion—psychophysical and computational approaches. Psychol Bull 99(1):52–65
3. Andersen GJ, Braunstein ML (1985) Induced self-motion in central vision. J Exp Psychol Hum Percept Perform 11(2):122–132
4. Ash A, Palmisano S, Allison RS (2012) Vection in depth during treadmill locomotion. J Vis. 12(9):181–181
5. Ash A, Palmisano S, Govan DG, Kim J (2011) Display lag and gain effects on vection experienced by active observers. Aviat Space Environ Med 82(8):763–769. doi:10.3357/ASEM. 3026.2011
6. Ash A, Palmisano S, Kim J (2011) Vection in depth during consistent and inconsistent multisensory stimulation. Perception 40(2):155–174. doi:10.1068/p6837
7. Becker W, Nasios G, Raab S, Jürgens R (2002) Fusion of vestibular and podokinesthetic information during self-turning towards instructed targets. Exp Brain Res 144(4):458–474
8. Becker W, Raab S, Jürgens R (2002) Circular vection during voluntary suppression of optokinetic reflex. Exp Brain Res 144(4):554–557
9. Berger DR, Schulte-Pelkum J, Bülthoff HH (2010) Simulating believable forward accelerations on a stewart motion platform. ACM Trans Appl Percept 7(1):1–27. doi:10.1145/ 1658349.1658354
10. Berthoz A, Pavard B, Young LR (1975) Perception of linear horizontal self-motion induced by peripheral vision (linearvection)—basic characteristics and visual-vestibular interactions. Exp Brain Res 23(5):471–489
11. Bles W, Bos JE, de Graaf B, Groen E, Wertheim AH (1998) Motion sickness: Only one provocative conflict? Brain Res Bull 47(5):481–487
12. Bles W, Dejong JMBV, Rasmussens JJ (1984) Postural and Oculomotor Signs in labyrinthine-defective subjects. Acta Oto-Laryngologica 96(s406):101–104
13. Bles W (1981) Stepping around: circular vection and Coriolis effects. In: Long J, Baddeley A (eds) Attention and performance IX. Erlbaum, Hillsdale, NJ, pp 47–61
14. Bles W, Kapteyn TS (1977) Circular vection and human posture .1. Does proprioceptive system play a role. Agressologie 18:325–328
15. Brandt T, Büchele W, Arnold F (1977) Arthrokinetic nystagmus and ego-motion sensation. Exp Brain Res 30(2):331–338. doi:10.1007/BF00237260
16. Brandt T, Dichgans J, Koenig E (1973) Differential effects of central versus peripheral vision on egocentric and exocentric motion perception. Exp Brain Res 16:476–491
17. Brandt T, Wist ER, Dichgans J (1975) Foreground and Background in Dynamic Spatial Orientation. Percept Psychophysics 17(5):497–503
18. Bruggeman H, Piuneu VS, Rieser JJ, Pick HLJ (2009) Biomechanical versus inertial information: Stable individual differences in perception of self-rotation. J Exp Psychol Hum Percept Perform 35(5):1472–1480. doi:10.1037/a0015782
19. Bubka A, Bonato F (2010) Natural visual-field features enhance vection. Perception 39(5):627–635. doi:10.1068/p6315
20. Campos JL, Siegle JH, Mohler BJ, Bülthoff HH, Loomis JM (2009) Imagined self-motion differs from perceived self-motion: evidence from a novel continuous pointing method. PLoS ONE 4(11):e7793. doi:10.1371/journal.pone.0007793

21. Cheung BSK, Howard IP, Money KE (1990) Visually-induced tilt during parabolic flights. Exp Brain Res 81(2):391–397. doi:10.1007/BF00228131
22. Cheung BSK, Howard IP, Nedzelski JM, Landolt JP (1989) Circularvection about earth-horizontal axes in bilateral labyrinthine-defective subjects. Acta Oto-Laryngologica 108(5):336. doi:10.3109/00016488909125537
23. Dichgans J, Brandt T (1978) Visual-vestibular interaction: effects on self-motion perception and postural control. Perception, handbook of sensory physiology vol VIII. Springer, Berlin, pp 756–804
24. DiZio P, Lackner JR (2002) Proprioceptive adaptation and aftereffects. In: Stanney KM (ed) Handbook of virtual environments. Lawrence Erlbaum, New York, pp 751–771
25. Durgin FH, Gigone K, Scott R (2005) Perception of visual speed while moving. J Exp Psychol Hum Percept Perform 31(2):339–353. doi:10.1037/0096-1523.31.2.339
26. Durgin FH, Pelah A, Fox LF, Lewis JY, Kane R, Walley KA (2005) Self-motion perception during locomotor recalibration: more than meets the eye. J Exp Psychol Hum Percept Perform 31(3):398–419
27. Ernst MO, Bülthoff HH (2004) Merging the senses into a robust percept. Trends Cogn Sci 8(4):162–169
28. Fischer MH, Kornmüller AE (1930) Optokinetisch ausgelöste Bewegungswahrnehmung und optokinetischer Nystagmus [Optokinetically induced motion perception and optokinetic nystagmus]. J für Psychol und Neurol 41:273–308
29. Frissen I, Campos JL, Souman JL, Ernst MO (2011) Integration of vestibular and proprioceptive signals for spatial updating. Exp Brain Res 212(2):163–176. doi:10.1007/s00221-011-2717-9
30. Giannopulu I, Lepecq JC (1998) Linear-vection chronometry along spinal and sagittal axes in erect man. Perception 27(3):363–372
31. Harm DL (2002) Motion sickness neurophysiology, physiological correlates, and treatment.In: Stanney KM (ed) Handbook of virtual environments. Lawrence Erlbaum, New York, pp 637–661
32. Held R, Dichgans J, Bauer J (1975) Characteristics of moving visual scenes influencing spatial orientation. Vis Res 15(3):357–365. IN1. doi:10.1016/0042-6989(75)90083-8
33. von Helmholtz H (1866) Handbuch der physiologischen Optik. Voss, Leipzig, Germany
34. Hettinger LJ (2002) Illusory self-motion in virtual environments.In: Stanney KM (ed) Handbook of virtual environments. Lawrence Erlbaum, New York, pp 471–492
35. Hettinger LJ, Berbaum KS, Kennedy RS, Dunlap WP, Nolan MD (1990) Vection and simulator sickness. Mil Psychol 2(3):171–181. doi:10.1207/s15327876mp0203_4
36. von der Heyde M, Riecke BE (2002) Embedding presence-related terminology in a logical and functional model (pp. 37–52). Presented at the Presence. Retrieved from http://edoc.mpg.de/39355
37. Hollerbach JM (2002) Locomotion interfaces. In: Stanney KM (ed) Handbook of virtual environments. Lawrence Erlbaum, New York, pp 239–254
38. Howard IP (1982) Human visual orientation. Wiley, Chichester, New York
39. Howard IP (1986) The perception of posture, self motion, and the visual vertical. In: Boff KR, Kaufman L, Thomas JP (eds) Sensory processes and perception, Handbook of human perception and performance. vol 1. Wiley, New York, pp 18.1–18.62
40. Howard IP, Childerson L (1994) The contribution of motion, the visual frame, and visual polarity to sensations of body tilt. Perception 23(7):753–762. doi:10.1068/p230753
41. Howard IP, Heckmann T (1989) Circular vection as a function of the relative sizes, distances, and positions of two competing visual displays. Perception 18(5):657–665. doi:10.1068/p180657
42. Howard IP, Howard A (1994) Vection–the contributions of absolute and relative visual motion. Perception 23(7):745–751
43. Howard IP, Jenkin HL, Hu G (2000) Visually-induced reorientation illusions as a function of age. Aviat Space Environ Med 71(9 Suppl):A87–91

44. Howard IP, Zacher JE, Allison RS (1998) Post-rotatory nystagmus and turning sensations after active and passive turning. J Vestib Res 8(4):299–312
45. Ito H, Shibata I (2005) Self-motion perception from expanding and contracting optical flows overlapped with binocular disparity. Vis Res 45(4):397–402. doi:10.1016/j.visres.2004.11.009
46. Ji JTT, So RHY, Cheung RTF (2009) Isolating the effects of vection and optokinetic nystagmus on optokinetic rotation-induced motion sickness. Hum Factors 51(5):739–751. doi:10.1177/0018720809349708
47. Johansson G (1977) Studies on visual-perception of locomotion. Perception 6(4):365–376. doi:10.1068/p060365
48. Johnson WH, Sunahara FA, Landolt JP (1999) Importance of the vestibular system in visually induced nausea and self-vection. J Vestib Res Equilibr Orientation 9(2):83–87
49. Jürgens R, Becker W (2006) Perception of angular displacement without landmarks: evidence for Bayesian fusion of vestibular, optokinetic, podokinesthetic, and cognitive information. Exp Brain Res 174(3):528–543. doi:10.1007/s00221-006-0486-7
50. Kennedy RS, Drexler JM, Compton DE, Stanney KM, Lanham DS, Harm DL (2003) Configural scoring of simulator sickness, cybersickness, and space adaptation syndrome: similarities and differences. In: Lawrence J Hettinger, Micheal W Hwas (eds) Virtual and adaptive environments: applications, implications, and human performance, issues, Lawrence Erlbaum Associates, USA, pp 247–278. http://www.amazon.com/Virtual-Adaptive-Environments-Applications-Implications/dp/080583107X/ref=sr_1_2?ie=UTF8&qid=1359862979&sr=8-2&keywords=Configural+scoring+of+simulator+sickness%2C+cybersickness%2C+and+space+adaptation+syndrome%3A#reader_080583107X
51. Kitazaki M, Murata A, Onimaru S, Sato T (2008) Vection during walking: effects of vision-action direction congruency and visual jitter. Poster presented at the International Multisensory Research Forum, Hamburg, Germany
52. Kitazaki M, Onimaru S, Sato T (2010) Vection and action are incompatible. Presented at the 2nd IEEE VR 2010 workshop on perveptual illusions in virtual environments (PIVE). Waltham, USA, pp 22–23
53. Kitazaki M, Sato T (2003) Attentional modulation of self-motion perception. Perception 32(4):475–484. doi:10.1068/p5037
54. Lackner JR (1977) Induction of illusory self-rotation and nystagmus by a rotating sound-field. Aviat Space Environ Med 48(2):129–131
55. Lackner JR, DiZio P (1988) Visual stimulation affects the perception of voluntary leg movements during walking. Perception 17(1):71–80. doi:10.1068/p170071
56. Larsson P, Västfjäll D, Kleiner M (2004) Perception of self-motion and presence in auditory virtual environments. In: Proceedings of 7th annual workshop of presence, pp 252–258
57. Lepecq JC, Giannopulu I, Baudonniere PM (1995) Cognitive effects on visually induced body motion in children. Perception 24(4):435–449
58. Lepecq JC, Jouen F, Dubon D (1993) The effect of linear vection on manual aiming at memorized directions of stationary targets. Perception 22(1):49–60
59. Loomis J, Knapp J (2003) Visual perception of egocentric distance in real and virtual environments. In: Hettinger LJ, Haas MW (eds) Virtual and adaptive environments: applications, implications, and human performance issues. Lawrence Erlbaum, Mahwah, USA, pp 21–46
60. Mach E (1875) Grundlinien der Lehre von der Bewegungsempfindung. Engelmann, Leipzig, Germany
61. Mergner T, Becker W (1990) Perception of horizontal self-rotation: Multisensory and cognitive aspects. In: Warren R, Wertheim AH (eds) Perception and control of self-motion. Erlbaum, London, pp 219–263
62. Nakamura S (2006) Effects of depth, eccentricity and size of additional static stimulus on visually induced self-motion perception. Vis Res 46(15):2344–2353. doi:10.1016/j.visres.2006.01.016
63. Nakamura S (2008) Effects of stimulus eccentricity on vection reevaluated with a binocularly defined depth. Jpn Psychol Res 50(2):77–86. doi:10.1111/j.1468-5884.2008.00363.x

64. Nakamura S (2010) Additional oscillation can facilitate visually induced self-motion perception: the effects of its coherence and amplitude gradient. Perception 39(3):320–329. doi:10.1068/p6534
65. Nakamura S, Shimojo S (1999) Critical role of foreground stimuli in perceiving visually induced self-motion (vection). Perception 28(7):893–902
66. Nichols S, Patel H (2002) Health and safety implications of virtual reality: a review of empirical evidence. Appl Ergonomics 33(3):251–271. doi:10.1016/S0003-6870(02)00020-0
67. Ohmi M, Howard IP (1988) Effect of stationary objects on illusory forward self-motion induced by a looming display. Perception 17(1):5–12. doi:10.1068/p170005
68. Ohmi M, Howard IP, Landolt JP (1987) Circular vection as a function of foreground-background relationships. Perception 16(1):17–22
69. Onimaru S, Sato T, Kitazaki M (2010) Veridical walking inhibits vection perception. J Vis 10(7):860. doi:10.1167/10.7.860
70. Palmisano S (1996) Perceiving self-motion in depth: The role of stereoscopic motion and changing-size cues. Percept Psychophysics 58(8):1168–1176
71. Palmisano S, Allison RS, Howard IP (2006) Illusory scene distortion occurs during perceived self-rotation in roll. Vis Res 46(23):4048–4058. doi:10.1016/j.visres.2006.07.020
72. Palmisano S, Allison RS, Kim J, Bonato F (2011) Simulated viewpoint jitter shakes sensory conflict accounts of vection. Seeing Perceiving 24(2):173–200. doi:10.1163/187847511X570817
73. Palmisano S, Bonato F, Bubka A, Folder J (2007) Vertical display oscillation effects on forward vection and simulator sickness. Aviat Space Environ Med 78(10):951–956
74. Palmisano S, Burke D, Allison RS (2003) Coherent perspective jitter induces visual illusions of self- motion. Perception 32(1):97–110
75. Palmisano S, Chan AYC (2004) Jitter and size effects on vection are immune to experimental instructions and demands. Perception 33(8):987–1000
76. Palmisano S, Gillam B (1998) Stimulus eccentricity and spatial frequency interact to determine circular vection. Perception 27(9):1067–1077
77. Palmisano S, Gillam BJ, Blackburn SG (2000) Global-perspective jitter improves vection in central vision. Perception 29(1):57–67
78. Palmisano S, Keane S (2004) Effect of visual jitter on visual-vestibular interaction during vection. Aust J Psychol 56(Suppl. S):213
79. Post RB (1988) Circular vection is independent of stimulus eccentricity. Perception 17(6):737–744. doi:10.1068/p170737
80. Presson CC, Montello DR (1994) Updating after rotational and translational body movements: coordinate structure of perspective space. Perception 23(12):1447–1455. doi:10.1068/p231447
81. Prothero JD, Hoffman HG, Parker DE, Furness TA, Wells MJ (1995) Foreground/background manipulations affect presence. Proc Hum Factors Ergon Soc Ann Meet 39(21):1410–1414. doi:10.1177/154193129503902111
82. Prothero JD, Parker DE (2003) A unified approach to presence and motion sickness. In: Hettinger LJ, Haas MW (eds) Virtual and adaptive environments: applications, implications, and human performance issues. Lawrence Erlbaum, New York, pp 47–66
83. Riecke BE (2003) How far can we get with just visual information? Path integration and spatial updating studies in virtual reality. vol. 8, Berlin http://www.logos-verlag.de/cgi-bin/buch/isbn/0440
84. Riecke BE (2006) Simple user-generated motion cueing can enhance self-motion perception (Vection) in virtual reality. In: Proceedings of the ACM symposium on virtual reality software and technology (VRST) Limassol, ACM, Cyprus, pp 104–107. doi:10.1145/1180495.1180517
85. Riecke BE (2009) Cognitive and higher-level contributions to illusory self-motion perception ("vection"): does the possibility of actual motion affect vection? Jpn J Psychon Sci 28(1):135–139

86. Riecke BE (2011) Compelling self-motion through virtual environments without actual self-motion–using self-motion illusions ("Vection") to improve user experience in VR. In: Kim J-J (ed) Virtual reality, pp 149–176. doi:10.5772/13150. InTech. http://www.intechopen.com/articles/show/title/compelling-self-motion-through-virtual-environments-without-actual-self-motion-using-self-motion-ill
87. Riecke BE, Feuereissen D (2012) To move or not to move: can active control and user-driven motion cueing enhance self-motion perception ("Vection") in virtual reality? ACM symposium on applied perception SAP (accepted full paper) ACM. Los Angeles, USA, pp 1–8
88. Riecke BE, Feuereissen D, Rieser JJ (2009) Auditory self-motion simulation is facilitated by haptic and vibrational cues suggesting the possibility of actual motion. ACM Trans Appl Percept 6(3):1–22. doi:10.1145/1577755.1577763
89. Riecke BE, Feuereissen D, Rieser JJ, McNamara TP (2011) Spatialized sound enhances biomechanically-induced self-motion illusion (vection). Proceedings of the 2011 annual conference on Human factors in computing systems, CHI '11. ACM SIG.CHI, Vancouver, Canada, pp 2799–2802. http://doi.acm.org/10.1145/1978942.1979356
90. Riecke BE, Feuereissen D, Rieser JJ, McNamara TP (2012) Self-motion illusions (Vection) in VR–are they good for anything? IEEE virtual reality 2012. Orange County, USA, pp 35–38: doi:10.1109/VR.2012.6180875
91. Riecke BE, Schulte-Pelkum J, Avraamides MN, Bülthoff HH (2004) Enhancing the visually induced self-motion illusion (Vection) under natural viewing conditions in virtual reality. Proceedings of 7th annual workshop presence 2004, pp 125–132. doi:10.1.1.122.5636
92. Riecke BE, Schulte-Pelkum J, Avraamides MN, Heyde MVD, Bülthoff HH (2006) Cognitive factors can influence self-motion perception (vection) in virtual reality. ACM Trans Appl Percept (TAP) 3(3):194–216. doi:10.1145/1166087.1166091
93. Riecke BE, Schulte-Pelkum J,Caniard F (2006)Visually induced linear vection is enhanced by small physical accelerations. 7th international multisensory research forum (IMRF). Dublin, Ireland
94. Riecke BE, Schulte-Pelkum J, Caniard F, Bülthoff HH (2005a) Towards lean and elegant self-motion simulation in virtual reality. Proceedings of IEEE virtual reality 2005. Bonn, Germany, pp 131–138. http://doi.ieeecomputersociety.org/10.1109/VR.2005.83
95. Riecke BE, Väljamäe A, Schulte-Pelkum J (2009) Moving sounds enhance the visually-induced self-motion illusion (circular vection) in virtual reality. ACM Trans Appl Percept 6(2):1–27
96. Riecke BE, Västfjäll D, Larsson P, Schulte-Pelkum J (2005) Top-down and multi-modal influences on self-motion perception in virtual reality. Proceedings of HCI international 2005. Las Vegas, USA, pp 1–10. http://en.scientificcommons.org/20596227
97. Rieser JJ (1989) Access to knowledge of spatial structure at novel points of observation. J Exp Psychol Learn Mem Cogn 15(6):1157–1165. doi:10.1037/0278-7393.15.6.1157
98. Rieser JJ, Pick HL, Ashmead D, Garing AE (1995) Calibration of human locomotion and models of perceptual-motor organization. J Exp Psychol Hum Percept Perform 21(3):480–497
99. Sasaki K, Seno T, Yamada Y, Miura K (2012) Emotional sounds influence vertical vection. Perception 41(7):875–877
100. Schulte-Pelkum J (2007) Perception of self-motion: vection experiments in multi-sensory virtual environments (PhD thesis). Ruhr-Universität Bochum. http://www-brs.ub.ruhr-uni-bochum.de/netahtml/HSS/Diss/SchultePelkumJoerg/
101. Schulte-Pelkum J, Riecke BE, von der Heyde M, Bülthoff HH (2003) Circular vection is facilitated by a consistent photorealistic scene. Talk presented at the presence 2003 Conference. Aalborg, Denmark
102. Seno T, Hasuo E, Ito H, Nakajima Y (2012) Perceptually plausible sounds facilitate visually induced self-motion perception (vection). Perception 41(5):577–593
103. Seno T, Ito H, Sunaga S (2009) The object and background hypothesis for vection. Vis Res 49(24):2973–2982. doi:10.1016/j.visres.2009.09.017

104. Seno T, Ito H, Sunaga S (2011a) Inconsistent locomotion inhibits vection. Perception 40(6):747
105. Seno T, Ito H, Sunaga S (2011b) Attentional load inhibits vection. Attention Percept Psychophysics 73(5):1467–1476. doi:10.3758/s13414-011-0129-3
106. Seno T, Ogawa M, Ito H, Sunaga S (2011) Consistent air flow to the face facilitates vection. Perception 40(10):1237–1240
107. Seno T, Palmisano S, Ito H, Sunaga S (2012) Vection can be induced without global-motion awareness. Perception 41(4):493–497. doi:10.1068/p7206
108. Seno T, Yamada Y, lhaya, K (2011) Narcissistic people cannot be moved easily by visual stimulation. Perception 40(11):1390–1392. doi:10.1068/p7062
109. Siegle JH, Campos JL, Mohler BJ, Loomis JM, Bülthoff HH (2009) Measurement of instantaneous perceived self-motion using continuous pointing. Exp Brain Res 195(3):429–444. doi:10.1007/s00221-009-1805-6
110. Slater M, Steed A, McCarthy J, Maringelli F (1998) The influence of body movement on subjective presence in virtual environments. Hum Factors 40(3):469–477
111. Steinicke F, Bruder G, Hinrichs K, Jerald J, Frenz H, Lappe M (2009) Real walking through virtual environments by redirection techniques. J Virtual Reality Broadcast 6(2)
112. Trutoiu LC, Mohler BJ, Schulte-Pelkum J, Bülthoff HH (2009) Circular, linear, and curvilinear vection in a large-screen virtual environment with floor projection. Comput Graph 33(1):47–58. doi:10.1016/j.cag.2008.11.008
113. Trutoiu LC, Streuber S, Mohler BJ, Schulte-Pelkum J, Bülthoff HH (2008) Tricking people into feeling like they are moving when they are not paying attention. Applied perception in graphics and visualization (APGV) 190–190. http://doi.acm.org/10.1145/1394281.1394319
114. Urbantschitsch V (1897) Über Störungen des Gleichgewichtes und Scheinbewegungen. Z Ohrenheilk 31:234–294
115. van der Steen FAM (1998) Self-motion perception (PhD thesis). Technical University Delft, Netherlands
116. van der Steen FAM, Brockhoff PTM (2000) Induction and impairment of saturated yaw and surge vection. Percept Psychophysics 62(1):89–99
117. Väljamäe A (2007) Sound for multisensory motion simulators (PhD thesis). Chalmers University of Technology, Gothenburg, Sweden
118. Väljamäe A (2009) Auditorily-induced illusory self-motion: A review. Brain Res Rev 61(2):240–255. doi:10.1016/j.brainresrev.2009.07.001
119. Väljamäe A, Alliprandini PMZ, Alais D, Kleiner M (2009) Auditory landmarks enhance circular vection in multimodal virtual reality. J Audio Eng Soc 57(3):111–120
120. Väljamäe A, Larsson P, Västfjäll D, Kleiner M (2006) Vibrotactile enhancement of auditory induced self-motion and spatial presence. J Acoust Eng Soc 54(10):954–963
121. Wallach H (1940) The role of head movements and vestibular and visual cues in sound localization. J Exp Psychol 27(4):339–368
122. Warren HC (1895) Sensations of rotation. Psychol Rev 2(3):273–276. doi:10.1037/h0074437
123. Warren R, Wertheim AH (eds) (1990) Perception and control of self-motion. Erlbaum, London
124. Wertheim AH, Reymond G (2007) Neural noise distorts perceived motion: the special case of the freezing illusion and the Pavard and Berthoz effect. Exp Brain Res 180:569–576. doi:10.1007/s00221-007-0887-2
125. Wolpert L (1990) Field-of-view information for self-motion perception. In: Warren R, Wertheim AH (eds) Perception and control of self-motion. Erlbaum, Hillsdale, pp 101–126
126. Wong SCP, Frost BJ (1981) The effect of visual-vestibular conflict on the latency of steady-state visually induced subjective rotation. Percept Psychophysics 30(3):228–236
127. Wood RW (1895) The "Haunted Swing" illusion. Psychol Rev 2(3):277–278. doi:10.1037/h0073333
128. Wright WG (2009) Linear vection in virtual environments can be strengthened by discordant inertial input. 31st annual international conference of the IEEE EMBS (Engineering in Medicine and Biology Society). Minneapolis, USA, pp 1157–1160. doi:10.1109/IEMBS.2009.5333425

129. Wright WG, DiZio P, Lackner JR (2005) Vertical linear self-motion perception during visual andinertial motion: More than weighted summation of sensory inputs. J Vestib Res Equilibr Orientation 15(4):185–195
130. Wright WG, DiZio P, Lackner JR (2006) Perceived self-motion in two visual contexts: dissociable mechanisms underlie perception. J Vestib Res 16(1–2):23–28
131. Yabe Y, Taga G (2008) Treadmill locomotion captures visual perception of apparent motion. Exp Brain Res 191(4):487–494. doi:10.1007/s00221-008-1541-3
132. Yabe Y, Watanabe H, Taga G (2011) Treadmill experience alters treadmill effects on perceived visual motion. PLoS ONE 6(7):e21642. doi:10.1371/journal.pone.0021642
133. Young LR, Oman CM, Dichgans JM (1975) Influence of head orientation on visually induced pitch and roll sensation. Aviat Space Environ Med 46(3):264–268
134. Young LR, Shelhamer M (1990) Weightlessness enhances the relative contribution of visually-induced self-motion. In: Warren R, Wertheim AH (eds) Perception and control of self-motion: resources for ecological psychology. Erlbaum, Hillsdale

Chapter 3
Biomechanics of Walking in Real World: Naturalness we Wish to Reach in Virtual Reality

Franck Multon and Anne-Hélène Olivier

Abstract In most virtual reality (VR) simulations the virtual world is larger than the real walking workspace. The workspace is often bounded by the tracking area or the display devices. Hence, many researchers have proposed technical solutions to make people walk through large virtual spaces using various types of metaphors and multisensory feedback. To achieve this goal it is necessary to understand how people walk in real life. This chapter reports biomechanical data describing human walking including kinematics, dynamics and energetics knowledge for straight line and nonlinear walking. Reference and normative values are provided for most of these variables, which could help developers and researchers improve the naturalness of walking in large virtual environments, or to propose evaluation metrics. For each section of this chapter, we will provide some potential applications in VR. On the one hand, this type of knowledge could be used to design more natural interaction devices such as omnidirectional treadmills, walk-in-place methods, or other facilities. A specific section is dedicated to comparisons between treadmill and ground walking as it is one of the most popular approaches in VR. On the other hand, this knowledge could also be useful to improve the quality of multisensory feedback when walking, such as adding sounds, vibrations, or more natural camera control.

3.1 Introduction

Walking is one of the most used motions in everyday life and has consequently been widely studied from antiquity. Exclusive biped walking is a special characteristic of human locomotion that cannot be found in any other primate. Hence, it is the

F. Multon (✉)
University Rennes2, Campus La Harpe, Av. Charles Tillon, CS24414, F35044 Rennes, France
e-mail: franck.multon@univ-rennes2.fr

A. Olivier
Campus Universitaire de Beaulieu, F35042 Rennes, France

F. Steinicke et al. (eds.), *Human Walking in Virtual Environments*,
DOI: 10.1007/978-1-4419-8432-6_3,

result of a complex evolutionary strategy that has tended to make humans walk on two feet, which is a challenge for balance maintenance. It is impossible to provide exhaustive knowledge about human walking and many very relevant states of the art exist in many domains. This chapter aims at providing relevant information about gait kinematics and dynamics, which should help developers and scientists design more natural walking facilities in large virtual environments. In most VR simulations the virtual world is larger than the real walking workspace. The workspace is often bounded by the tracking area or the display devices. Hence, researchers in VR have proposed a wide set of systems to make people navigate in wide virtual environments while maintaining them in restricted real areas. One of the most popular approaches consists in designing metaphors. Metaphors consist in using one thing to describe another thing, such as moving an arm laterally to describe a translation of a 3D object placed in virtual environment.

Another problem is to deliver multisensory feedback that has positive effects on presence and immersion when navigating in virtual environments. The goal of this chapter is to provide information to address these two problems.

The first section of this chapter describes kinematics of human walking: global parameters and joint angles. This section provides the reader with normative reference data of human gait in straight line and nonlinear walking. It also addresses how these data change according to various parameters such as gender, age, speed and orientation. The second part of this chapter provides some knowledge about dynamics of human walking, including the pattern of external forces, their influence on balance and the muscle activation patterns. Again, reference values and their relation to other parameters are provided in this section. The last section deals with the comparison between ground and treadmill walking. Treadmills are widely used in many immersive applications that require natural navigation in immersive environments. However, the way people use this device and perceive the feedback associated with this device is still unclear. This section tries to provide a synthesis of the knowledge and discussions about this point. In all the sections, the authors provide an example of scientific contributions in VR that use this type of knowledge. The goal of this chapter is thus twofold. Firstly, it provides complementary information to imagine new systems or improvements of previous works. Secondly, it should help people better understand the advantages and limits of these previous works.

3.2 Kinematics of Human Walking

Kinematics of human walking has been studied for a long time. In this section, we give an overview of the most commonly studied kinematic parameters of human walking. Information about joint kinematics could be used:

- to create more realistic camera displacements when navigating in VR [70] so that classical linear camera displacements could be replaced by more natural motions, in order to enhance the feeling of naturally walking in VR,

- to animate avatars of the user with only little information, such as gait speed [12], by computing the corresponding joint angles, so that the user could embody the virtual human which is walking in VR,
- and to compute relevant multisensory feedback such as forces, sounds, vibrations... in accordance with the current walking state; the goal is to improve immersion and the feeling of being there in the virtual environment.

3.2.1 Global Description

3.2.1.1 Linear Walking

A gait cycle is defined as the time interval between two identical movement events during gait. The heel contact to the floor (heel strike) is widely accepted as the starting point and the next contact of the same heel as the end of a gait cycle. Similarly, a stride is defined as a period of time between two successive heel contacts of the same foot. A step is defined as a period between successive heel contacts of the opposite feet during double support. Bipedal walking is defined as a quasi-cyclic motion and is generally characterized by a sequence of single (one foot in contact with the ground) and double support (two feet on the ground) phases.

Within a stride, the double-support phase represents approximately 20 % of the total time [3, 31] . For a given leg, the stride is composed of a contact phase (60 % of the total time) and a swing phase (40 % of the total time). The ratio between the contact and the swing phase changes according to speed. Indeed, the duration of the contact phase decreases when speed increases. When the duration of the contact phase decreases down to 50 %, the double-support phase disappears, which corresponds to the walk-run transition [46, 76]. Intra-individual variability of these durations is low (below 4 %), while inter-individual variability is greater than 10 % [69].

The leg walking cycle could be decomposed in a more accurate manner according to relevant events: left (resp. right) foot-strike, right (resp. left) toe-off, right (resp. left) foot-strike and left (resp. right) toe-off.

The walking cycle is globally represented by three quantitative parameters: the step length (L), the step frequency (F) and the walking speed (V), which are linked through the relation:

$$V = L * F$$

In order to increase walking speed (V), humans jointly increase the step length and frequency. When continuously increasing walking speed on a treadmill, people generally increase step length first until a limit is reached and then increase step frequency [74]. However the results change when analyzing ground walking.

While people are walking at a freely chosen step-rate it has been reported that step-length divided by step-rate (called 'the walk ratio' [45, 62]) does not vary over wide ranges of speed [5, 62, 88] during ground walking. The walk ratio seems to be smaller in older adults than in younger adults, and tends to decrease with age [45]. Sekiya and Nagasaki [64] have shown that the walk ratio is a reliable measure for evaluating pathological and aging walking patterns. The average step length, step rate, velocity and walk ratio of healthy male and female subjects is reported in Table 3.1. The walk ratio does not differ significantly between males and females. It is also globally invariant except at low speeds in females [63].

Knowing this walk ratio for a user, it is possible to automatically estimate his/her step length and frequency depending on his/her current speed. As a result, it might be possible to more accurately adapt the trajectory of the camera in real-time [70], or to animate avatars simply driven by walking speed [12]. Conversely, it could be possible to deduce walking speed using step frequency or step length. Hence, knowing the frequency of the head oscillations (related to the step frequency) of the user with a simple webcam, it is possible to compute the corresponding walking speed of the avatar or of the camera [71].

It is also widely acknowledged that stride length shows a moderate inverse correlation with the age of the walker ($r = -0.68$, $P < 0.001$) as well as velocity, but the correlation is not as strong ($r = -0.53$, $P < 0.05$) [34]. Hence, the walk ratio seems to depend on both gender and age, which is an interesting property. In VR the trajectory of the camera and avatar motions can be customized with only few input parameters (such as step frequency or gait speed).

In everyday life, we usually walk at our own moderate pace. It is well known that casual walking involves optimization of the energy costs for moving. Briefly, optimal human locomotion is achieved when people move freely in their own selected manner which could be viewed as a definition of casual walking [1, 2, 8, 9]. In casual walking it seems that people always walk at a constant speed over a prescribed distance. Walking speed has thus been implicitly assumed to be a dominant parameter of casual walking [4, 42, 84]. Men walk generally faster than women but some authors have shown that it is mainly true for high speeds [64] and it also depends on age [6]. One has also to notice that walking speed decreases with age in both genders [6]. Casual walking speed starts to decrease during the sixth decade for men and during the seventh decade for women, as reported in Table 3.2.

Some authors [35] have demonstrated that cycle duration is more stable than stride length and walking speed in casual ground walking.

All this knowledge about casual walking can help developers in VR introduce variability in camera motion and avatar control thanks to only very little information. Hence, elderly and young avatars, male and female avatars, would each lead to different camera motions in the virtual environment in a very simple manner. The method described in [70] would be very easy to adapt in order to take this type of information into account.

Kirtley et al. [34] have also highlighted significant correlations between step frequency (expressed in steps per minute) and four gait parameters. The regression equations between these parameters are given in Table 3.3. Again, this type of information would help camera controllers to adapt more accurately to the situation. Such type of knowledge has been used in the past to animate virtual humans with only few control parameters [12].

Table 3.1 Reference values for step length, step frequency, speed and walk ratio for male and female subjects walking at three speeds (slowest to fastest speeds)

	Male				Female			
	Step length (m)	Step freq. (steps/s)	Speed (m/s)	Walk ratio (M/steps per s)	Step length (m)	Step freq. (steps/s)	Speed (m/s)	Walk ratio (M/steps per s)
Slowest (std deviation)	0.552 (0.114)	1.390 (0.145)	0.770 (0.188)	0.397 (0.084)	0.582 (0.051)	1.276 (0.130)	0.743 (0.098)	0.462 (0.072)
Preferred (std dev.)	0.664 (0.048)	1.808 (0.127)	1.201 (0.115)	0.366 (0.042)	0.695 (0.058)	1.782 (0.133)	1.242 (0.178)	0.390 (0.030)
Fastest (std dev.)	0.882 (0.810)	2.217 (0.142)	1.958 (0.273)	0.396 (0.030)	0.846 (0.056)	2.253 (0.198)	1.908 (0.233)	0.378 (0.036)

Adapted from [64]

Table 3.2 Mean and S.D. per decade for gait variables for men and women

Age (year)	Gender	Walking speed	Stride freq.	Stride length
		m/s	Hz	m
	Male	1.59 (0.13)	0.97 (0.06)	1.65 (0.12)
20–29	Female	1.54 (0.12)	1.03 (0.06)	1.49 (0.10)
	Male	1.54 (0.12)	0.98 (0.06)	1.57 (0.11)
30–39	Female	1.56 (0.11)	1.04 (0.06)	1.50 (0.09)
	Male	1.63 (0.15)	1.00 (0.06)	1.64 (0.15)
40–49	Female	1.50 (0.10)	1.06 (0.04)	1.42 (0.08)
	Male	1.42 (0.08)	0.96 (0.05)	1.49 (0.07)
50–59	Female	1.48 (0.12)	1.05 (0.08)	1.41 (0.07)
	Male	1.47(0.11)	0.96 (0.04)	1.53 (0.09)
60–69	Female	1.35 (0.09)	1.03 (0.04)	1.32 (0.08)
	Male	1.32 (0.12)	0.95 (0.06)	1.38 (0.08)
>70	Female	1.26 (0.19)	1.00 (0.08)	1.25 (0.15)

Adapted from [6]

Table 3.3 Correlations between step frequency SF (steps per minute) and four gait parameters

Parameters	Units	Correlation coefficient	Regression equation
Stride length	m	0.81	0.0088 SF + 0.58
Velocity	$m{\cdot}s^{-1}$	0.95	0.021 SF − 0.79
Stance phase duration	% cycle	−0.68	−0.073 SF + 67.0
Double support time	% cycle	−0.57	−0.058 SF + 15.6

Adapted from [34]

3.2.1.2 Nonlinear Walking

All of the above results have generally been obtained for straight line walking. For curved paths that involve rotations, the problem is more complex. Definitions of step length and width in such nonlinear walking are still being discussed. A modern definition is given in Fig. 3.1 [30].

In both linear and non-linear walking the foot placement on the floor seems to be constrained in order to minimize the risk of falling, to maximize the possibility of fast change of directions, and to ensure continuity of the walking trajectory [50]. When turning, two main strategies [24] arise for the feet placement. In the "step-turn" strategy, the change of direction is opposite to the contact foot (such as turning left while the right foot is in contact with the ground). This strategy is very similar to the one used in straight line walking and tends to enlarge the base of support (area delimited with the border of the two feet), which minimizes the risk of falling. In the "spin-turn" strategy, the change of direction and the contact foot are in the same size (such as turning left during a left contact phase). This strategy decreases the size of the base of support leading to more unstable, but also faster rotational displacements. Some authors have shown a high preference for using the "step turn" strategy compared to the "spin turn" strategy [23].

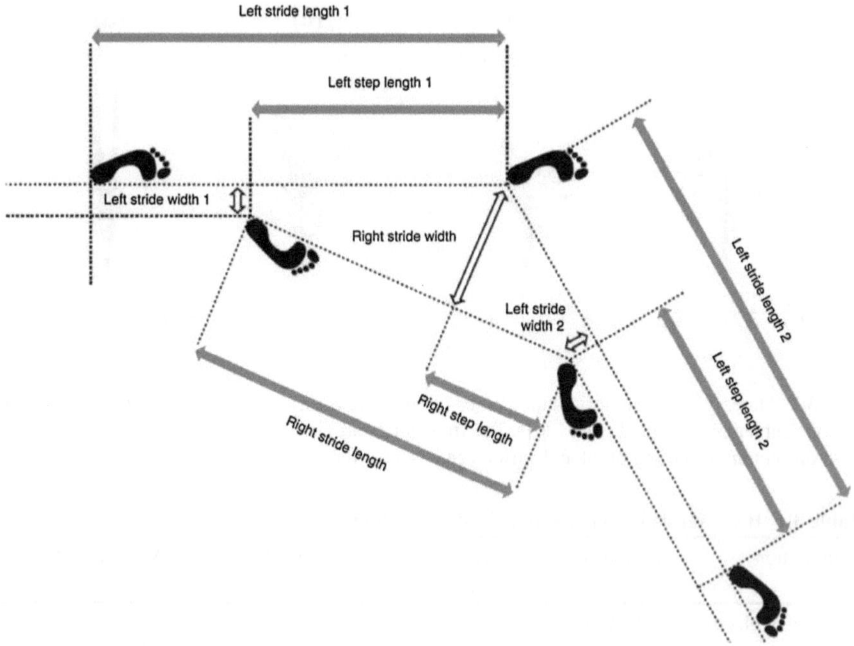

Fig. 3.1 Definition of step length and width in curved walking. Adapted from [30]

Whatever the strategy is, some authors have shown that the formation of trajectories during goal-oriented locomotion in humans demonstrates several general properties [26]. This stereotype appears in both the trajectory and the velocity profile for goal-oriented locomotion (such as starting from one point with a given orientation and reaching another point with an imposed final orientation). This stereotype is shown in Fig. 3.2.

This stereotype seems to be associated with a maximization of the smoothness of the trajectory [53]. Hence humans seem to minimize the Jerk (derivative of the acceleration) [22] and the snap (second derivative of the acceleration) [57]. Hence, these authors [57] suggested to design optimal controllers based on Jerk minimization to compute these stereotyped trajectories for various conditions.

Similarly to arms movements [79], it seems that trajectories in natural locomotion [25, 77] obey the power law:

$$V(t) = K \cdot R(t)^p$$

where K is a constant, V(t) is the instantaneous velocity, R(t) is the instantaneous radius of gyration and p is a real value. For elliptic motions, p has been identified to be close to 1/3, but it seems to change according to the shape of the trajectory.

All of these results tend to show that there is a control of the trajectory instead of a control of the foot placement. This could be an important issue when simulating human navigation in virtual environments. Some approaches tend to control camera

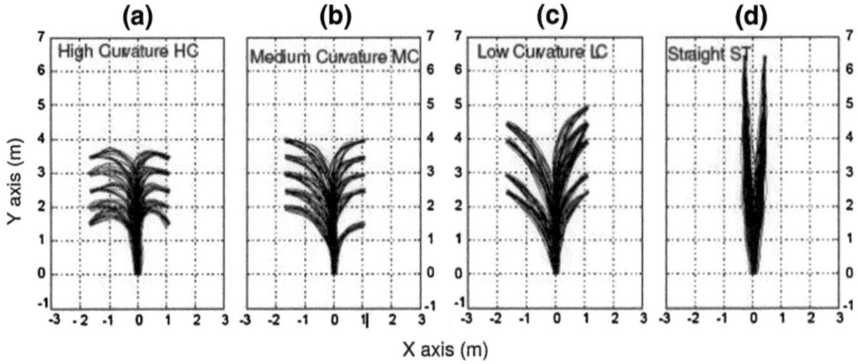

Fig. 3.2 Stereotype of the horizontal Center of Mass trajectory in goal-directed locomotion for various curvatures (adapted from [26]). The trajectories are plotted in horizontal plan (XY plane) with one common starting point and various ending positions

Table 3.4 Head translation and rotation for different human locomotion

Locomotion	Amplitude (m)	Velocity ($m{\cdot}s^{-1}$)	Acceleration (g)	Vertical Rot. (deg)
Walking in place	(0.01–0.025)	(0.15–0.20)	(0.2–0.3)	7.5 ± 3
Free walking	(0.05–0.09)	(0.25–0.35)	(0.3–0.7)	8.5 ± 2.5
Running	(0.07–0.16)	(0.7–1.5)	(1.4–2.6)	13.5 ± 4
Hopping	(0.15–0.25)	(1–1.8)	(1.2–2.2)	13.5 ± 4.5

Adapted from [55]

motion through the displacement of the head of the user. This "Shake-Your-Head" metaphor [71] consists of revisiting the Walking-In-Place technique to match a larger set of configurations and apply it notably to the context of desktop VR. Previous works [55] in neuroscience have shown stabilization of the head in human locomotion and provides us with reference data for the head behavior (see Table 3.4) to tune the parameters of such a method.

3.2.2 *Joint Kinematics*

The previous section focused on global parameters, but it is sometimes relevant to get information about joint kinematics. It could help to either simulate human walking in VR with little information or to analyze outputs of various sensors placed over body parts to deduce global information about gait cycle. However, many parameters may affect joint kinematics, such as walking up and down stairs, or locally avoiding obstacles. In this section, we describe some of the relevant knowledge about joint kinematics in human walking that could be reused to animate avatars or to interpret local body segment information to drive navigation in immersive environments.

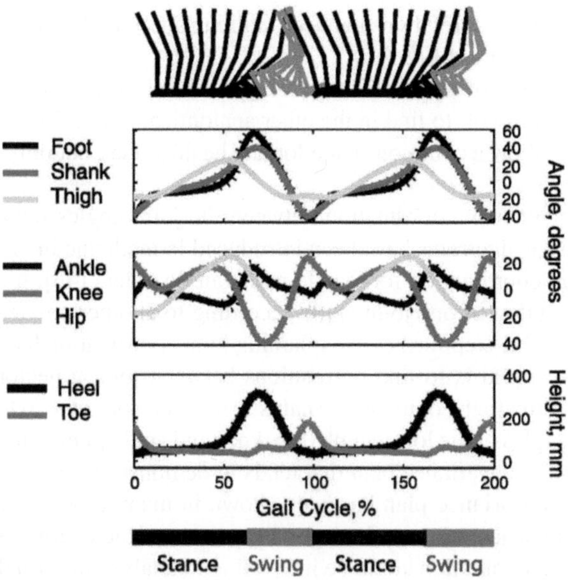

Fig. 3.3 Description of a stereotyped gait cycle with averaged trajectories for the joint angles (ankle, knee and hip joints), the body segment orientations (foot, shank and thigh), and the height of heels and toes (adapted from [11]). The *bottom* part of the figure depicts how the stance and swing phases occur in the gait cycle

Human walking is generally described using joint angle trajectories, which exhibit a typical shape as shown in Fig. 3.3 [11, 85].

Some authors have highlighted statistical significant correlations between global parameters, such as step frequency, length and speed, and knee angles [34]. Knee flexion during the stance phase could be extrapolated from step frequency SF (expressed in steps per minutes) with the following equation:

Stance Phase Flexion (deg) $= 0.27\ \text{SF} - 5.50\ (R = 0.74)$

Knee flexion during swing phase is given by:

Swing Phase Flexion (deg) $= 0.19\ \text{SF} + 41.4\ (R = 0.68)$

The same kind of correlations can be expressed between stride length SL (m) and these knee angles:

Stance Phase Flexion (deg) $= 26.2\ \text{SL} - 16.8\ (R = 0.79)$

Swing Phase Flexion (deg) $= 16.0\ \text{SL} + 37.5\ (R = 0.62)$

In the same manner speed S ($\text{m}\cdot\text{s}^{-1}$) is correlated with the same angles:

Stance Phase Flexion (deg) $= 13.0\ \text{S} + 4.7\ (R = 0.78)$

Swing Phase Flexion (deg) $= 8.6\ \text{S} + 49.6\ (R = 0.66)$

For more detailed correlations between walking speed and joint angles the reader is encouraged to refer to [39].

The typical shape of joint trajectories can be mostly found in the sagittal plane and seems more difficult to find in the other anatomical plans. This means that joint angles seem to follow a common shape for any healthy user, but only for flexion and extension.

Let us consider now coordination between the joint angles involved in human locomotion. Angle diagrams have been introduced in medicine in order to highlight changes in joint coordination for people with gait disorders [43]. It consists of displaying the evolution of one joint angle according to another one leading to a kind of 2D signature. One step further in explaining joint coordination has been proposed by [11]. The key idea is to find correlations between joint trajectories using data reduction with Principal Component Analysis. If we consider the 3D space based on the three joint angles of the lower-limb (hip, knee and ankle joints), many researchers have highlighted a covariance plan that tends to demonstrate coordination between the joints. This covariance plan has been shown in many types of locomotion and its orientation in space seems to be a good predictor of the energy expenditure and style [10]. This coordination could be interesting to analyze in virtual environments to check if joint coordination is affected by immersion. It would thus participate in defining a metric to measure the quality of the walking cycle in immersive environments compared to real ground walking.

In nonlinear walking, all of these variables are affected [48]. Self-selected walking speed while turning was reduced compared to walking straight ahead. It decreased from $1.00 \pm 0.02\,\mathrm{m{\cdot}s^{-1}}$ for straight line walking down to $0.91 \pm 0.01\,\mathrm{m{\cdot}s^{-1}}$ and $0.87{\pm}0.01\,\mathrm{m{\cdot}s^{-1}}$ for spin and step turns respectively. Table 3.5 reports other relations between turning strategies and gait parameters.

Hence, nonlinear walking not only changes the orientation of the body, but also affects global gait variables and joint trajectories. It may be interesting to evaluate to what extent taking this knowledge into account in camera motion control would improve immersion in navigation tasks. To our knowledge this type of evaluation has not been carried-out yet.

3.3 Dynamics of Human Walking

Joint angles and overall kinematic parameters are strongly linked to dynamic constraints such as external forces and muscle activation patterns. Hence, one has to

Table 3.5 Reference gait parameters values for straight line walking compared to spin turns and step turns

Variable	Straight	Spin	Step
Stride length (m)	1.30 ± 0.02	1.14 ± 0.03	1.36 ± 0.03
Peak ankle dorsiflexion (°)	14.4 ± 1.3	15.6 ± 1.4	10.6 ± 1.7
Peak knee extension (°)	-1.65 ± 1.11	1.21 ± 1.25	-3.49 ± 1.48

Adapted from [48]

fight against gravity and produce external forces to propel his body while maintaining balance. To this end, one has to produce forces and torques at each joint (resulting from the muscle activation patterns) in order to satisfy the external constraints. In this section, we provide a brief description of the available knowledge about these points. Many devices have been introduced in order to measure dynamic constraints in walking such as force plates and could be used to control avatars or cameras. The Wii balance board from Nintendo has been used to immerse users in videogames by measuring the instantaneous ground reaction force and the center of pressure of the user. This type of device is strongly associated with the idea of balance control which is also a key point in walking. We thus provide the reader with some knowledge about balance control in walking and give some examples of how it could be used in VR. We also address the problem of energy expenditure which is a key point in bipedal locomotion and which could be used to evaluate the relevance of immersive systems for navigating in virtual environments.

3.3.1 Forces and Torques Description

From the mechanical point of view, the external forces and torques exerted on the body while walking are due to gravity g and ground reaction force GRF. It comes:

$$\mathrm{BW} + \mathrm{GRF} = \mathrm{m}\,\gamma$$

where BW is body weight, m is the total mass and γ is the center of mass acceleration. Hence, GRF is very important in order to understand motor strategies and to analyze human walking. It is generally analyzed using a force plate through the three main axes. Even if GRFs vary [5] depending on speed, style, mechanical properties of the ground, the type of shoes worn by the subject, etc., it exhibits a stereotyped shape [87], as shown in Figure 3.4.

During walking, the vertical component of GRF exhibits two peaks, which correspond to heel-trike (associated with the absorption of the downward COM velocity) and toe-off (associated with propulsion of the body forward and upward). These two peaks reach higher values than body weight. From the muscle point of view, "absorption" is linked to gathering elastic potential energy and "propulsion" is associated with release of this elastic potential energy.

Females have a higher vertical ground reaction force (expressed as a percentage of body weight) in heel-strike and toe-off stages than males [15, 16]. Indeed females tend to increase their walking speed by increasing step length instead of frequency, which results in a higher vertical ground reaction force than males. Walking speed also has an effect on vertical ground reaction force so that in VR it could be possible to adapt feedback associated with force according to speed and gender.

Some authors have used information about the shape of the ground reaction force to provide vibration feedback to the user [73] and sound feedback could also be tuned according to this type of information [37]. This type of feedback has positive effects

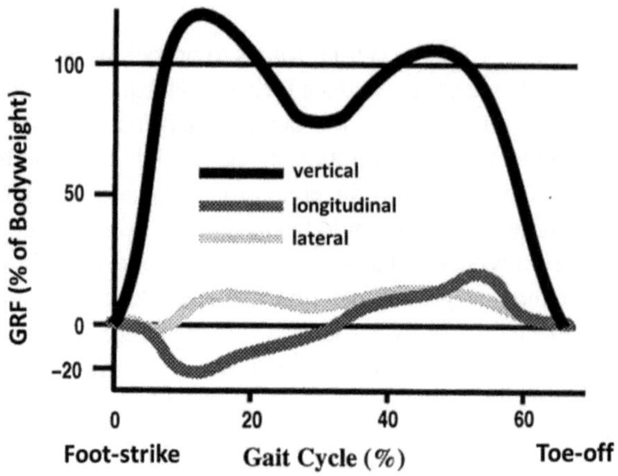

Fig. 3.4 Ground reaction force (GRF) measured below a foot while walking. Adapted from [87]

on presence. But it could be possible to go further by adapting the force to speed and any other parameter. It could also be possible to analyze these forces instantaneously on an equipped treadmill (with force sensors) to predict changes of speed and gait parameters of the user. See Chung and Wang [16] for detailed correlations of the vertical ground reaction force and speed and gender.
However, some authors have shown that GRF differs when walking on a treadmill compared to the ground [84] because the treadmill does not perfectly apply a constant velocity, especially at foot-strike, which is associated with a high friction force of the treadmill surface. Moreover, the horizontal components of GRF are strongly affected by the fact that the contact foot is pulled over the treadmill surface compared to normal ground walking where the user voluntarily applies a horizontal component of force.

If no force device is available in the system, joint torques and forces can be deduced using inverse dynamics, as described in Part 2, Chap. 3 "sensing human walking" of this book. This method is very sensitive to the anthropometric data used to model the body, as shown by [51, 56]. Despite these differences, there exists a general pattern for these joint torques. Many researchers in biomechanics have studied how torque is adapted to different populations (elderly, young, obese or disabled people).

Force might be an important sensory feedback when walking. The Walking-In-Place technique [65] enables a real physical movement (similar to some extent to natural walking), which leads to real contacts with the ground. The user has to consciously walk in place while motions of his body are tracked and analyzed [21, 72, 83]. In this case, the feedback associated with GRF is not simulated, but intrinsically delivered by the motion performed by the user.

Force information can also be used in VR to compute other sensory feedback when walking in virtual environments. For example, shoe-based devices can be associated with real time audio simulation [49]. Contact sensors embedded in shoes are used to detect footsteps and both vibro-tactile and auditory feedback is provided to match a specific virtual ground surface. Tactile tiles can also be used with spatialized audio in a CAVE to provide a complete simulation using the haptic, auditory and visual modalities [78]. All of these approaches tend to deliver various types of feedback that are more or less linked to dynamic properties.

As no device exists to directly measure joint torques, this type of information seems difficult to apply for navigation in virtual environments, except maybe for evaluation purposes. However, muscle activation can be monitored thanks to electromyographic sensors and could thus be used either as an input signal or an output for evaluation purposes. Indeed, electromyograms (EMG) (see Part 2, Chap. 3 "sensing human walking" of this book for an introduction of this measure) provide us with muscle coordination or muscle activation patterns and can give interesting information about walking performance. Figure 3.5 shows the muscle activation pattern of the most relevant muscles when walking [59]. It clearly shows that despite large inter-individual variations, a stereotyped shape appears for each muscle. It could thus be possible to apply signal processing and pattern recognition to determine in which phase of the walking cycle the subject is when activating the muscles. Some authors in biomechanics have used such type of information to simulate phases of a gait in musculoskeletal models [33]. Other authors were able to estimate the joint angle velocities depending on EMG signals [19].

EMG is very difficult to measure with good accuracy and it is sometimes difficult to correctly deduce the muscle activation. However, the main idea here could be to analyze EMG as people do with Electroencephalography (EEG) signals used in Brain Computer Interfaces [40]. We could imagine that EMG electrodes placed over relevant muscles of the lower-limb may detect activation patterns that could be an input for the navigation system. They can also be used for control if the proposed walking interface generates natural muscle activation patterns when navigating in virtual environments. However, it seems that this modality has not been used yet to control navigation in immersive environments.

3.3.2 Energetics of Human Walking

Many researchers have shown that locomotion is a very economic motion from the energetic point of view, especially for naturally selected speeds [17, 18]. People are able to walk for a long time and along very long distances. However, people have to support and propel their body mass, which seems to be a costly task from the energetic point of view. The reported minimal energy expenditure is thus the result of a positive coordination pattern.

Firstly, there exists an anti-phase coordination between the global potential and kinetic energies [13, 75], which enables us to maintain an almost constant total amount of energy without introducing huge mechanical work (see Fig. 3.6).

The anti-phase coordination pattern between kinetic and potential energies favors transfers between these two energies. This phenomenon can be observed in an inverted pendulum, which is a very good approximation of the human body during a step, especially during casual walking [13]. This energy transfer is less efficient when the gait parameters diverge from those observed during casual walking [17, 18].

Fig. 3.5 Muscle excitation patterns for walking: group average EMG linear envelopes for GMAX (gluteus maximus, adductor magnus), GMED (anterior and posterior portion of gluteus medius), IL (iliacus, psoas), HAM (biceps femoris long head, medial hamstrings), VAS (three-component vasti), RF (rectus femoris), BFsh (biceps femoris short head), TA (tibialis anterior), GAS (medial and lateral gastrocnemius) and SOL (soleus), (adapted from [59])

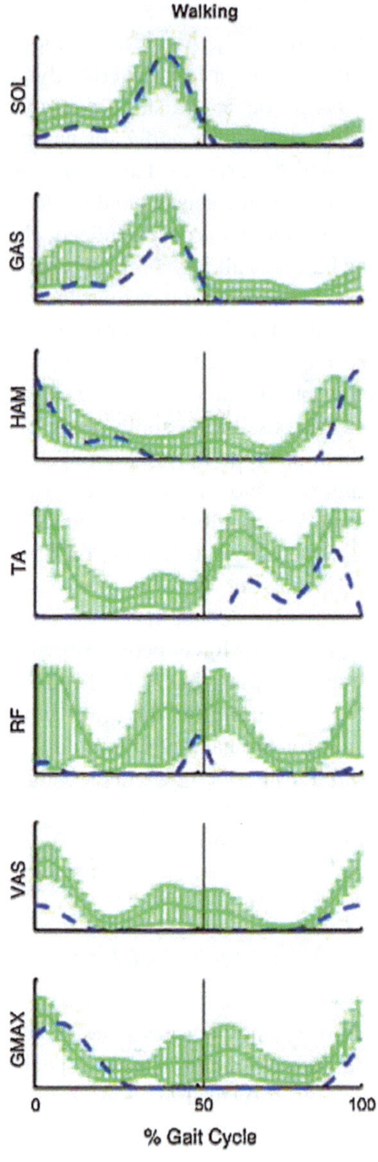

Energy expenditure is generally evaluated through the measurement of oxygen consumption. We could imagine using oxygen consumption in a virtual environment in order to evaluate the performance of the user's actions in navigation tasks. However, it involves wearing a mask, which can be too invasive for some users. Other devices such as heart rate monitors can be used to indirectly evaluate energy expenditure but they are very inaccurate unless some very restrictive protocols are used.

Fig. 3.6 Energy transfer between kinetic (*bold curve*) and potential energy (*dashed curve*) which demonstrate an anti-phase coordination (adapted from [13])

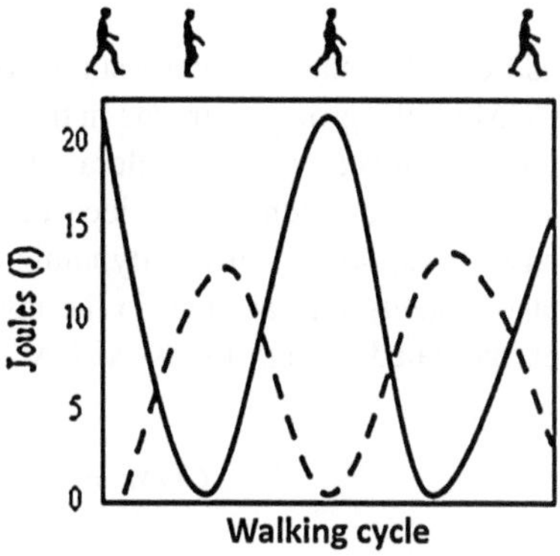

Another solution to indirectly evaluate energy expenditure without these types of devices consists of computing the mechanical internal work. Using the Angular Momentum Theorem and knowing joints kinematics it is possible to approximate the internal mechanical work W_{int} [86]:

$$W_{int} = \Delta\left[\sum_{i=1}^{n} \frac{1}{2} m_i \dot{G}_i^2 + \frac{1}{2} I_i \omega_i^2 + m_i g h_i\right]$$

where m_i is the mass of the ith segment, G_i its center of mass position, I_i its inertia, h_i its height and ω_i its angular velocity. g stands for gravity. The absolute values of the instantaneous work (computed at each time) is known to be well correlated with the energy expenditure measured using measures of oxygen consumption [86], although there exist some debate about this relation. It could thus be used in VR with motion capture devices to evaluate the difficult of or the performance on a navigation task.

3.3.3 Balance

In bipedal locomotion balance is a key point as the supported area on the ground is very small compared to the body surface. As stated above, energy expenditure is a key point in human walking because of transfers between kinetic and potential energy suggesting that walking could be viewed as a sequence of forward falls. Hence, it must have an impact on several multisensory inputs, especially on vestibular information. It has been shown that changes in walking conditions, such as walking on a treadmill, could affect stability and balance, compared to ground walking [29]. To maintain balance people increase step width and generally reduce step length for a given speed. Hence, balance status is used by some researchers to evaluate the naturalness of treadmill walking in immersive environments [61].

To address balance in walking, the human body is generally considered as an inverted pendulum with the contact point being the ankle joint and for which the total mass is placed on the COM. This problem of unstable balance is generally addressed in static or quasi-static conditions by assuming that the projection of the COM on the ground remains in the contact area. However, this assumption becomes false when velocity or accelerations cannot be neglected, such as in walking. Some authors have extended the idea of maintaining the projection of COM inside the base of support to such a dynamic situation [27] (see Fig. 3.7). The key idea is to take velocity into account in the inverted pendulum model. The resulting so-called extrapolated center of mass xCOM is thus defined by:

$$xCOM = COM + \frac{\dot{COM}}{\omega_0} \text{ with } \omega_0 = \sqrt{\frac{g}{l}}$$

where l stands for the length of the pendulum (the distance between the ankle joint and the COM).

This point has been used to demonstrate a simple control strategy to deal with balance while walking [28] and could thus be evaluated in immersive environments. This control strategy states that the COP should be placed at a certain distance behind and outward of the xCOM when foot-strike occurs. Hence, a disturbance in COM velocity change Δv can be compensated by a change in COP position equal to $\Delta v / \omega_0$. This control strategy is also available when starting, stopping, or turning. This point is now widely used in biomechanics and neurosciences, but it has not been used in VR protocols to analyze the balance status of users who have to navigate in virtual environments.

However, understanding dynamic balance in walking is still a challenge. Indeed, the inverted-pendulum model which globally can generally explain energy exchanges in walking (as explained in the previous section), could demonstrate forward falling phases while walking, which naturally leads to stepping forward to prevent the user from falling. Falling forward would also be a good representation of what is happening in step initiation. This idea has been reused recently in VR. The "Joyman"

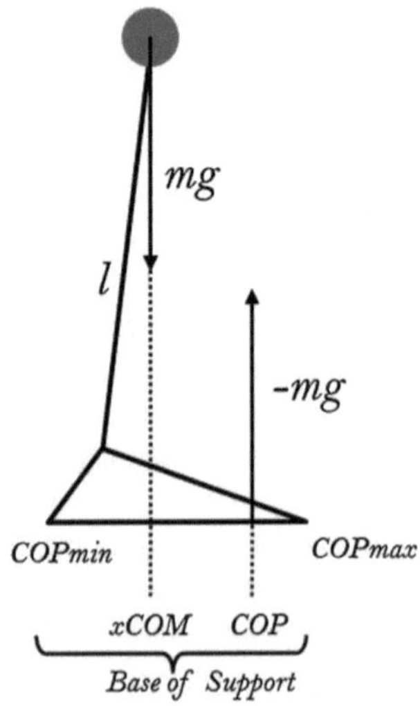

Fig. 3.7 Inverted pendulum model in static condition, with length l (distance between the ankle joint and the COM), mass m. COPmin and COP-max stand for the minimal and maximal limits that the COP can take to remain in the base of support. In the static condition, xCOM is simply the projection of the COM on the ground. Adapted from [27]

device [41, 52] is based on the idea that the user has to lean his whole body to indicate direction and speed of the navigation task in virtual environments. The key idea is to naturally stimulate the vestibular system by simply leaning the body in order to maintain the perception-action coupling that occurs when walking.

3.4 Comparison Between Ground and Treadmill Walking

One of the most popular methods used to walk through virtual environments is to use treadmills [32, 67]. It physically engages the user in a very natural manner and the proprioceptive and sound feedbacks are naturally generated through walking. However, it is important to check if treadmill walking is really similar to ground walking to ensure that users are engaged in a natural walking experience in VR.
Some authors [66] suggest that walking on the CyberWalk, a large omnidirectional treadmill, is very close to normal walking, especially after some initial familiarization. This statement was mainly based on comparing kinematic gait parameters, but it would be interesting to evaluate to what extent this statement is true from the dynamic point of view and if this difference in ground reaction force could affect presence in navigation tasks. Some authors have demonstrated that even if kinematic gait parameters could become similar in ground and treadmill walking after some training, the joint torques and muscular activity remains different [38].

The main difference between ground and treadmill walking is that some variations of the belt speed occur in treadmill walking (some researchers reported a 5 % change in treadmill speed within a walking cycle [58]). These variations seem to be correlated with the subject's mass, gait speed, and the intrinsic mechanical properties of the treadmill itself. The general consensus in the literature is thattreadmill walking leads to higher step frequency and lower step length [14, 80]. It seems that these variations lead to an increase in variability during walking [60] and a decrease of the casual speed compared to ground walking [7]. Globally, variations of the treadmill speed lead to instability with a longer stance phase compared to ground walking [14, 44]. Perception of walking speed is also affected and vision is globally perturbed [81]. Recent studies have confirmed that one's perception of speed is influenced by the treadmill, such that individuals were unable to match their corresponding self-selected ground running speed [36]. The unmatched perception of speed is likely due to the distortion of normal visual inputs resulting from the discrepancy between observed and expected optic flow [36]. Some authors have artificially induced greater step variability through visual perturbations from a VR display [47]. Perturbations generally induced greater variability in both step width and (to a lesser degree) step length.

However, it seems that most of the reported differences decrease with some training period. Four to six minutes of training seem to significantly decrease kinematic differences between ground and treadmill walking [68], but the adaptation period differs from one user to another. For example, elderly people have more difficulty adapting to treadmill walking compared to younger adults [61, 82].

However, there is no real consensus about differences between ground and treadmill walking. When walking in virtual environments it seems positive to offer natural interfaces and multisensory feedback, although it is not always possible to physically walk in as large an environment as displayed in the virtual environment. Hence, all of the reported differences may lead to perceptual perturbations that could affect presence. Durgin et al. [20] found that perceived speed in a richly structured near environment differ by only about 10 % in treadmill compared to wide-areawalking. They also found that trial-to-trial variations in step frequency predicted changes in perceived locomotor speed. It has been reported that traveled distance is under-estimated in virtual environments. However, some authors were able to separate the component of real walking from other sources that possibly affect distance estimation [54].

As belt speed variation seems to be a key point to explain these differences, some authors proposed an innovative treadmill speed controller that compensates for possible perturbations in real-time [66]. It seems that this kind of controller does not disturb immersiveness, but the results are very sensitive to the gains of the belt speed controller.

To summarize, it seems that treadmill walking may lead to similar kinematic data to ground walking. This is especially true for treadmills that offer accurate belt speed control and when users are trained to use treadmills (4–6 min training should be enough to reach this objective). However, perceptual studies seem to demonstrate that treadmill walking affects distance and speed estimation that could lead to instability. This is particularly true in immersive environments where other authors reported that distance evaluation is affected in any type of immersive application. One has to notice that kinematic data in treadmill walking may result in values similar to those in ground walking, but dynamic parameters seem to remain different (especially joint torques and muscle activation patterns). Hence, proprioception in both situations may be different. This problem is still open and further studies will be necessary to clearly understand the real advantages and limitations of using treadmills in VR.

3.5 Conclusion

In virtual environments we potentially know everything about the environment and it is thus possible to imagine infinite possibilities to navigate. However, if we wish to improve the quality of immersion it is important to notice how people pick-up information and select the most appropriate action. This perception-action coupling seems to be very relevant to allow for realistic navigation in VR. One must acknowledge that any additional cognitive load could significantly change things for the user; using metaphors may add cognitive load that may affect gait quality.

Hence, biomechanical knowledge reported in this chapter could be used in three main directions. Firstly, it could help to measure the user's navigation wishes using few biomechanical signals and more natural metaphors. Treadmills have been widely studied to directly measure speed and direction, but could lead to some gait perturbations that could affect many dynamic parameters including balance, ground reaction force and muscle activation. Walk-in-place metaphors and adaptations could be tuned according to biomechanical knowledge in order to enrich the variability and commands when navigating in immersive environments. Data reported in this chapter provide information for designing navigation systems that could adapt to gender, age, size, and direction. Some authors have tried to implement such knowledge in their controllers but it could be extended to more complex variations. Other types of gait parameters have not been used yet such as EMG signals. Brain computer interfaces use EEG signals coming from the brain to drive simulations in VR, but we could imagine combining EEG and EMG to enhance the performance of the classifiers in order to navigate in virtual environments. Such type of work has been briefly addressed in biomechanics but might be relevant for VR applications.

Secondly, it can help to introduce multisensory feedback such as vibrations, sounds, camera motion, etc., which should increase the naturalness of the navigation task, as described in some recent papers. Adapting sound and vibrators' oscillations to step length, frequency and weight, and to external parameters such as the type of ground, is a promising issue. It has been explored in some recent papers and seems to improve the quality of the navigation task in VR. In the same way, modifying camera motions to avoid linear displacements which are not natural seems to be well appreciated by users. Thanks to biomechanical knowledge it is possible to adapt camera motions for many different parameters. For example, it would be possible to recognize female and male camera motions. Indeed, it is well known that only a few kinematic inputs enable people to recognize male and female gaits. Is it true for camera motion?

Thirdly, it provides us with validation criterions that could help designers and scientists evaluate the naturalness of navigation tasks in VR. Indeed, when navigating in virtual environments the constraints of the real environment may disappear such as trying to minimize metabolic energy, tiredness, jerk, etc. Moreover, interfaces, such as using joysticks or pads, may lead to some binary commands. The resulting trajectory could be composed of straight lines with few redirections which is very different from real trajectories. Biomechanical knowledge reported in this chapter and the related papers could help designers and scientists design a cost metrics function to evaluate the performance of various navigation systems without requiring huge experiments in real navigation tasks for comparison to their system.

References

1. Alexander RM (1986) Optimization and gaits in the locomotion of vertebrates. Physiol Rev 69:1199–1227
2. Alexander RM (2002) Energetics and optimization of human walking and running: the 2000 Raymond Pearl memorial lecture. Am J Hum Biol 14:641–648
3. Alexander RM (2003) Principles of animal locomotion. Princeton University Press, Princeton, p 384
4. Alton F, Baldey L, Caplan S, Morrissey MC (1998) A kinematic comparison of overground and treadmill walking. Clin Biomech 13:434–440
5. Andriacchi T, Ogle J, Galante J (1977) Walking speed as a basis for normal and abnormal gait measurements. J Biomech 10(4):261–268

6. Auvinet B, Berrut G, Touzard C, Moutel L, Collet N, Chaleil D, Barrey E (2002) Reference data for normal subjects obtained with an accelerometric device. Gait Posture 16(2):124–134
7. Bayat R, Barbeau H, Lamontagne A (2005) Speed and temporal distance adaptations during treadmill and overground walking following stroke. Neurorehabil Neural Repair 19(2):115–124
8. Bertram JE, Ruina A (2001) Multiple walking speed-frequency relations are predicted by constrained optimization. J Theor Biol 209:445–453
9. Bertram JE (2005) Constrained optimization in human walking: cost minimization and gait plasticity. J Exp Biol 208:979–991
10. Bianchi L, Angelini D, Orani GP, Lacquaniti F (1998) Kinematic coordination in human gait: relation to mechanical energy cost. J Neurophysiol 79:2155–2170
11. Borghese NA, Bianchi L, Lacquaniti F (1996) Kinematic determinants of human locomotion. J Physiol 494(3):863–879
12. Boulic R, Magnenat-Thalmann N, Thalmann D (1990) A global human walking model with real-time kinematic personication. Vis Comput 6(6):344–358
13. Cavagna GA, Kaneko M (1977) Mechanical work and efficiency in level walking and running. J Physiol 268:467–481
14. Chatterley F, Chockalingam N, Greenhalgh A (2007) Comparison of pelvic complex kinematics during normal locomotion in treadmill vs. overground walking. J Biomech 40(S2):S510–S511
15. Chiu MC, Wang MJ (2007) The effect of gait speed and gender on perceived exertion, muscle activity, joint motion of lower extremity, ground reaction force and heart rate during normal walking. Gait Posture 25:385–392
16. Chung M, Wang MJ (2010) The change of gait parameters during walking at different percentage of preferred walking speed for healthy adults aged 20–60 years. Gait Posture 31:131–135
17. Cotes JE, Meade F (1960) The energy expenditure and mechanical energy demand in walking. Ergonomics 3:97–119
18. Dill DB (1965) Oxygen used in horizontal and grade walking and running on the treadmill. J Appl Physiol 20:19–22
19. Ding Q, Zhao X, Xiong A, Han J (2011) A novel motion estimate method of human joint with EMG-Driven model. In: International conference on bioinformatics and biomedical engineering (iCBBE), pp 1–5
20. Durgin FH, Reed C, Tigue C (2007) Step frequency and perceived self-motion. ACM Trans Appl Percept 4(1):article 5
21. Feasel J, Whitton MC, Wendt J (2008) Llcm-wip: low-latency, continuous-motion walking-in-place. In: Proceedings of the IEEE symposium on 3D user interfaces, pp 97–104
22. Flash T, Hogan N (1985) The coordination of arm movements: an experimentally confirmed mathematical model. J Neurosci 5(7):1688–1703
23. Glaister B, Bernatz G, Klute G, Orendurff M (2007) Video task analysis of turning during activities of daily living. Gait Posture 25(2):289–294
24. Hase K, Stein R (1999) Turning strategies during human walking. J Neurophysiol 81(6):2914–2922
25. Hicheur H, Berthoz A (2005) How do humans turn? Head and body movements for the steering of locomotion. In: Proceedings of the 5th IEEE-RAS international conference on humanoid robots, pp 265–270
26. Hicheur H, Pham Q, Arechavaleta G, Laumond J, Berthoz A (2007) The formation of trajectories during goal-oriented locomotion in humans. I. A stereotyped behaviour. Eur J Neurosci 26(8):2376–2390
27. Hof AL, Gazendam MG, Sinke WE (2005) The condition for dynamic stability. J Biomech 38:1–8
28. Hof AL (2008) The 'extrapolated center of mass' concept suggests a simple control of balance in walking. Hum Mov Sci 27:112–125
29. Hollman JH Does walking in a virtual environment induce unstable gait? Gait Posture 26:289–294
30. Huxham F, Gong J, Baker R, Morris M, Iansek R (2006) Defining spatial parameters for non-linear walking. Gait Posture 23(2):159–163

31. Inman VT, Ralston HJ, Todd F (1981) Human walking. Williams & Wilkins, Baltimore
32. Iwata H, Yoshida Y (1999) Path reproduction tests using a torus treadmill. Presence Teleoper Virt Environ 8(6):587–597
33. Jonkers I, Spaepen A, Papaioannou G, Stewart C (2002) An EMG-based, muscle driven forward simulation of single support phase of gait. J Biomech 35(5):609–619
34. Kirtley C, Whittle MW, Jefferson RJ (1985) Influence of walking speed on gait parameters. J Biomed Eng 7(4):282–288
35. Kito T, Yoneda T (2006) Dominance of gait cycle duration in casual walking. Hum Mov Sci 25:383–392
36. Kong PW, Koh TMC, Tan WCR, Wang YS (2012) Unmatched perception of speed when running overground and on a treadmill. Gait Posture 36(1):46–48
37. Lécuyer A, Marchal M, Hamelin A, Wolinski D, Fontana F, Civolani M, Papetti S, Serafin S (2011) Shoes-your-style: changing sound of footsteps to create new walking experiences. In: Proceedings of workshop on sound and music computing for human-computer interaction (CHItaly), Alghero, Italy, 13–16 Sept 2011
38. Lee SJ, Hidler J (2007) Biomechanics of overground versus treadmill walking in healthy individuals. J Appl Physiol 104:01380
39. Lelas JL, Merriman GJ, Riley PO, Kerrigan DC (2003) Predicting peak kinematic and kinetic parameters from gait speed. Gait Posture 17:106–112
40. Lotte F, Van Langhenhove A, Lamarche F, Ernest R, Renard Y, Arnaldi B, Lécuyer A (2010) Exploring large virtual environments by thoughts using a brain-computer interface based on motor imagery and high-level commands. Presence Teleoper Virtual Env 19:1 (MIT Press, Cambridge)
41. Marchal M, Pettré J, Lécuyer A (2011) Joyman: a human-scale joystick for navigating in virtual worlds. In: Proceedings of IEEE symposium on 3D user interface (3DUI'11), Singapour, 19–20 March 2011, pp 19–26
42. Masani K, Kouzaki M, Fukunaga T (2002) Variability of ground reaction forces during treadmill walking. J Appl Physiol 92:1885–1890
43. Milner M, Dall D, McConnel VA, Brennan PK, Hershler C (1973) Angle diagrams in the assessment of locomotor function. S.A Med J 47:951–957
44. Murray MP, Spurr GB, Sepic SB, Gardner GM, Mollinger LA (1985) Treadmill vs. foor walking: kinematics, electromyogram, and heart rate. J Appl Physiol 59(1):87–91
45. Nagasaki H, Ito H, Hashizurne K, Furuna T, Maruyama H, Kinugasa T (1996) Walking patterns and finger rhythm of older adults. Percept Mot Skills 82:435–447
46. Nilsson J, Thorstensson A, Halbertsam J (1985) Changes in leg movements and muscle activity with speed of locomotion and mode of progression in humans. Acta Physiol Scand 123(4):457–475
47. O'Connor SM, Xub HZ, Kuo AD (2012) Energetic cost of walking with increased step variability. Gait Posture 36(1):102–107
48. Orendurff MS, Segal AD, Berge JS, Flick KC, Spanier D, Klute GK (2006) The kinematics and kinetics of turning: limb asymmetries associated with walking a circular path. Gait Posture 23:106–111
49. Papetti S, Fontana F, Civolani M, Berrezag A, Hayward V (2010) Audio-tactile display of ground properties using interactive shoes. In: Proceedings of the haptic and audio interaction design, pp 117–128
50. Patla A, Prentice S, Rietdyk S, Allard F, Martin C (1999) What guides the selection of alternate foot placement during locomotion in humans. Exp Brain Res 128:441–450
51. Pearsall DJ, Costigan PA (1999) The effect of segment parameter error on gait analysis results. Gait Posture 9:173–183
52. Pettré J, Marchal M, Siret O, de la Rivière J-B, Lécuyer A (2011) Joyman: an immersive and entertaining interface for virtual locomotion. ACM SIGGRAPH Asia Emerging Technologies, Hong-Kong
53. Pham Q, Hicheur H, Arechavaleta G, Laumond J, Berthoz A (2007) The formation of trajectories during goal-oriented locomotion in humans. II. A maximum smoothness model. Eur J Neurosci 26(8):2391–2403

54. Popp MM, Platzer E, Eichner M, Schade M (2004) Walking with and without walking: perception of distance in large-scale urban areas in reality and in virtual reality. Presence Teleoper Virtual Env 13(1): 61–76
55. Pozzo T, Berthoz A, Lefort L (1990) Head stabilization during various tasks in humans. I. Normal subjects. Exp Brain Res 82(1):97–106
56. Rao G, Amarantini D, Berton E, Favier D (2006) Infuence of body segments' parameters estimation models on inverse dynamics solutions during gait. J Biomech 39:1531–1536
57. Richardson M, Flash T (2002) Comparing smooth arm movements with the two-thirds power law and the related segmented-control hypothesis. J Neurosci 22(18):8201–8211
58. Riley PO, Paolini G, Croce UD, Paylo KW, Kerrigan DC (2007) A kinematic and kinetic comparison of overground and treadmill walking in healthy subjects. Gait Posture 26(1):17–24
59. Sasaki K, Neptune RR (2005) Differences in muscle function during walking and running at the same speed. J Biomech 39(11):2005–2013
60. Savelberg HHCM, Vorstenbosch MATM, Kamman EH, van de Weijer JGW, Schambardt HC (1998) Intra-stride belt-speed variation affects treadmill locomotion. Gait Posture 7(1):26–34
61. Schellenbach M, Lövdén M, Verrel J, Krüger A, Lindenberger U (2010) Adult age differences in familiarization to treadmill walking within virtual environments. Gait Posture 31:295–299
62. Sekiya N, Nagasaki H, Ito H, Furuna T (1996) The invariant relationship between step length and step rate during free walking. J Hum Mov Stud 30:241–257
63. Sekiya N, Nagasaki H, Ito H, Furuna T (1997) Optimal walking in terms of variability in step-length. J Orthop Sports Phys Ther 26:266–272
64. Sekiya N, Nagasaki H (1998) Reproducibility of the walking patterns of normal young adults: test-retest reliability of the walk ratio (step-length: step-rate). Gait Posture 7:225–227
65. Slater M, Usohg M, Steed A (1995) Taking steps: the influence of a walking technique on presence in virtual reality. ACM Trans Comput Hum Interact 2(3):201–219
66. Souman JL, Giordano PR, Frissen I, De Luca A, Ernst MO (2010) Making virtual walking real: perceptual evaluation of a new treadmill control algorithm. ACM Trans Appl Percept 7(2):Article11
67. Stanney KM (2002) Handbook of virtual environments: design, implementation, and applications. Lawrence Erlbaum Associates, Mahwah
68. Taylor N, Evans O, Goldie P (1996) Angular movements of the lumbar spine and pelvis can be reliably measured after 4 minutes of treadmill walking. Clin Biomech 11(8):484–486
69. Terrier P, Schutz Y (2003) Variability of gait patterns during unconstrained walking assessed by satellite positioning (GPS). Eur J Appl Physiol 90(5):554–561
70. Terziman L, Lécuyer A, Hillaire S, Wiener J (2009) Can camera motions improve the perception of traveled distance in virtual environments? In: IEEE international conference on virtual reality (IEEE VR), Lafayette, US
71. Terziman L, Marchal M, Emily M, Multon F, Arnaldi B, Lecuyer A (2010) Shake-your-head: revisiting walking-in-place for desktop virtual reality. In: ACM symposium on virtual reality software and technology (ACM VRST), Hong-Kong, China
72. Terziman L, Marchal M, Multon F, Arnaldi B, Lécuyer A (2011) Comparing virtual trajectories made in slalom using walking-in-place and joystick techniques. In: Proceedings of joint virtual reality conference (Joint eurographics symposium on virtual environments-Euro VR conference, JVRC), Nottingham UK, 20–21 Sept 2011
73. Terziman L, Marchal M, Multon F, Arnaldi B, Lécuyer A (2012) The king-kong effects: improving sensation of walking in VR with visual and tactile vibrations at each step. In: Proceedings of the IEEE symposium on 3D user interfaces 2012
74. van Emmerik R, Wagenaar R (1996) Effects of walking velocity on relative phase dynamics in the trunk in human walking. J Biomech 29(9):1175–1184
75. Vaughan KR (1984) Biomechanics of running gait. Crit Rev Eng 12(1):1–48
76. Vaughan CL, Davis BL, O'Connor J (1992) Dynamics of human gait. Human Kinetics Publishers, Champaign
77. Vieilledent S, Kerlirzin Y, Dalbera S, Berthoz A (2001) Relationship between velocity and curvature of a human locomotor trajectory. Neurosci Lett 305(1):65–69

78. Visell Y, Law A, Ip J, Smith S, Cooperstock J (2010) (2010) Interaction capture in immersive virtual environments via an intelligent floor surface. In: Proceedings of IEEE virtual reality, pp 313–314
79. Viviani P, Terzuolo C (1982) Trajectory determines movement dynamics. Neuroscience 7:431–437
80. Warabi T, Kato M, Kiriyama K, Yoshida T, Kobayashi N (2005) Treadmill walking and overground walking of human subjects compared by recording sole-floor reaction force. Neurosci Res 53(3):343–348
81. Warren WHJ, Kay BA, Zosh WD, Duchon AP, Sahuc S (2001) Optic flow is used to control human walking. Nat Neurosci 4(2):213–216
82. Wass E, Taylor NF, Matsas A (2005) Familiarisation to treadmill walking in unimpaired older people. Gait Posture 21(1):72–79
83. Wendt J, Whitton MC, Brooks F (2010) Gud wip: gait-understanding-driven walking-in-place. In: Proceedings of IEEE virtual reality, pp 51–58
84. White SC, Yack HJ, Tucker CA, Lin HY (1998) Comparison of vertical ground reaction forces during overground and treadmill walking. Med Sci Sports Exercise 30:1537–1542
85. Whittle M (1991) Gait Analysis: an introduction. Butterworth-Heinemann, Oxford
86. Winter DA (1979) A new definition of mechanical work done in human movement. J Appl Physiol Respirat Environ Exercise Physiol 46(1):79–83
87. Winter DA (1991) The biomechanics and motor control of human gait: normal, elderly and pathological. University of Waterloo press, Ontario
88. Zijlstra W, Rutgers AWF, Hof AL, Van Weerden TW (1995) Voluntary and involuntary adaptation of walking to temporal and spatial constraints. Gait Posture 3:13–18

Chapter 4
Affordance Perception and the Visual Control of Locomotion

Brett R. Fajen

Abstract When people navigate through complex, dynamic environments, they select actions and guide locomotion in ways that take into account not only the environment but also their body dimensions and locomotor capabilities. For example, when stepping off a curb, a pedestrian may need to decide whether to go now ahead of an approaching vehicle or wait until it passes. Similarly, a child playing a game of tag may need to decide whether to go to the left or right around a stationary obstacle to intercept another player. In such situations, the possible actions (i.e., affordances) are partly determined by the person's body dimensions and locomotor capabilities. From an ecological perspective, the ability to take these factors into account begins with the perception of affordances. The aim of this chapter is to review recent theoretical developments and empirical research on affordance perception and its role in the visual control of locomotion, including basic locomotor tasks such as avoiding stationary and moving obstacles, walking to targets, and selecting routes through complex scenes. The focus will be on studies conducted in virtual environments, which have created new and exciting opportunities to investigate how people perceive affordances, guide locomotion, and adapt to changes in body dimensions and locomotor capabilities.

4.1 Introduction

To successfully interact with one's environment (real or virtual), it is necessary to select actions and guide movements in ways that take one's body dimensions and movement capabilities into account. To illustrate this point, consider what would happen if the designer of a virtual environment (VE) were to dramatically alter the dimensions or dynamics of the user's virtual body. In a VE, one's virtual body could

B. R. Fajen (✉)
Department of Cognitive Science, Rensselaer Polytechnic Institute,
Troy, NY, USA
e-mail: fajenb@rpi.edu

F. Steinicke et al. (eds.), *Human Walking in Virtual Environments*,
DOI: 10.1007/978-1-4419-8432-6_4,

be made taller, shorter, leaner, or stockier; one's arms or legs could be stretched or compressed; one could be made faster, slower, stronger, or weaker. Before the user has a chance to adapt, he or she may attempt actions that have no chance of success, pass up opportunities to perform actions that would lead to beneficial outcomes, follow suboptimal routes, or inadvertently collide with objects in the virtual environment.

Such behavior can be observed in real environments as well. Infants, upon first learning to walk, have difficulty gauging their actions to their new movement capabilities [1]. As such, they often attempt to descend sloped surfaces that are impossibly steep or cross gaps that are impossibly wide. Similarly, when older children ride bicycles, they have difficulty taking into account how long it takes to initiate movement [27]. This puts them at greater risk when crossing busy streets because it leaves them less time to reach the other side before approaching cars arrive. In both cases, one's ability to select appropriate actions is impaired by changes that affect one's action capabilities.

In most situations, however, people are remarkably good at choosing actions that are appropriate given their body dimensions and movement capabilities (see [9] for a review). They know whether an aperture between two stationary obstacles is wide enough to permit safe passage [35], whether a gap in the ground surface is small enough to jump [22], and whether to pass in front of or behind a moving obstacle [6].

The primary aim of this chapter is to consider how people take their body dimensions and movement capabilities into account when interacting with real and virtual environments. The specific focus will be on attempts, motivated largely by the ecological approach to perception and action [14], to explain how people take their body dimensions and dynamics into account without appealing to internal models of the body. I will begin by reviewing some classic work on the perception of affordances (i.e., possibilities for action), their specification by eyeheight-scaled visual information, and why affordance perception offers a starting point for addressing this problem. I will then present more recent research aimed at generalizing the affordance-based approach to account for behavior in a wider variety of circumstances.

The second aim of this chapter is to show how VEs have been used to study affordance perception and the visual control of locomotion. Much of the research presented in this chapter takes advantage of VEs and could not be conducted in the real world. In short, VEs have created new and exciting opportunities to investigate the way in which people perceive affordances, guide locomotion, and adapt to changes in their body dimensions and locomotor capabilities.

4.2 Taking Body Dimensions and Movement Capabilities into Account

4.2.1 Theoretical Approach

How do people choose actions and guide locomotion in a way that takes their body dimensions and locomotor capabilities into account? By some accounts, the starting point is to assume some form of knowledge of the dimensions and dynamics of the body [20]. For proponents of the computational approach to motor control, it is axiomatic that the brain builds and exploits internal models of the body and environment [30]. Within this framework, there is an abundance of studies aimed at understanding how the motor system makes use of detailed internal models of the dimensions of the body [15], the dynamics of the body [29], and motor variability [5, 32].

In this chapter, I will consider an alternative approach—one that is inspired by the ecological approach to perception and action [14]. Ecological psychologists have long appreciated the importance of the "fit" between the organism (its dimensions and movement capabilities) and the environment. Indeed, the idea that the environment is perceived in relation to the body is one of the conceptual pillars of the ecological approach. One of the goals of this chapter is to consider how people take their body dimensions and dynamics into account by appealing to what is provided for free from the environment in the form of perceptual information.

4.2.2 *Affordance Perception and the Control of Locomotion*

From an ecological perspective, the ability to take one's body dimensions and movement capabilities into account begins with the perception of affordances—that is, possibilities for action provided by the environment for an animal with particular dimensions and capabilities [14, 33]. Affordances differ from conventional spatial properties in that they reflect the fit between the animal and its environment. For example, the passability of a gap between a tree and a boulder depends not only on the physical size of the gap but also on the actor's body dimensions. Similarly, the leapability of a stream depends not only on the width of the stream but also on the actor's leaping capabilities. Because affordances reflect the fit between the actor and the environment, the perception of affordances makes it possible to choose actions that are appropriately gauged to one's body dimensions and dynamics.

There is a vast amount of empirical research assessing the accuracy with which affordances can be perceived. A review of this literature is beyond the scope of this chapter (but see [9] for a recent review). Instead, the focus will be on how affordances are perceived, the available information that makes it possible to perceive affordances, and the role of affordance perception in the control of locomotion.

4.2.3 *Eyeheight-Scaled Information*

Although the main focus of this chapter is on recent developments in the study of affordance perception, it is worth taking a brief detour to consider a classic study by Warren and Whang [35] on the perception of passability. This study is a good place to start because it offers an instructive example of the ecological approach to the problem of how people choose actions that are appropriately gauged to their body dimensions. In other words, this study provides an example of the kind of solution that one might seek in attempting to understand the more general problem of how people take their body dimensions and movement capabilities into account.

Among the most commonly encountered potential impediments to forward locomotion is a narrow opening (or aperture) between obstacles, such as a doorway or the space between a stationary object and a wall. To select safe and efficient routes through environments, one must be able to perceive whether such apertures are sufficiently wide to allow safe passage. Further, it must be possible to perceive passability in advance—for trial-and-error is neither a safe nor efficient option. Of course, whether or not an aperture is passable depends not only on the size of the aperture but also on the size of the observer's body. Therefore, the decision about whether to pass through or circumvent the aperture must be made in a way that takes into account the size of the aperture in relation to the size of the body.

The availability of *eyeheight-scaled information* [28] offers a potential solution to this problem. As illustrated in Fig. 4.1, the width of the aperture (G) is optically specified in units of eyeheight (E) by $[2\tan(\alpha/2)]/\tan\gamma$, where α is the angle subtended by the inside edges of the obstacle and γ is the angle of declination of the base of the obstacle (see [35] for derivation). The fact that aperture width is specified in units of eyeheight is important because it means that dimensions of the environment are specified in the same units as dimensions of the body. Further, because body width (W) is a fixed proportion of standing eyeheight, such information also specifies aperture width in relation to body width, which is sufficient for perceiving the passability of the aperture. Consistent with the hypothesis that passability is perceived on the basis of eyeheight-scaled information, subtle increases in the height of the ground surface beneath the obstacles make apertures appear more passable [35].

Aperture width is not the only dimension that is specified by eyeheight-scaled information. The horizontal and vertical dimensions of any visible surface that is

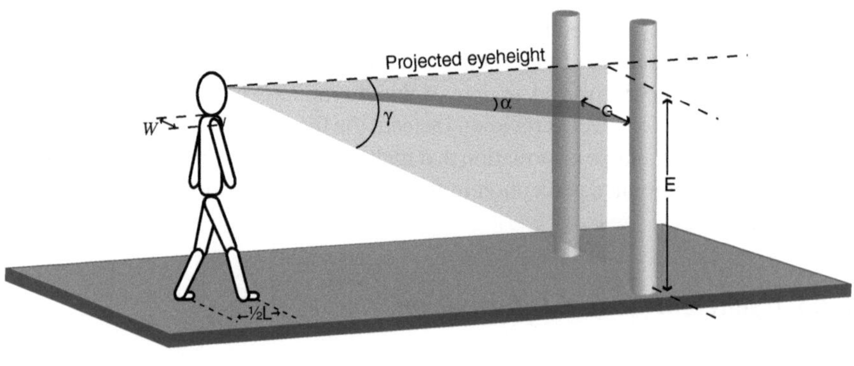

Fig. 4.1 Optical specification of aperture size by eyeheight-scaled information. α is the angle subtended by the inside edges of the obstacle, γ is the angle of declination of the base of the obstacle, G is the size of the gap, E, W, and L are the observer's eyeheight, body width, and stride length, respectively

perpendicular to the line of sight and resting on the ground plane are specified in intrinsic units by eyeheight-scaled information. Thus, affordances such as the climbability and sit-on-ability of horizontal surfaces can be directly perceived using eyeheight-scaled information [21, 37]. Distances along the ground plane are also optically specified in units of eyeheight [28].

Nonetheless, eyeheight-scaled information alone is not sufficient for a general affordance-based account of visually guided locomotion. There are other important aspects of locomotion that require people to take their body dimensions and action capabilities into account but cannot be attributed to the use of eyeheight-scaled information alone.

First, eyeheight-scaled information cannot be the only means by which dimensions of the environment are perceived in relation to dimensions of the body. For such information to reliably specify the size of an object, that object must be resting on the ground plane. Although objects are not normally suspended in the air, it is not uncommon for parts of objects to extend outward from their base of support. Thus, the size of an aperture formed by an overhanging tree branch or a kitchen countertop that juts out into an opening cannot be perceived on the basis of eyeheight-scaled information. Moreover, when the environment contains slopes, stairs, or tiers of level ground surfaces, relying on eyeheight-scaled information can lead to biases in perceived size [35, 37]. Are there sources of information other than eyeheight-scaled information that specify dimensions of the environment in relation to dimensions of the body? Second, the affordances that I have discussed in this section are *body-scaled affordances* in that they reflect relations between dimensions of the environment and dimensions of the actor's body. There are also affordances that are defined by the actor's movement capabilities, which are referred to as *action-scaled affordances.* For example, if an aperture is bounded by a pair of moving obstacles that are converging toward each other to form a shrinking gap, then passability depends not only on body dimensions but also on locomotor capabilities, such as how quickly the person can move. Can we account for the perception of action-scaled affordances without invoking internal models of the dynamics of the body? Third, once an appropriate action is selected, reaching the goal often involves guiding movement based on continuously available information, which also requires one to take one's body dimensions and movement capabilities into account. The role of affordance perception in selecting actions has been widely investigated but its role in on-line control is less well understood. In the remainder of this chapter, I will discuss recent attempts to extend the theory of affordances to address the three aforementioned problems.

4.3 Perceiving Body-Scaled Affordances

Although the role of eyeheight-scaled information is well established, there are other sources of information that specify dimensions of objects, including those that are not resting on the ground surface. In other words, eyeheight is not the only yardstick by which properties of the world are optically specified in intrinsic units. In fact, such metrics need not be defined by dimensions of the body. As anticipated by Lee [19, 34], dimensions of the actor's movements provide other possible yardsticks. For example, legged locomotion is normally accompanied by side-to-side head movements (i.e., lateral head sway). As a person approaches an aperture, the size of the aperture is optically specified in units of lateral head sway amplitude (A) by a higher-order variable involving the optical expansion rates ($\dot{\phi}$) and drift rates ($\dot{\theta}$) of the obstacles on the left and right sides (Fig. 4.2; see also, [11] for the full derivation). In other words, just as there is information about aperture size in units of eyeheight [35], there is also information about aperture size in units of lateral head sway amplitude. To the degree that lateral head sway amplitude remains roughly constant during normal locomotion, its relation to body width is relatively stable. Therefore, just as the perceptual system can calibrate itself to the relation between eyeheight and body width, it can also calibrate itself to the relation between head sway amplitude and body width. Head-sway-scaled information constitutes an alternative source of information for perceiving aperture size in units of body width.

Another dimension of the observer's movement that can serve as a yardstick for scaling dimensions of the environment is stride length. In his 1980 paper, Lee showed that the size of an approached object, such as an aperture, is optically specified in units of stride length (L) by $\alpha\tau_\alpha/t_s$, where τ_α is the ratio of α to the first temporal derivative of α and t_s is stride duration (see Fig. 4.1 for definition of α).

Whereas eyeheight-scaled information is static (i.e., available even when the observer is stationary), head-sway-scaled and stride-length-scaled information are dynamic in that the observer must be moving. However, both dynamic sources of information specify aperture width in body-scaled units regardless of whether the

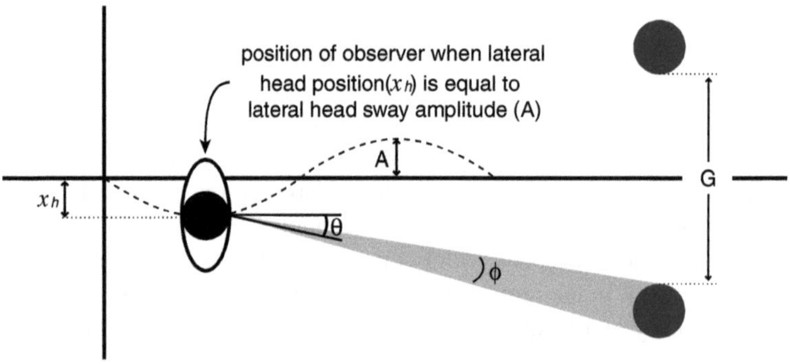

$$\frac{G}{A} \approx -\tfrac{1}{2}\left(\frac{\dot{\theta}_R(t)}{\dot{\phi}_R(t)} - \frac{\dot{\theta}_L(t)}{\dot{\phi}_L(t)} - 1\right) \Bigg/ \left(\frac{\dot{\theta}_R(t_{x_h=A})}{\dot{\phi}_R(t_{x_h=A})} + \frac{\dot{\theta}_L(t_{x_h=A})}{\dot{\phi}_L(t_{x_h=A})}\right)$$

Fig. 4.2 Optical specification of aperture size by stride-length-scaled information illustrated from a top-down view. As the observer on the *left* approaches the aperture between two obstacles (*dark gray circles*) on the *right*, the size of the gap (G) is optically specified in units of lateral head sway amplitude (A) by a higher-order variable involving the optical expansion rates ($\dot{\phi}$) and drift rates ($\dot{\theta}$) of the obstacles on the *left* (L) and *right* (R) sides base of the objects that define the gap is resting on the ground surface. Thus, when an aperture is bounded by objects that are not directly anchored to the ground surface, eyeheight-scaled information is not available but both stride-length-scaled and head-sway-scaled information are. Likewise, when an aperture is resting on a ground surface that is higher or lower than the ground surface upon which the observer is standing, eyeheight-scaled information is not reliable but both stride-length-scaled and head-sway-scaled information are.

The sufficiency of these two dynamic sources of information was recently tested in an experiment conducted in an immersive VE [11]. The VE was monocularly viewed through a head-mounted display. Subjects approached and walked through narrow openings between virtual obstacles, rotating their shoulders as necessary, while head and shoulder position and orientation were tracked. The task was performed in three VEs (Post, Tall Post, and Wall) that differed in terms of the availability of eyeheight-scaled, head-sway-scaled, and stride-length-scaled information.

In the Post condition, the aperture was an opening between a pair of cylindrical obstacles resting on a textured ground surface (see left column of Fig. 4.3). As subjects approached the obstacles, all three sources of information were available. In the Tall Post condition, the ground surface was absent and the cylindrical obstacles on either side of the aperture spanned the entire visual field from bottom to top (see middle column of Fig. 4.3). Both head-sway-scaled and stride-length-scaled information were available in the Tall Post condition but because the ground plane was absent and the cylinders had no visible base, eyeheight-scaled information was not available. In the Wall condition, the aperture was an opening between two untextured walls that spanned the visual field from bottom to top and from the edge of the aperture to the periphery (see right column of Fig. 4.3). The only source of information about aperture size that was available in this condition was stride-length-scaled information, which is based on the visual angle subtended by the aperture. Eyeheight-scaled information was unavailable because the ground surface was absent and head-sway-scaled information was unavailable because the walls were textureless and extended from left to right, making it impossible to detect the local optical expansion (i.e., $\dot{\phi}$ in Fig. 4.2).

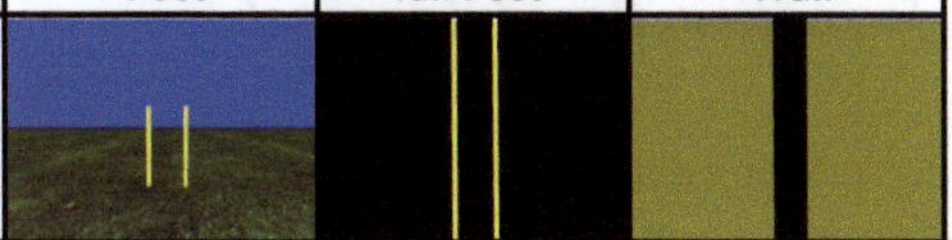

		Post	Tall Post	Wall
Information	EH	✓		
	HS	✓	✓	
	SL	✓	✓	✓

Fig. 4.3 Three VEs used in Fath and Fajen [11] and information available in each. EH, HS, and SL correspond to eyeheight-scaled, head-sway-scaled, and stride-length-scaled information, respectively

The main finding was that subjects performed the task successfully in all three conditions, with only a minor degradation in performance in the Wall condition. That is, subjects successfully timed and appropriately scaled the amplitude of shoulder rotation to fit through the aperture in all three conditions. These findings suggest that visual information other than eyeheight-scaled information can be used to guide locomotion through apertures. In particular, both stride-length-scaled and head-sway-amplitude-scaled information are sufficient to perceive aperture size in relation to shoulder width, supporting Lee's [19] assertions that dimensions of the world can be perceived in units related to dimensions of the actor's movements. Thus, for dimensions of the world that are not specified by eyeheight-scaled information, the availability of other body-scaled information makes it possible to perceive the environment in relation to one's body dimensions.

4.4 Perceiving Action-Scaled Affordances

The overwhelming majority of research on affordance perception has focused on the perception of affordances that are constrained by the observer's body dimensions. There is an entirely different class of affordances that are defined by one's movement capabilities—that is, *action-scaled affordances*. Most of the few studies on action-scaled affordances focus on the accuracy with which such affordances are perceived (e.g., [26]). Much less is known about the means by which such affordances are perceived and their specification by information.

Earlier in the chapter, I described how the availability of body-scaled information makes it possible to perceive affordances such as passability without knowledge of body dimensions. In this section, I will attempt to develop an analogous account of action-scaled affordances. That is, I will show how action-scaled affordances can be perceived based on information in optic flow without resorting to internal models of the body.

4.4.1 The Information-Based Approach

I will begin by considering a first attempt by ecological psychologists to explain how action-scaled affordances could be perceived using visual information alone without relying on knowledge of action capabilities. Because this approach relies on visual information alone, I will refer to it as the *information-based approach*. Consider a person walking toward a shrinking gap between obstacles such as an elevator with its doors beginning to close. Whether or not the shrinking gap is passable depends not only on the person's body dimensions but also his or her locomotor capabilities. In such situations, there exists visual information that specifies whether the person's current speed of locomotion is sufficient to safely pass through the gap before it closes [36]. That is, such information informs the person whether he or she will pass through the gap if current locomotor speed is maintained. Bootsma [3] referred to this as one's *current future* because it concerns one's future assuming that current conditions (e.g., current locomotor speed) are maintained. For the purposes of this discussion, the particular optical variable that specifies the sufficiency of one's current speed is less relevant than the fact that such information could be used to perceive whether the gap is passable. For example, if the person is moving fast enough to safely pass through the gap before it closes, then the aforementioned information specifies that the gap is passable (assuming current speed can be maintained). On the other hand, if the person is moving as quickly as possible but will not reach the gap before it closes, then the aforementioned information specifies that the gap is impassable.

A similar approach was proposed by Oudejans et al. [25] to explain how outfielders in baseball perceive whether a fly ball is catchable. In that case, the relevant information specifies whether the fielder's current running speed is sufficient to reach the landing location in time to catch the ball [4, 23, 24]. Thus, if an outfielder is running fast enough to reach the landing location in time to catch the ball and he or she can maintain that speed, such information can be used to perceive that the ball is catchable. Likewise, if the outfielder is running as fast as possible but will not reach the landing location in time, such information can be used to perceive that the ball is uncatchable.

For the present purposes, the important point is that the availability of information about the sufficiency of one's current state makes it possible to perceive action-scaled affordances such as passability and catchability based on visual information alone without having to rely on knowledge of one's running capabilities. The same would apply to other tasks for which there is information about one's current future, such as intercepting a moving target [10] and braking to avoid a collision [18, 38].

4.5 Testing the Information-Based Approach

An important corollary of this hypothesis is that affordances are specified only when the observer is actually moving. When the observer is stationary, the information specifies only that his or her current speed (which is zero if the observer is stationary) is not sufficient. The information does not specify whether the action is still within the person's capabilities once movement is initiated. This leads to the testable prediction that movement is necessary to perceive action-scaled affordances; that is, if action-scaled affordances are perceived on the basis of information about the sufficiency of one's current state, then because such information is available only when observers are moving, stationary observers should perceive such affordances less accurately [25].

This prediction was recently tested in a set of experiments conducted in an immersive VE [6]. The task was to judge whether it was possible to safely pass through a gap between a pair of converging obstacles before the gap closed. The initial distance to the point of convergence and the rate of closure varied across trials, yielding a range of conditions that varied from easily passable at slow walking speeds to impassable at fast walking speeds. In one condition, which I will refer to as the Move condition, subjects began walking from a designated home location at the same time that a pair of cylindrical obstacles began converging toward a point along their future path. The cylinders disappeared 1 s after they began moving and subjects were instructed to press one of two buttons on a handheld remote mouse to indicate whether they could have safely passed through the gap before it closed. In another condition, which I will refer to as the Delayed Move condition, subjects waited at the home location for 1 s after the cylinders began moving. After 1 s, an auditory "go" signal was presented instructing participants to begin walking. At the same time as the go signal, the cylinders disappeared and subjects were instructed to judge whether they could have safely passed through the gap before it closed.

This manipulation allowed [6] to test the prediction of the information-based hypothesis that subjects can accurately perceive action-scaled affordances only while moving. In the Move condition, subjects were allowed to move for 1 s while viewing the cylinders before making a judgment. In the Delayed Move condition, subjects were stationary for the entire 1 s during which the moving cylinders were visible and did not begin moving until the cylinders disappeared. Therefore, if the information-based hypothesis is correct, judgments about the passability of the gap should be accurate in the Move condition but not in the Delayed Move condition.

To measure the accuracy with which subjects perceived passability in the Move and Delayed Move conditions, judgments were compared with behavior on two other sets of trials in which the obstacles remained visible and subjects actually attempted to pass through the gap. That is, for both the Move and Delayed Move conditions, there was a corresponding set of trials in which the cylinder remained visible beyond 1 s and subjects attempted to walk through the gap. When judgments were compared to actual behavior, there was no evidence that subjects either overestimated or underestimated their ability to safely pass through the gap. That is, subjects tended to judge that the gap was passable in conditions in which they were able to pass through the gap and impassable in conditions in which they were unable to pass through the gap. Furthermore, this was true in both the Move and Delayed Move conditions. Thus, regardless of whether subjects made the judgment while moving or while stationary, they were able to accurately perceive passability. Such findings do not support the information-based account.[1]

To summarize, the information-based approach explains how action-scaled affordances, such as the passability of a shrinking gap, could be perceived by relying entirely on visual information without appealing to stored knowledge of movement capabilities. However, because the visual information upon which actors are thought to rely is only available "on the fly," the information-based approach predicts that action-scaled affordances cannot be accurately perceived while stationary [25].

[1] In an earlier study, Oudejans et al. [25] found that judgments were more accurate when subjects were allowed to move. However, Fajen et al. [6] pointed out several methodological problems that explain their findings and showed that once these problems were corrected, judgments were equally accurate regardless of whether subjects were stationary or moving.

The results of Fajen et al. [6] do not support this prediction and demonstrate that the range of conditions across which people can accurately perceive action-scaled affordances is broader than one would expect based on the information-based account. This should not come as a surprise, for it is not enough to know whether an action is within or beyond one's limits only when moving.

4.5.1 An Alternative Account

If people do not rely on information about their current future to perceive action-scaled affordances, then how are such affordances perceived? Does the fact that people reliably perceive such affordances even when stationary mean that they must rely on knowledge of their locomotor capabilities (i.e., an internal model)? In this section, I will introduce an alternative account that still bypasses the need for knowledge of locomotor capabilities but better accounts for the range of conditions across which action-scaled affordances are perceived. I will illustrate this approach using the same shrinking gap task that was described above.

A shrinking gap is passable if the minimum locomotor speed needed to safely pass through the gap ($\nu_{\min}$) is less than the actor's maximum possible locomotor speed. In terms of spatial variables, $\nu_{\min}$ is equal to the minimum distance that the observer must travel to pass between the obstacles divided by the amount of time remaining until the size of the gap is equal to the width of the observer's body; that is,

$$\nu_{\min}(t) = \left[z_m(t^*) - z_o(t)\right] / (t^* - t) \tag{4.1}$$

where z_m and z_o are the positions along the z-axis of the moving obstacle and the observer, respectively, and t^* is the time at which the size of the gap is equal to the width of the observer's body (see Fig. 4.4a). This is equivalent to:

$$\frac{\nu_{\min}(t)}{E/(t^* - t)} = \frac{[z_m(t) - z_o(t)]}{E} + \frac{\dot{z}_m}{E} \times TTC \times \left[1 - k\left(\frac{E}{g(t)}\right)\right] \tag{4.2}$$

where E is the observer's eyeheight, $\dot{z}_m$ is approach speed of the obstacle, *TTC* (time-to-contact) is the amount of time remaining until the obstacle reaches the locomotor axis, k is a constant equal to W/E, and g is the spatial gap between the inside edge of the obstacle and the z-axis (see Fig. 4.4a). As shown in Fig. 4.4b, each component of Eq. 4.2 (and therefore $\nu_{\min}$ itself) is optically specified. Therefore, one can perceive the minimum locomotor speed needed to safely pass through the shrinking gap. (See [7] for the full derivation of Eq. 4.2 and its optical specification).

There are four important points to be made about the information for $\nu_{\min}$ in Fig. 4.4. First, $\nu_{\min}$ is specified in a way that takes into account the physical sizes of the obstacles and the observer's body. Therefore, by detecting such information, the passability of the shrinking gap can be perceived in a way that takes these properties into account. Second, $\nu_{\min}$ is specified in units of $\left[E/(t^* - t)\right]$, which is the number of

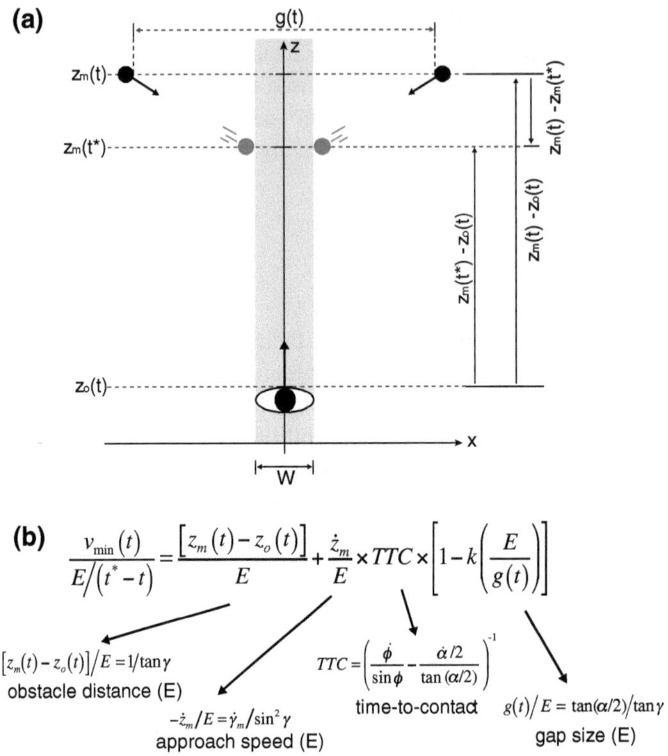

Fig. 4.4 **a** *Top-down* view of observer and a pair of converging obstacles at time t (*black circles*) and time t* (*gray circles*). t* is the time at which the size of the gap (g) between obstacles is equal to the observer's body width (W). **b** Optical specification of minimum walking speed ($\nu_{\min}$)

eyeheights that must be covered in the amount of time remaining until the obstacle reaches the locomotor path. Because such units are intrinsic rather than extrinsic, information about $\nu_{\min}$ can be calibrated, allowing one to perceive $\nu_{\min}$ in relation to maximum locomotor speed, in the same way that eyeheight-scaled information about aperture size can be calibrated to allow for the perception of aperture size in relation to body width [35]. Third, information about $\nu_{\min}$ is available regardless of whether the observer is stationary or moving. Therefore, unlike the information-based approach described in the previous section, this approach accounts for the fact that stationary and moving observers can perceive passability equally well [6]. Putting these first three points together, detecting and calibrating information about $\nu_{\min}$ allows for the direct perception of the passability of a shrinking gap by stationary and moving observers, taking into account both the width of the observer's body and his or her locomotor capabilities.

A fourth point is that detecting information about $\nu_{\min}$ requires the visual system to recover the object-motion component of optic flow independent of self-motion. When a person moves in the presence of other moving objects, the optic flow field is

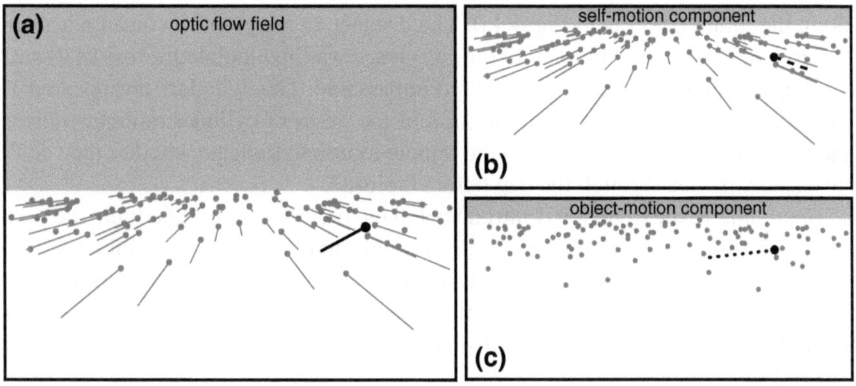

Fig. 4.5 **a** Optic flow field generated by combined motion of observer and object (*black dot*). **b** The component of optic flow due to self-motion independent of object motion. **c** The component of optic flow due to object motion independent of self-motion. The optic flow field (**a**) is the vector sum of the self-motion (**b**) and object-motion (**c**) components

the vector sum of the object-motion component and the self-motion component (see Fig. 4.5). Information about $\nu_{\min}$ is found in the object-motion component of optic flow. This is because the optical specification of $\dot{z}_m/E$ involves $\dot{\gamma}_m$ (see Fig. 4.4b), which is the component of $\dot{\gamma}$ that is due to the motion of the object independent of the observer's self-motion. Recall that γ is the visual angle between eye level and the base of the moving object (see Fig. 4.1). $\dot{\gamma}$, which is the rate of change of γ, is influenced by the movement of both the observer and the object. Specifically, $\dot{\gamma}$ is the sum of $\dot{\gamma}_o$ (the rate of change of γ due to the observer's self-motion) and $\dot{\gamma}_m$ (the rate of change of γ due to object motion). Because the optical specification of $\dot{z}_m/E$ involves $\dot{\gamma}_m$, detecting information about $\nu_{\min}$ while moving requires the visual system to factor out the influence of self-motion.

In principle, factoring out the self-motion component of optic flow could be achieved using visual information about self-motion, non-visual information about self-motion, or some combination of both. This leads to the prediction that manipulations of visual and/or non-visual self-motion information should influence the detection of information about $\nu_{\min}$ and any actions that are selected on the basis of such information. This prediction was recently tested using a VE to manipulate self-motion information (Fajen and Matthis, in press; [8]).

4.6 Testing the Affordance-Based Approach

The task required subjects to judge whether they could safely walk through a shrinking gap between a pair of converging obstacles before the gap closed. On each trial, subjects were walking along a path at steady state when two cylindrical obstacles in the VE positioned symmetrically about the path began to converge toward a point along the future path (as in Fig. 4.4a). The distance to the point of convergence and the amount of time remaining until the gap closed was manipulated across trials such that gaps ranged from easily passable to impassable. The cylinders disappeared 1 s after they began moving and within 1.2 s of the onset of cylinder motion, subjects had to press one of two buttons on the remote mouse to indicate whether they could have safely passed through the gap before it closed.

According to the hypothesis introduced in the previous section, judging the passability of a shrinking gap requires detecting information about the minimum locomotor speed needed to safely pass through the gap (ν_{min}), which in turn requires the visual system to factor out the influence of self-motion. Therefore, manipulations of visual and/or non-visual self-motion information should affect judgments of passability.

To investigate the influence of non-visual self-motion information (e.g., proprioception), [8] took advantage of the fact that such information, to be useful, must be continually recalibrated. This is because the relation between non-visual self-motion information and self-motion is dependent upon factors such as surface compliance and load that can vary from situation to situation. For such information to be useful for the purposes of perceiving self-motion, it must be possible to recalibrate when conditions change. In a VE, recalibration can be brought about by increasing or decreasing the speed with which subjects move through the VE relative to the physical world (i.e., the visual gain). Fajen and Matthis increased visual gain to $1.5\times$, which means that subjects moved through the VE 50 % faster than they moved through the physical world. The manipulation of visual gain affects the relation between non-visual self-motion information and self-motion in the VE. Therefore, as subjects moved around the VE with the increased visual gain, non-visual self-motion information became recalibrated—that is, subjects learn to attribute more optic flow to their own actively generated self-motion. If subjects rely on non-visual self-motion information to factor out the influence of self-motion, then the component of optic flow that is attributed to self-motion should be greater when subjects are calibrated to the faster-than-normal visual gain (see Fig. 4.6). The remaining component (i.e., the component that is attributed to object motion) should be less. Therefore, subjects should perceive that the obstacles will converge toward a point that is farther away along the locomotor axis and should be less likely to perceive that the gap is passable. This prediction was tested by comparing passability judgments when subjects were recalibrated to the faster-than-normal visual gain versus when they were calibrated to a normal visual gain. As predicted, subjects were less likely to perceive the gap as passable when they were calibrated to a faster-than-normal visual gain. Such findings indicate that non-visual self-motion information plays a role in recovering the object-motion component of optic flow, which is required to detect information about ν_{min}. In a follow-up experiment, Fajen and Matthis (in press) demonstrated that visual self-motion information also plays a role.

To summarize, when people move in the presence of other moving objects, the optic flow field is the vector sum of the self-motion and object-motion components. Visual information that is relevant to perceiving affordances such as the passability of a shrinking gap is found in the object-motion component of optic

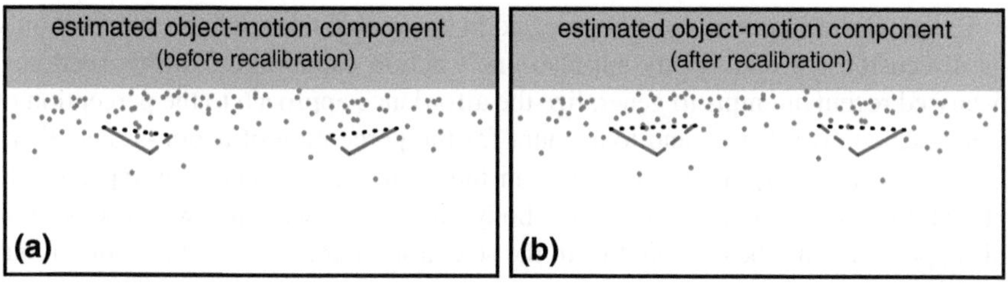

Fig. 4.6 Estimated object motion component (*dotted line*) after self-motion component (*dashed line*) is factored out from the optic flow field, which includes both the self-motion and object-motion components (*solid line*). The magnitude of optic flow attributed to self-motion (i.e., the component that is factored out) is greater in (**b**) than (**a**) due to recalibration of non-visual self-motion information

flow. To detect such information, people must factor out the influence of self-motion. The findings of Fajen and Matthis (in press [8]) show that both visual and non-visual self-motion information can be used to factor out the influence of self-motion, allowing for the perception of passability.

4.7 Extensions of the Affordance-Based Approach

In the shrinking gap example above, it was assumed that there are two obstacles positioned symmetrically about the locomotor axis and converging toward a common point at the same speed. We can relax this assumption and assume that there are multiple obstacles at different depths that will cross the locomotor axis in different places and at different times. In such situations, there are many possible routes because each obstacle introduces a choice point at which the actor must decide whether to pass in front or behind. Further, the decision must be made in a way that takes into account the physical sizes of the obstacles and the observer's body, as well as the observer's locomotor capabilities. The source of information identified in the previous section provides a basis for selecting actions in such situations. In the same way that such information can be used to perceive $v_{\min}$ for the shrinking gap, the same information can also be used to perceive $v_{\min}$ for each obstacle in the scene—that is, how fast one would need to move to pass in front of that obstacle. In addition, by changing the reference point from the leading edge of the obstacle to its trailing edge, the same information can also be used to perceive the maximum speed at which one could move to pass behind each obstacle. Thus, one can perceive the range of speeds at which not to move (i.e., because doing so would result in a collision) as well as how fast one would need to move to pass in front of or behind each obstacle. Such information could be used to decide which route to follow. Further, this applies regardless of whether there is one, two, or many obstacles in the environment. Therefore, such information could be used to select routes in arbitrarily complex environments with multiple moving obstacles.

To summarize, actions must be selected in a way that takes into account not only the dimensions of one's body but also one's action capabilities. In this section, I discussed recent attempts to generalize the affordance approach to the perception of action-scaled affordances and to account for the perception of action-scaled affordances based on information available in the optic array rather than appealing to internal models of the dynamics of the body. In the next section, I will discuss how this approach could be extended from the selection of action to on-line, continuous control.

4.8 Affordance Perception and the Continuous Control of Locomotion

Up to this point, the focus has been on the role of affordance perception in the selection of appropriate actions. Once an action is selected, it is necessary to move to the goal, which often involves guiding one's movements based on continuously available information. Just as selecting appropriate actions requires one to take body dimensions and locomotor capabilities into account, so does the on-line guidance of action. In this section, I will briefly discuss how the sources of information described in the previous section could also be used for the purposes of continuous control, using the task of intercepting moving targets as an example.

When intercepting a moving target, one needs to know how fast to move. To properly coordinate speed and direction during interception, it must be possible to perceive required speed as a function of the direction of locomotion. Thus, one could intercept the target quickly by turning toward the target, but doing so would require moving faster (see Fig. 4.7). Likewise, one could intercept the target while moving slower, but at the cost of taking more time. In addition, one must know the range of directions for which interception is not possible due to the fact that the speed required to reach the target exceeds the actor's maximum possible speed (see black region in Fig. 4.7). Thus, speed and direction of locomotion must be coordinated in a way that takes into account the actor's locomotor capabilities [2].

A variant of the information described in the previous section can also be used to guide interception of moving targets. Suppose that Fig. 4.4a contained a single object (a moving target) rather than two. A variant of the information in Fig. 4.4b specifies how fast the observer would need to move along the z-axis to intercept the target. In fact, required speed is specified for any arbitrary direction of locomotion. In Fig. 4.4, the reference frame within which visual angles are measured is aligned with the direction of locomotion. If the visual angles are measured in a reference frame that is rotated, then the value of required speed that is specified will equal that which is needed to intercept the target by moving in that direction. Thus, the speed required to intercept the target is specified for any arbitrary direction of locomotion. In principle, such information could be used to coordinate locomotor speed and direction during interception. At this point, the actual contribution of this information in visually

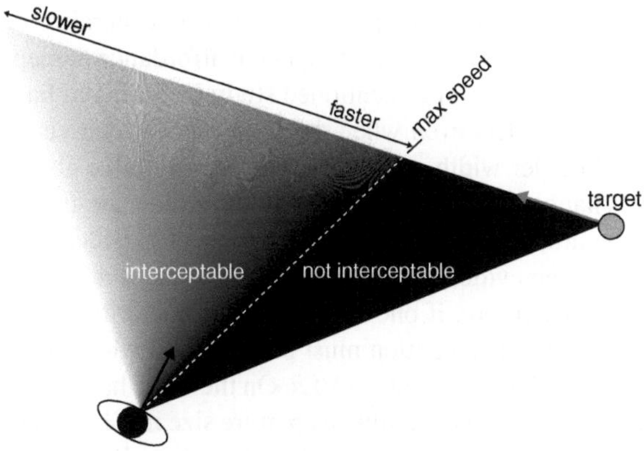

Fig. 4.7 *Top-down* view of an observer intercepting a moving target. The gradient illustrates how the speed needed to intercept the target varies with direction of locomotion. The *dashed line* indicates the direction in which the observer would need to move to intercept the target as quickly as possible. This line also partitions the space into directions in which the target is interceptable and directions in which the target is not interceptable due to limits on how fast the observer can move

guided interception remains unknown. Further experiments in VEs are needed to investigate the information and control strategies used to coordinate steering and speed during interception.

To summarize, continuously controlled visually guided actions such as intercepting a moving target on foot, require actors to guide their movements in ways that take their locomotor capabilities into account. This means that affordance perception may play a role not only in the selection of appropriate actions but also in the continuous guidance of locomotion. In this section, I described how information that specifies affordances for action selection (i.e., passability) may also be used to guide locomotion during interception.

4.9 Conclusions

Virtual environments have already proven to be a powerful tool for studying many aspects of perception and action [39]. In this chapter, I illustrated how VEs have also been used to study the perception of affordances and their role in the control of locomotion. This research has provided insights into the means by which people select actions and guide movements in ways that take their body dimensions and locomotor capabilities into account. The findings build upon previous research inspired by the ecological approach and contribute to a more general, affordance-based approach to visually guided locomotion.

Given the well-documented differences in perception and motor control in virtual versus real environments (e.g., [12, 16, 17, 31]), it is reasonable to ask whether findings from studies of affordance perception in VEs generalize to the real world. The answer most likely depends on what aspect of affordance perception one is interested in. For example, in the aforementioned study by Fath and Fajen [11], subjects often rotated their shoulders even when the distance between virtual obstacles was more than 1.3× shoulder width, which was the critical value above which subjects in Warren and Whang's [35] real world study did not rotate their shoulders. Such behavior could be attributed to an underestimation of exocentric distances in VEs (although there is some evidence that exocentric distances are accurately perceived in VEs; e.g., [13]). Therefore, if one is interested in the critical value for gap passability in the real world, then caution must be exercised when drawing conclusions based on passability judgments made in VEs. On the other hand, there is no reason to believe that the tendency to underestimate aperture size in VEs should affect subjects differently in the three conditions used in that study (i.e., Post, Tall Post, and Wall). Therefore, the perceptual bias does not present any problems for testing the contributions of eyeheight-scaled, head-sway-scaled, and stride-length-scaled information, which was the primary goal of that study.

Another issue that arises in studying affordance perception in VEs is that people's motor capabilities in VEs may differ from those in the real world. For example, subjects are unlikely to move as quickly in VEs due to the added weight of the HMD, greater postural instability, or fear of colliding with walls or other objects in the laboratory. Interestingly, Fajen et al. [6] found that subjects' judgments of their ability to pass through a shrinking gap between a pair of moving obstacles closely matched their ability to actually pass through shrinking gaps. Given that subjects' locomotor capabilities in the VE differ from those in the real world, the accuracy of judgments may initially seem surprising. Even in the real world, however, locomotor capabilities are not fixed. People's ability to move is continually affected by factors such as fatigue, injury, or carrying a heavy load. Therefore, the ability to adapt to altered locomotor capabilities in VEs may reflect a well-developed ability to adapt to changes in the real world. To summarize, although there are differences in perception and motor control in virtual versus real environments, experiments can be designed in such a way as to minimize the significance of these differences, thereby allowing researchers to take advantage of manipulations that are not possible in the real world.

Just as our understanding of affordance perception in the real world has benefited from VEs, so might our understanding of perception and action in VEs benefit from the study of affordance perception. The need to take one's body dimensions and locomotor capabilities into account applies in VEs just as it does in the real world. Attempts to understand how people cope with the many sensorimotor rearrangements and disruptions that are encountered in VEs often focus on the perception (or misperception) of conventional spatial properties. If the goal is to understand how people move in and interact with VEs, then more research on the perception of action-relevant properties (i.e., affordances) in VEs could be beneficial.

Acknowledgments Preparation of this chapter was supported by a grant from the National Institutes of Health (1R01EY019317).

References

1. Adolph KE (2008) The growing body in action: what infant locomotion tells us about perceptually guided action. In: Klatzky MBR, MacWhinney B (ed) Embodiment, ego-space, and action. Erlbaum, Mahwah, pp 275–321
2. Bastin J, Fajen BR, Montagne G (2010) Controlling speed and direction during interception: an affordance-based approach. Exp Brain Res 201(4):763–780
3. Bootsma RJ (2009) Lee's 1976 paper. Perception 38(6):851
4. Chapman S (1968) Catching a baseball. Am J Phys 36:368–370
5. Chhabra M, Jacobs RA (2006) Near-optimal human adaptive control across different noise environments. J Neurosci 26(42):10883–10887
6. Fajen BR, Diaz G, Cramer C (2011) Reconsidering the role of movement in perceiving action-scaled affordances. Hum Mov Sci 30(3):504–533
7. Fajen BR, Matthis JM (in press). Visual and non-visual contributions to the perception of object motion during self-motion. PLoS ONE
8. Fajen BR, Matthis JS (2011) Direct perception of action-scaled affordances: the shrinking gap problem. J Exp Psychol Hum Percept Perform 37(5):1442–1457
9. Fajen BR, Riley MA, Turvey MT (2009) Information, affordances, and the control of action in sport. Int J Sport Psychol 40(1):79–107
10. Fajen BR, Warren WH (2004) Visual guidance of intercepting a moving target on foot. Perception 33(6):689–715
11. Fath AJ, Fajen BR (2011) Static and dynamic visual information about the size and passability of an aperture. Perception 40:887–904
12. Fink PW, Foo PS., Warren WH (2007) Obstacle avoidance during walking in real and virtual environments. ACM Trans Appl Percept 4(1):1–18 (Article 2)
13. Geuss MN, Stefanucci JK, Creem-Regehr SH, Thompson WB (2012) Effect of viewing plane on perceived distances in real and virtual environments. J Exp Psychol Hum Percept Perform
14. Gibson JJ (1979) The ecological approach to visual perception. Houghton Mifflin, Boston
15. Ivanenko YP, Dominici N, Daprati E, Nico D, Cappellini G, Lacquaniti F (2011) Locomotor body scheme. Hum Mov Sci 30(2):341–351
16. Kelly JW, Riecke B, Loomis JM, Beall AC (2008) Visual control of posture in real and virtual environments. Percept Psychophys 70(1):158–165
17. Knapp JM, Loomis JM (2004) Limited field of view of head-mounted displays is not the cause of distance underestimation in virtual environments. Presence Teleoper Virtual Environ 13(5):572–577
18. Lee DN (1976) A theory of visual control of braking based on information about time-to-collision. Perception 5(4):437–459
19. Lee DN (1980) Visuo-motor coordination in space-time. In: Stelmach GE, Requin J (eds) Tutorials in motor behavior. North Holland, Amsterdam, pp 281–295
20. Loomis JM, Beall AC (2004) Model-based control of perception-action. In: Vaina LM, Beardsley SA, Rushton SK (eds) Optic flow and beyond. Kluwer, Dordrecht
21. Mark LS (1987) Eyeheight-scaled information about affordances: a study of sitting and stair climbing. J Exp Psychol Hum Percept Perform 13:361–370
22. Mark LS, Jiang Y, King SS, Paasche J (1999) The impact of visual exploration of judgments of whether a gap is crossable. J Exp Psychol Hum Percept Perform 25(1):287–295
23. McLeod P, Reed N, Dienes Z (2006) The generalized optic acceleration cancellation theory of catching. J Exp Psychol Hum Percept Perform 32(1):139–148
24. Michaels CF, Oudejans RR (1992) The optics and actions of catching fly balls: zeroing out optical acceleration. Ecol Psychol 4:199–222
25. Oudejans RD, Michaels CF, Bakker FC, Dolné MA (1996) The relevance of action in perceiving affordances: perception of catchableness of fly balls. J Exp Psychol Hum Percept Perform 22(4):879–891
26. Pepping G, Li FX (2000) Changing action capabilities and the perception of affordances. J Hum Mov Stud 39:115–140

27. Plumert JM, Kearney JK, Cremer JF (2004) Children's perception of gap affordances: bicycling across traffic-filled intersections in an immersive virtual environment. Child Dev 75(4):1243–1253
28. Sedgwick HA (1980) The geometry of spatial layout in pictorial representation. In: Hagen M (ed) The perception of pictures VI. Academic Press, New York, pp 33–90
29. Shadmehr R, Mussa-Ivaldi FA (1994) Adaptive representation of dynamics during learning of a motor task. J Neurosci 14(5 Pt 2):3208–3224
30. Shadmehr R, Smith MA, Krakauer JW (2010) Error correction, sensory prediction, and adaptation in motor control. Annu Rev Neurosci 33:89–108
31. Thompson WB, Willemsen P, Gooch AA, Creem-Regehr SH, Loomis JM, Beall AC (2004) Does the quality of the computer graphics matter when judging distances in visually immersive environments? Presence Teleoper Virtual Environ 13(5):560–571
32. Trommershauser J, Maloney LT, Landy MS (2008) Decision making, movement planning and statistical decision theory. Trends Cogn Sci 12(8):291–297
33. Warren WH (1984) Perceiving affordances: visual guidance of stair climbing. J Exp Psychol Hum Percept Perform 10:683–703
34. Warren WH (2007) Action-scaled information for the visual control of locomotion. In: Pepping G, Grealy MA (eds) Closing the gap: the scientific writings of David N. Lee. Lawrence Erlbaum, Mahwah
35. Warren WH, Whang S (1987) Visual guidance of walking through apertures: body-scaled information for affordances. J Exp Psychol Hum Percept Perform 13(3):371–383
36. Watson G, Brault S, Kulpa R, Bideau B, Butterfield J, Craig C (2011) Judging the 'passability' of dynamic gaps in a virtual rugby environment. Hum Mov Sci 30(5):942–956
37. Wraga M (1999) The role of eye height in perceiving affordances and object dimensions. Percept Psychophys 61(3):490–507
38. Yilmaz EH, Warren WH Jr (1995) Visual control of braking: a test of the tau hypothesis. J Exp Psychol Hum Percept Perform 21(5):996–1014
39. Zaal FTJM, Bootsma RJ (2001) Virtual reality as a tool for the study of perception-action: the case of running to catch fly balls. Presence 20(1):93–103

Chapter 5
The Effect of Translational and Rotational Body-Based Information on Navigation

Roy A. Ruddle

Abstract Physical locomotion provides internal (body-based) sensory information about the translational and rotational components of movement. This chapter starts by summarizing the characteristics of model-, small- and large-scale VE applications, and attributes of ecological validity that are important for the application of navigation research. The type of navigation participants performed, the scale and spatial extent of the environment, and the richness of the visual scene are used to provide a framework for a review of research into the effect of body-based information on navigation. The review resolves contradictions between previous studies' findings, identifies types of navigation interface that are suited to different applications, and highlights areas in which further research is needed. Applications that take place in small-scale environments, where maneuvering is the most demanding aspect of navigation, will benefit from full-walking interfaces. However, collision detection may not be needed because users avoid obstacles even when they are below eye-level. Applications that involve large-scale spaces (e.g., buildings or cities) just need to provide the translational component of body-based information, because it is only in unusual scenarios that the rotational component of body-based information produces any significant benefit. This opens up the opportunity of combining linear treadmill and walking-in-place interfaces with projection displays that provide a wide field of view.

5.1 Introduction

Navigation is central to many types of virtual environment (VE) applications. However, with only a few exceptions (mostly in military training), these applications use abstract navigation interfaces. That is, users press buttons and manipulate

Roy A. Ruddle (✉)
University of Leeds, School of Computing, Leeds, UK
e-mail: royr@comp.leeds.ac.uk

Roy A. Ruddle
Max Planck Institute for Biological Cybernetics, Tübingen, Germany

F. Steinicke et al. (eds.), *Human Walking in Virtual Environments*,
DOI: 10.1007/978-1-4419-8432-6_5, © Springer Science+Business Media New York 2013

devices such as joysticks and mice to travel through a VE and look around, and are provided with minimal *body-based sensory information* about their movement. Given the difficulty that many users encounter when trying to learn spatial layouts in desktop VEs [1], which only provide visual information, it is likely that "walking" interfaces could have a widespread and beneficial impact on VE applications.

This chapter is divided into four main parts. The first summarizes the characteristics of VE applications from a navigational perspective, by mapping them onto different scales of environment (model vs. small vs. large). The second identifies attributes of ecological validity that should be considered when applying the results of navigation research to a given VE application. The third, and most substantive, part reviews experimental studies that have investigated the effect of body-based information on navigation, focusing on studies that investigated the rotational and/or translational *components* of body-based information, rather that different *cues* (proprioception, vestibular and efference copy) [2]. These studies are categorized according to type of navigation participants performed while acquiring knowledge of the environment (single-route vs. whole-environment), the scale of the environment (small vs. large), the environment's spatial extent, and the richness of the visual scene. The chapter concludes by using these research results to identify the types of navigation interface that are suited to different applications, and highlight areas in which further research is needed.

5.2 Applications of Virtual Environments

From a navigational perspective, VE applications [3–5] may be divided into three broad categories (see Table 5.1). The categories are defined by the scale of the environment in spatial cognition terms [6].

In the first category are *model-scale* applications, where users look around while remaining in one position (model-scale spaces, which in the real world would be placed on a table top, can be seen and reached from a single place). Examples include designing the layout of the cockpit of a car and training communication between the pilot and winch-man of search and rescue helicopters. Head-mounted displays (HMDs) are ideal for these applications, because they allow users to look around naturally by turning their head, with positional changes lying within the bounds of low-cost tracking devices (say, a 1 m^3). This means that effective navigation interfaces for these applications do not require a walking interface, so they are not considered further until the Conclusions section of this chapter.

The second category is *small-scale* applications, where users can resolve all of the detail necessary for navigation from a single place (e.g., any position in a room), but have to travel through the VE during usage. Examples range from analyzing the ease with which an engine may be assembled, or a control room layout for visibility, to being a witness in a virtual identity parade (a courtroom lineup, conducted using avatars in a VE). In these applications it is typically straightforward for users to

Table 5.1 Examples of VE applications that, in spatial cognition terms, are model-, small-, or large-scale

Application theme	Environment scale		
	Model	Small	Large
Design	Cockpit layout	Engine assembly Control room layout	Chemical plant Architecture Retail shop layout
Training	Close-range naval weaponry Helicopter crew collaboration	–	Search building Learn evacuation route
Health	–	Motor rehabilitation	Post-traumatic stress disorder
Science	Molecular docking	–	Data visualization
Other	–	Identity parade	Heritage and tourism Computer games

determine where they wish to move, but it is often non-trivial to make the maneuvers that are necessary for that movement.

The third category is *large-scale* applications where users travel through a large environment (e.g., a building, city, forest or dataset) over an extended period of time, integrating sensory information obtained during their movement to maintain knowledge of their location in the environment and avoid getting lost. Sometimes, and as with small-scale applications, maneuvering is non-trivial (e.g., when training to search a building for the enemy [7, 8]), but typically it is the acquisition of spatial knowledge that is the greatest navigational challenge.

5.3 Ecological Validity

Experimental (and especially laboratory) studies of navigation use stimuli and tasks that have been chosen to investigate specific hypotheses, and are sometimes simplified to an extreme. To assess the relevance of an experiment's findings, it is important to balance the generality of those findings with the ecological validity of the stimuli and tasks for a given type of VE application. In particular, attention should be paid to the VE's scale, extent and visual scene, the paths users follow during navigation and how frequently they follow them, and how users' knowledge is assessed.

A VE may be model-, small-, or large-scale in spatial cognition terms (see above). The cognitive processes involved for navigation in each of these differ substantially, as does the difficulty of and time required for users to acquire accurate spatial knowledge. For example, in a few minutes users can learn an environment's layout from a map (the map is effectively a model-scale representation of the environment), but such knowledge takes orders of magnitude more time to learn by direct navigation in the environment itself, which is large-scale [9], although knowledge gained in the latter will be ultimately more detailed. Thus, particular caution should be taken when applying the findings of research conducted in model- or small-scale environments [10–12] to applications that require users to navigate large-scale VEs.

To overcome the above issues with the scale of a VE, it is common for studies of full walking interfaces to condense an environment so that it fits within the physical space of a laboratory. This leads to a situation where the environment is large-scale, but its *spatial extent* (physical size) is small (say, less than $10 \times 10\,m$) [13–15]. This is rather unrealistic (in both VE applications and the real world a large-scale environment is almost always also large in extent), but necessary for the purposes of the experiment. However, extent changes the time cost of traveling from one place to another and influences navigational behavior [16]. Few walking studies have actually investigated the effect of extent, but a notable exception is [17].

In early VE navigation studies it was rare for a visually rich environment to be used (a notable exception was [18]), but this richness is now more common, partly due to the ability of PC graphics cards to render complex scenes in real time. Real-world environments and modern VE applications typically contain a surfeit of visual cues, which compete to become landmarks and may be used in a different manner to landmarks in a visually impoverished setting [19]. Thus, apart from specialized applications such as training for evacuation during a fire, a rich visual scene is essential for ecological validity.

The paths people navigate in VEs and the real world often involve many navigational decisions. By contrast the paths used for some navigation research, particularly studies that investigate low-level mechanisms such as distance perception and path integration, are simplistic and so may engage different cognitive processes (e.g., working vs. long-term memory) and brain regions [20] than when users navigate in a real VE application. Most experimental studies only expose a participant once to an environment before testing, which has similarities with being a first-time visitor to a place, but is clearly different from settings that a user visits repeatedly and develops spatial knowledge of over an extended period of time. In those latter circumstances a user has more opportunity to learn the layout of the environment as a whole (survey knowledge).

Lastly, studies adopt a variety of measures, some of which are designed to assess specific aspects of users' route or survey knowledge, and others that are designed to assess to ephemeral concepts such as presence. These measures should be considered in the context of the tasks users perform in a given VE application before the relevance of research findings can be judged.

5.4 The Effect of Body-Based Information

This section reviews the findings of research into the effect of body-based information on navigation, and offers explanations for contradictions between some of the studies' findings. The review attempts to inform: (a) our basic understanding of how translational versus rotational body-based information affects human navigation, and (b) simplify the process of applying those findings to VE applications. In terms of scope, the review focuses on the effect of different components of body-based information (rotation vs. translation), rather than different cues, because during active walking users are provided with a full set of body-based cues. The criteria for inclusion in the review were that a study: (a) involved users changing both position and orientation as they navigated, and (b) investigated different components (rotational; translational) of body-based information, not just different cues (proprioception; vestibular; efference copy). Low-level studies that focused exclusively on rotational movement (e.g., [21]) or distance perception are omitted (e.g., [22]).

5.4.1 Review Framework

The studies that are reviewed are divided into four groups (see Table 5.2), which are dictated by the type of navigation participants performed while acquiring knowledge of the environment (single-route vs. whole-environment) and the scale of the environment (small vs. large). Single-route acquisition is where participants only navigated one specific route. Whole-environment acquisition is where participants either freely explored the environment or navigated to find target locations that were distributed around the environment, in specific but changing orders. The distinction between small- versus large-scale environments is explained above.

Spatial *extent* is classified as either small (room-sized; a maximum of approximately 10×10 m) or large (building-sized or greater). The richness of the *visual scene* is classified as low, medium or high. Low corresponds to environments where, apart from target landmarks, variations in the visual scene were just designed to provide optic flow. High corresponds to rich visual scenes that contained a surfeit of visual detail of deliberately added landmarks (e.g., at each junction in a building), and medium corresponds to scenes that did not belong clearly to either of the other categories.

The experimental results summarized in Table 5.2 are divided into *navigation* performance (time taken and distance traveled metrics that show how efficiently participants moved between places) and survey knowledge (*direction* estimates and straight line *distance* estimates). These survey metrics are the basic information people need if they are to know the location of places in relation to each other, or take shortcuts [30]. Absolute direction estimate errors were used, rather than signed errors that indicate response biases (e.g., see [31]) and, in all except the triangle completion studies (single-route acquisition/small-scale environments), the distance estimates were estimates of relative straight line distance, which are accurate if people have a well-developed cognitive map [9]. For a discussion of distance estimation methodologies, see [32].

Each type of results is subdivided into four columns: *Vis*, *Rot*, *Tran*, and *Full*. Vis is where participants were only provided with visual information (e.g., a desktop VE). Rot and Tran are where participants were provided with the rotational and translational component of body-based information, respectively, in addition to visual information. Full is where participants were physically walking through the

Table 5.2 Effect of body-based information on navigation performance and survey knowledge (direction and straight line distance estimates)

Study	Extent	Visual scene	Navigation				Direction estimates				Distance estimates			
			Vis	Rot	Tran	Full	Vis	Rot	Tran	Full	Vis	Rot	Tran	Full
Single-route acquisition and small-scale environments														
Klatzky et al. [23]	Small	Low					worst	best						
Kearns et al. [24]	Small	Low					best			worst	n.s.			n.s.
Peruch et al. [25]	Small	Low					worst	worst		best	worst	intermediate		best
Single-route acquisition and large-scale environments														
Chance et al. [13]	Small	Low					worst	intermediate		best				
Ruddle et al. (Expt. 2) [15]	Small	High		worst		best								
Suma et al. (Exp. 1) [14]	Small	High		n.s.		n.s.								
Witmer et al. [18]	Large	High		worst		best								
Grant et al. [26]	Large	High	worst			best	n.s.			n.s.				
Waller et al. [2]	Large	High					worst			best				
Waller et al. [27]	Large	High					n.s.			n.s.				
Whole-environment acquisition and small-scale environments														
Ruddle et al. (Expt. 2) [11]	Small	Low	worst			best								
Riecke et al. [12]	Small	Low	worst	best		best								
Ruddle et al. (Expt. 1) [11]	Small	Medium	worst	worst		best								
Whole-environment acquisition and large-scale environments														
Ruddle et al. (Expt. 1) [17]	Small	High	worst	worst		best	worst	intermediate		best	intermediate	worst		best
Ruddle et al. [28]	Large	Low	n.s.	n.s.			n.s.	n.s.			worst	best		
Ruddle et al. [29]	Large	Medium	n.s.	n.s.			n.s.	n.s.			best	worst		
Ruddle et al. (Expt. 2) [17]	Large	High	n.s.	n.s.	n.s.	n.s.	worst	worst	best	best	worst	worst	best	best

See *Review framework* for a detailed explanation

environment and, therefore, were provided with rotational and translational body-based information, as well as visual information. The terms Vis, Rot, Tran and Full are used as generic group names in the remainder of this article.

In Table 5.2, cells are blank if the relevant metric or category of body-based information was not investigated in a given study. For example, Chance et al. [13] only investigated direction estimates for Vis, Rot and Full conditions. If there was no significant difference for a given metric then all the categories of body-based information that were investigated in the study are marked as "n.s." (e.g., Vis and Full, for distance estimates in Kearns et al. [24]). Where a study reported statistically significant differences, shading shows the worst, inermediate and best performing conditions. The logic used to determine the shading is best explained using examples. Chance et al. [13] uses all three levels of shading because there was a main effect of direction estimate accuracy, pair-wise comparisons showed that the Full group performed significantly better than the Vis group, and performance of the Rot group was clearly intermediate. Ruddle et al. (Expt. 1) [17] uses the "worst" and "best" shading levels because there was a main effect of navigation performance, and pair-wise comparisons showed that the Full group performed significantly better than either of the other groups. The Riecke et al. [12] data refers to the number of revisits metric, which is more sensitive than the percentage of perfect trials. There was a main effect, and pair-wise comparisons indicated that the Full and Rot groups were equivalent, but there was a marginally significant difference between the Rot and Vis groups.

Caution should be taken when drawing conclusions from results that were statistically not significant. Sometimes this is due to there being no underlying difference. On other occasions it is due to a lack of statistical power, and this is particularly true in navigation studies, which often have large individual differences.

5.4.2 Studies Investigating the Effect of Body-Based Information

5.4.2.1 Single-Route Acquisition and Small-Scale Environments

All three of the studies included in this section [23–25] used a triangle completion paradigm (this involved being guided along two legs of a path and then being asked to point or return directly to the start point, which assesses a participant's ability to take short cuts). Klatzky et al. reported a step change in performance between Vis and Rot groups of participants, with the latter performing accurately and the former not. By contrast, Kearns et al. (Experiments 1 and 3) reported a small but significant difference between Vis and Full groups, with the Vis group performing more accurately. The difference between the studies' findings may be due to participants' mode of response, because Klatzky's pointed to where they would have to travel to return to a trial's start point, and the errors were assumed to occur because the Vis group failed to update their cognitive heading. Kearns' participants' responded by traveling to where they thought the start point was located, and while doing so may have corrected their cognitive heading. Some support for this explanation is provided by subsequent research, which showed that the errors reported by Klatzy et al. did not occur if participants responded verbally [33].

Péruch et al. reported that participants who walked (a Full group) performed best and those who were in a Vis group performed worst, in direct contrast to the findings of Kearns et al. However, Péruch's study combined responses from triangle completion trials with responses from trials in which participants had to reverse the two-leg path that had been traveled. In research by Ruddle et al. [15] substantially fewer errors were made by participants who physically walked and then reversed a path (a Full group) than participants who were in a Rot group. If a similar difference occurred in Péruch's study then the Full group's superior performance on path-reversal trials more than compensated for slightly inferior performance on the triangle completion trials, and that would explain the difference with Kearns et al's findings.

Each of the above studies was designed to investigate specific low-level perceptual and cognitive processes that are involved in navigation. To do so, participants were either blindfolded or presented with optic flow visual information. Triangle completion is trivial to perform accurately in rich visual scenes [34] which, together with the simple (two leg) paths that participants followed, means that the above studies had little ecological validity with the environments and tasks used in VE applications.

5.4.2.2 Single-Route Acquisition and Large-Scale Environments

Compared with the above studies, investigations of the effect of body-based information that used single-route acquisition tasks and large-scale environments have produced more consistent findings. Whenever the results were statistically significant, the Full group performed best, and the worst performing group was either the Vis group (if such a group was part of the study) or the Rot group (if the study had no Vis group).

In identifying the above consistency in the findings, a number of caveats should be noted. First, Witmer et al. [18] asked participants to learn a route through either a real building (Full group) or a high visual fidelity VE model of the building (Rot group), and then tested training transfer to the real building. The Full group was superior in both training and testing, but the difference could have been caused by various factors that were associated with performing the task in the real world, not just the addition of translational body-based information. Second, Grant and Magee [26] also performed a training transfer study. The Full and Vis groups both trained in the VE, but there was not a significant difference between the groups' direction estimate accuracy when tested in that environment. A significant difference only occurred when navigational performance was tested in the equivalent real-world environment. Third, although Waller et al. [27] found no significant main effect for the accuracy of direction estimates, for the most complex routes (6–8 turns) the Full group's estimates were significantly more accurate than the other groups' estimates. Fourth, in both of Waller's studies [2, 27] the Full group moved actively, but the Vis group passively viewed movement that had been recorded by a camera worn by a person who walked. Fifth, Suma et al. [14] reported significant effects, but these were due to the poor performance of participants who used a move-where-pointing interface (a Vis group). There was not a significant difference for the time taken between a physical walking (Full) group and another Vis group, who used a move-where-looking interface, for the number of collisions with the environment's walls, or in recall and recognition tests about objects that had been in the environment. That contrasts with another study, where participants who physically walked (a Full group) were significantly better at both recognizing and correctly recalling the order of objects that had been in the environment than participants who were in a Rot group [15].

In summary, these large-scale environment studies indicate that navigating a route with full body-based information improves both the route and survey knowledge of participants. There is some evidence that rotational body-based information produces an intermediate accuracy of spatial knowledge, but only one study included factors of Vis versus Rot versus Full [13]. From an ecological validity perspective, these large-scale studies somewhat inevitably required participants to navigate routes that were more complex, and hence ecologically more valid, than the small-scale environment studies that were described in the previous section. It is also notable that all but one of the large-scale studies used a high-fidelity visual scene, unlike their small-scale counterparts. Finally, the pattern of results is independent of the spatial extent of the environment that was used.

5.4.2.3 Whole-Environment Acquisition and Small-Scale Environments

The experiments that used a whole-environment acquisition task in a small-scale environment have both consistencies and differences between their findings. A notable consistency is that participants tended to maneuver around objects in a VE when provided with a physical walking interface (a Full group), but collided with them when provided with interfaces that had less body-based information (Rot or Vis groups) [10, 11]. Participants' paths were also qualitatively different—curved with a walking interface but straight for participants in Rot and Vis groups.

Zanbaka et al. [10] gathered subjective responses from participants and measured their ability to maneuver. The other experiments quantified participants' ability to remember where they had traveled, and showed that participants in Full groups performed significantly better than those who were in Vis groups. However, there was an inconsistency in the findings for participants in Rot groups. When the environment was square those participants performed as poorly as participants who had no body-based information (a Vis group) [11], but when the environment was circular the Rot group's performance was comparable with that of a Full group [12]. Contrary to assertions made by the authors of the latter study, a likely explanation is that rotational body-based information is important when external (visual) orientation cues are absent (see also [35]).

5.4.2.4 Whole-Environment Acquisition and Large-Scale Environments

At first glance Table 5.2 appears to highlight several contradictions between the findings of this fourth category of experiment, but the following explanations make the underlying pattern of results more consistent. First, consider differences between Rot and Vis groups. In none of the studies did a statistical test show a significant difference between these groups for navigational performance. For survey knowledge, the differences between these groups appears to be metric-dependent in Experiment 1 of [17] (Table 5.2 indicates that direction estimates were worst for the Rot group, but distance estimates were worst for the Vis group), but this is due to post-hoc tests showing that the Full group differed significantly from the Vis group (direction estimates) and Rot group (distance estimates). There are indications that, with greater statistical power, other posthoc comparisons may also have been significant, which would have led to the same pattern of results as for navigational performance in that experiment, and all metrics for Experiment 2 of that study. However, the contradiction between the findings for distance estimates [28, 29] remains unexplained (if the findings had been the opposite way around then they could have been explained by environment layout, which was orthogonal [28] versus oblique [29]; see previous section).

Both Full and Trans groups have the benefit of translational body-based information, which accounts for the significantly more accurate direction and distance estimates made by those groups than Vis and Rot groups in both experiments of Ruddle et al. [17], and the significantly better navigational performance of the Full group in Experiment 1 of that study. The lack of an effect of translational body-based information in Experiment 2 may be because the increased (and ecologically more valid) spatial extent meant that participants had considerably more time to process visual information as they navigated, so body-based information made less contribution to their development of spatial knowledge. Alternatively, it is possible that the environment was not complex enough for a statistically significant effect of translational body-based information to occur (a ceiling effect).

Lastly, Suma et al's Experiment 2 also used whole-environment acquisition and a large-scale environment [14]. The Full group collided with the VE's walls less often than the Vis group, but the difference in the distance the groups traveled (greater for the Full group) in the time that each participant was given may have been due to either an inbuilt speed restriction or insufficient practice with the virtual travel interfaces that were provided for the Vis and Rot groups. Metrics involving a cognition questionnaire and a map placement test produced non-significant results between the groups, which is common for these tests' lack of sensitivity.

5.5 Summary and Conclusions for VE Applications

So how can the findings of experimental studies of body-based information inform the design of navigation interfaces for VE applications? Table 5.3 summarizes the answer to this question from a navigation perspective, taking into account the need for users to maneuver and develop of spatial knowledge, but does not attempt to consider other factors such as cost.

5.5.1 Model-Scale Environments

In applications that use environments which in spatial cognition terms are model-scale then users need to be able to look around, but make only localized adjust-

Table 5.3 Navigation interfaces that would benefit different types of VE applications

Navigation interface	Environment scale: Model	Small	Large
Abstract (Vis group)	–	–	Data visualization
Orientation-tracked (Rot group)	Cockpit layout Close-range naval weaponry Helicopter crew collaboration Molecular docking	–	–
Linear treadmill (Trans group)	–	Motor rehabilitation	Chemical plant Architecture Retail shop layout Learn evacuation route Heritage and tourism Computer games
Walking (Full group)	–	Engine assembly Control room layout Identity parade	Search building Post-traumatic stress disorder

ments to their position. Therefore, orientation-tracking (the type of interface used by Rot groups in the studies described above) is sufficient and there is no need for a full walking interface, as evidenced by a number of successful, military training applications [3].

5.5.2 Small-Scale Environments

In VE applications that utilize a small-scale environment, users generally know where they wish to travel, so the primary navigational challenge is maneuvering. The studies show a qualitative difference in people's maneuverability with Full (walking) interfaces, compared with Rot and Vis (abstract) interfaces, coupled with objective data that show that users collide with objects within the environment significantly less often when a walking interface is used [10, 11, 14]. This reduction takes place without the need to implement collision detection/feedback in the VE software, because users have a natural tendency to avoid objects so long as the interface provides sufficient maneuverability. Therefore, there is a clear indication that a walking interface is beneficial for applications such as engine assembly design, control room layout design and virtual identity parades, though cheaper, carefully designed desktop alternatives should also be considered (e.g., see [8]).

Control room layout combines the same requirement for maneuverability with the need for users to be spatially aware of the environment they are designing, which also benefits significantly from the provision of a walking interface [11, 12]. As with other small-scale applications, the spatial extent of the environments means that it is technically feasible to implement walking by tracking users in an empty room that contains the environment on a 1:1 scale, but a hybrid real/virtual walking interface may prove more practical [10].

Motor rehabilitation applications are concerned with a patient's gait, and so only require translational body-based information. This may be provided via a linear treadmill or a specialist exoskeleton-based device, but questions remain about the medical benefits of integrating such devices within a VE [36].

5.5.3 Large-Scale Environments

For applications that use large-scale VEs the consensus result is that a full walking interface is required [13, 15, 26], and probably necessary in applications that also require maneuverability (e.g., military training for searching a building; but see also [7, 8]). However, in a study that was unique in including a Trans condition, that condition was as effective as a Full condition in allowing participants to acquire spatial knowledge [17]. This highlights an opportunity for preserving the benefit to users while simplifying the technology used for walking interfaces. For example, although omnidirectional treadmills can be constructed [37], linear treadmills are simpler to design and so are smaller, cheaper and more reliable.

An exception is likely to be data visualization applications, because the scale involved (e.g., in genomics) is several orders of magnitude greater than other large-scale applications. Given that "magic" interfaces (interfaces that allow users to make movements that would be impossible in the real world, e.g., jump between widely separated places) [4] will always be needed if users are to move rapidly and precisely between levels of detail such as chromosome $\rightarrow$ base pair, such applications are likely to remain based on abstract navigation interfaces.

5.5.4 Further Research

The main area that requires further research into navigation interfaces is applications that use large-scale environments. One priority is to thoroughly evaluate techniques that allow the navigation of large spatial extents via walking movements made within a much smaller locality (treadmill, walking-in-place [38], and redirected walking [39]). Such interfaces are currently unproven, and we need to understand their effect on participants' navigational performance and the rate at which they develop route- and survey-type spatial knowledge. A second priority is to evaluate these interfaces with projection displays, because they hold advantages over HMDs in terms of image resolution and field of view.

Acknowledgments This research was supported by the award of an Alexander von Humboldt Foundation for Experienced Researchers to RAR, the University of Leeds, and the Max Planck Society.

References

1. Ruddle RA (2001) Navigation: am I really lost or virtually there? In: Harris D (ed) Engineering psychology and cognitive ergonomics. Ashgate, Burlington, pp 135–142
2. Waller D, Loomis JM, Haun DBM (2004) Body-based senses enhance knowledge of directions in large-scale environments. Psychon Bull Rev 11:157–163
3. Stone RJ (2002) Applications of virtual environments: an overview. In: Stanney KM (ed) Handbook of virtual environments. Lawrence Erlbaum, Mahwah, pp 827–756
4. Bowman DA et al (2004) 3D user interfaces: theory and practice. Addison-Wesley, London
5. Blascovich J, Bailenson JN (2011) Infinite reality. William Morrow, New York
6. Weatherford DL (1985) Representing and manipulating spatial information from different environments: models to neighborhoods. In: Cohen R (ed) The development of spatial cognition. Erlbaum, Hillsdale, pp 41–70
7. Whitton MC et al (2005) Comparing VE locomotion interfaces. In: Proceedings of IEEE virtual reality, IEEE, Los Alamitos, pp 123–130
8. Templeman JN et al (2007) Pointman: a new control for simulating tactical infantry movements. In: Proceedings of IEEE virtual reality, IEEE, Los Alamitos, pp 285–286
9. Ruddle RA, Payne SJ, Jones DM (1997) Navigating buildings in "desk-top" virtual environments: experimental investigations using extended navigational experience. J Exp Psychol Appl 3:143–159
10. Zanbaka C et al (2005) Comparison of path visualizations and cognitive measures relative to travel techniques in a virtual environment. IEEE Trans Vis Comput Graphics 11:694–705
11. Ruddle RA, S Lessels (2009) The benefits of using a walking interface to navigate virtual environments. ACM Trans Comp-Hum Int 16:article 5
12. Riecke BE et al (2010) Do we need to walk for effective virtual reality navigation? Physical rotations alone may suffice. In: Lecture notes in computer science, pp 234–247
13. Chance SS et al (1998) Locomotion mode affects the updating of objects encountered during travel: the contribution of vestibular and proprioceptive inputs to path integration. Presence-Teleoper Virtual 7:168–178
14. Suma EA et al (2010) Evaluation of the cognitive effects of travel technique in complex real and virtual environments. IEEE Trans Vis Comput Graph 16:690–702
15. Ruddle RA et al (2011) The effect of landmark and body-based sensory information on route learning. Mem Cognit 39:686–699
16. Ruddle RA et al (2000) The effects of hyperlinks on navigation in virtual environments. Int J Hum-Comput Stud 53:551–581
17. Ruddle RA, Volkova E, Bülthoff HH (2011) Walking improves your cognitive map in environments that are large-scale and large in extent. ACM Trans Comp-Hum Int 18(2):Article 10
18. Witmer B et al (1996) Virtual spaces and real-world places: transfer of route knowledge. Int J Hum-Comput Stud 45:413–428
19. Spiers H, Maguire E (2004) A 'landmark' study on the neural basis of navigation. Nat Neurosci 7:572–574
20. Burgess N, Maguire E, O'Keefe J (2002) The human hippocampus and spatial and episodic memory. Neuron 35:625–641
21. Lathrop WB, Kaiser MK (2002) Perceived orientation in physical and virtual environments: changes in perceived orientation as a function of idiothetic information available. Presence-Teleoper Virtual 11:19–32

22. Campos JL, Byme P, Sun HJ (2010) The brain weights body-based cues higher than vision when estimating walked distances. Eur J Neurosci 31:1889–1898
23. Klatzky RL et al (1998) Spatial updating of self-position and orientation during real, imagined, and virtual locomotion. Psychol Sci 9:293–298
24. Kearns MJ et al (2002) Path integration from optic flow and body senses in a homing task. Perception 31:349–374
25. Péruch P et al (2005) Direction and distance deficits in path integration after unilateral vestibular loss depend on task complexity. Cogn Brain Res 25:862–872
26. Grant SC, Magee LE (1998) Contributions of proprioception to navigation in virtual environments. Hum Factors 40:489–497
27. Waller D, Greenauer N (2007) The role of body-based sensory information in the acquisition of enduring spatial representations. Psychol Res 71:322–332
28. Ruddle RA, Payne SJ, Jones DM (1999) Navigating large-scale virtual environments: what differences occur between helmet-mounted and desk-top displays? Presence-Teleoper Virtual 8:157–168
29. Ruddle RA, Péruch P (2004) Effects of proprioceptive feedback and environmental characteristics on spatial learning in virtual environments. Int J Hum-Comput Stud 60:299–326
30. Thorndyke P, Hayes-Roth B (1982) Differences in spatial knowledge acquired from maps and navigation. Cognit Psychol 14:560–589
31. Loomis J et al (1999) Human navigation by path integration. In: Golledge R (ed) Wayfinding: cognitive mapping and other spatial processes. John Hopkins, Baltimore, pp 125–151
32. Montello DR (1991) The measurement of cognitive distance: methods and construct validity. J Environ Psychol 11:101–122
33. Avraamides MN et al (2004) Use of cognitive versus perceptual heading during imagined locomotion depends on the response mode. Psychol Sci 15:403–408
34. Riecke BE, van Veen HAHC, Bülthoff HH (2002) Visual homing is possible without landmarks: a path integration study in virtual reality. Presence-Teleoper Virtual 11:443–473
35. Souman JL et al (2009) Walking straight into circles. Curr Biol 19:1538–1542
36. Crosbie JH et al (2007) Virtual reality in stroke rehabilitation: still more virtual than real. Disabil Rehabil 29:1139–1146
37. De Luca A et al (2009) Control design and experimental evaluation of the 2D CyberWalk platform. In: Proceedings of the IEEE/RSJ international conference on intelligent robots and systems (IROS 2009), IEEE, Los Alamitos, pp 5051–5058
38. Feasel J, Whitton MC, Wendt JD (2008) LLCM-WIP: low-latency, continuous-motion walking-in-place. In: Proceedings of the 2008 IEEE symposium on 3D user interfaces, IEEE, Los Alamitos, pp 97–104
39. Peck TC, Fuchs H, Whitton MC (2009) Evaluation of reorientation techniques and distractors for walking in large virtual environments. IEEE Trans Vis Comput Graph 15:383–394

Chapter 6
Enabling Unconstrained Omnidirectional Walking Through Virtual Environments: An Overview of the CyberWalk Project

Ilja Frissen, Jennifer L. Campos, Manish Sreenivasa and Marc O.Ernst

Abstract The CyberWalk treadmill is the first truly omnidirectional treadmill of its size that allows for near natural walking through arbitrarily large Virtual Environments. The platform represents advances in treadmill and virtual reality technology and engineering, but it is also a major step towards having a single setup that allows the study of human locomotion and its many facets. This chapter focuses on the human behavioral research that was conducted to understand human locomotion from the perspective of specifying design criteria for the CyberWalk. The first part of this chapter describes research on the biomechanics of human walking, in partic-

I. Frissen (✉). · M. Sreenivasa · M. O. Ernst
Max Planck Institute for Biological Cybernetics, Multisensory Perception and Action Group, Tübingen, Germany
e-mail: iljafrissen@hotmail.com

I. Frissen
LUNAM Université, CNRS, Ecole Centrale de Nantes, IRCCyN (Institut de Recherche en Communications et Cybernétique de Nantes), 1 rue de la Noë, BP 92101, 44321 Nantes Cedex 3, France

J. L. Campos
Max Planck Institute for Biological Cybernetics, Multisensory Perception and Action Group, Tübingen, Germany
e-mail: Jennifer.Campos@uhn.ca

J. L. Campos
iDAPT, Toronto Rehabilitation Institute, Toronto, Ontario, Canada

J. L. Campos
Department of Psychology, University of Toronto, Toronto, Ontario, Canada

M. Sreenivasa
Nakamura Laboratory, Department of Mechano-Informatics, University of Tokyo, Tokyo, Japan
e-mail: manu@ynl.t.u-tokyo.ac.jp

M. O. Ernst
Department of Cognitive Neuroscience, University of Bielefeld, Bielefeld, Germany
e-mail: marc.ernst@uni-bielefeld.de

F. Steinicke et al. (eds.), *Human Walking in Virtual Environments*, DOI: 10.1007/978-1-4419-8432-6_6, © Springer Science+Business Media New York 2013 ular, the nature of natural unconstrained walking and the effects of treadmill walking on characteristics of gait. The second part of this chapter describes the multisensory nature of walking, with a focus on the integration of vestibular and proprioceptive information during walking. The third part of this chapter describes research on large-scale human navigation and identifies possible causes for the human tendency to veer from a straight path, and even walk in circles when no external references are made available. The chapter concludes with a summary description of the features of the CyberWalk platform that were informed by this collection of research findings and briefly highlights the current and future scientific potential for this platform.

6.1 Introduction

By far the most natural way to move through our environment is through locomotion. However, the seemingly effortless act of walking is an extremely complex process and a comprehensive understanding of this process involves scientific and clinical studies at different levels of analysis. Locomotion requires preparing the body posture before initiating locomotion, initiating and terminating locomotion, coordinating the rhythmic activation patterns of the muscles, of the limbs and of the trunk, and maintaining dynamic stability of the moving body [77]. There is also a need to modulate the speed of locomotion, to avoid obstacles, to select appropriate, stable foot placement, to accommodate different terrains, change the direction of locomotion, and guide locomotion towards endpoints that are not visible from the start. To this end, locomotion engages many different sensory systems, such as the visual, proprioceptive, auditory and vestibular systems, making it a particularly interesting multisensory problem. Importantly, these are also factors that must be considered when developing a realistic walking interface to be used with Virtual Reality (VR).

Although many of these aspects of locomotion have received extensive scientific attention, much of the earlier laboratory-based research, though highly valuable, has lacked ecological validity. Ultimately, scientific research should, when possible, evaluate human behaviors as they occur under natural, cue-rich, ecologically valid conditions. To this end, VR technology has been providing researchers with the opportunity to provide natural, yet tightly controlled, stimulus conditions, while also maintaining the capacity to create unique experimental scenarios that would (or could) not occur in the real world [16, 24, 68, 105]. An integral part of VR is to also allow participants to move through the Virtual Environments (VE) as naturally as possible. Until recently a very common way of having observers navigate through VEs was to have them manipulate unnatural control devices such as joysticks, computer mice, and keyboards. Despite having some advantages over mere visual stimulation, such rudimentary motion control devices are severely limited. While using such devices, the physical actions which drive self-motion are very different from the action of natural locomotion which they are intended to replace (e.g. clicking a mouse button to move forward versus stepping). Moreover, the sensory input is mainly visual and other important sensory information is lacking, notably proprioceptive feedback from the legs and vestibular feedback. Fortunately, more natural locomotion interfaces, such as bicycles, treadmills and fully-tracked free-walking spaces, are becoming more common (see [24] for a review). Although with these solutions locomotion is much closer to real life movements, they are still constrained in important ways. In the case of the bicycle, for instance, there is no absolute one-to-one relationship between the metrics of visual space and those of the proprioceptive movements because of the unknown scale of one pedal rotation (i.e., this would depend on the gear, for instance). Fully-tracked walking spaces are constrained by the size of the actual space within which they are contained. Treadmill setups are restrictive as most of them are rather small [94] and only allow walking in one direction. Indeed, in everyday navigational tasks, we rarely walk completely straight over extended periods of time. In short, today it is still difficult to allow people to freely walk through large scale VEs in an unconstrained manner.

It is this unsatisfactory situation that prompted some of the work reported in this volume and it likewise prompted the CyberWalk project. The goal of this project was the development of a novel, multimodal, omnidirectional walking interface, with at its core, a 4×4 m omnidirectional treadmill. The project encompassed an international consortium dedicated to both scientific and technological research. The CyberWalk platform is the first truly omnidirectional treadmill of its size that allows for natural walking in any direction through arbitrarily large Virtual Environments. It is a major step towards having a single setup that allows for the study of the many facets of human locomotion, ranging from the biomechanical to the cognitive processes involved in navigating large areas. The platform consists of segmented belts which are mounted on two large chains in the shape of a torus, which allows it to move the walking surface in both horizontal directions and thereby enables indefinite omnidirectional walking and turning (see Fig. 6.7). It is integrated with additional VR capabilities so that a virtual world is presented through a head-mounted display (HMD) and updated as a function of the movements of the user. The platform is described more fully in [95] and in Sect. 6.5 of this chapter. More detailed descriptions of specific technological and engineering aspects of the platform can be found elsewhere [29, 87–89, 94, 112].

The technological development of the platform had a strong human-centered approach and was guided by human gait and psychophysical research conducted at the Max Planck Institute for Biological Cybernetics (MPI), one of the consortium partners. Here we report on a selected number of these studies. Since a major objective was to develop a platform that enables natural walking, we studied basic gait parameters during natural unconstrained outdoor walking as a general reference. The CyberWalk platform has at its core a treadmill, and thus we investigated potential differences between normal overground walking and treadmill walking. Studies were also focused on the multisensory processes at play during human walking. While there is a wealth of research on the role of vision in locomotion, relatively little is known about the interaction between the different non-visual senses. Consequently, a series of studies was conducted to look at the interaction between vestibular and proprioceptive information during walking. Finally, a number of studies on human navigation were conducted on long-range navigational capabilities with and without the use of visual information.

6.2 Gait and Biomechanics

One of the major goals of the CyberWalk project was to enable natural and unconstrained walking on a treadmill based system. This original challenge introduced many questions and we highlight two of those here. First, in order to enable natural and unconstrained gait, a description of typical gait characteristics was needed. For instance, at what speed do people normally walk, how do they start and stop walking, how often and how much do they turn? Second, there is still a debate in the literature as to whether gait characteristics during treadmill walking are the same as during overground walking. Thus, we conducted a series of studies to address these questions. The results were intended to assign tangible constraints on a system intended to support natural walking (e.g., on the accelerations required and the size of the walking surface).

6.2.1 Natural Unconstrained Walking

There is, in fact, very little literature on natural unconstrained walking. One reason for this is a previous lack of measurement technologies suitable to capture gait with sufficient accuracy. In recent years, however, Global Positioning Systems (GPS) are providing a promising solution to this problem [70, 101]. For instance, Terrier and colleagues used a highly accurate GPS to show that inter- and intra-subject variability of gait characteristics can be measured outdoors [107, 108]. Moreover, GPS data can be combined with Inertial Measurement Unit (IMU) technologies to develop highly accurate measurement systems with high data rates [104]. Nevertheless, the few available studies that report GPS data are still highly constrained in a fashion reminiscent of laboratory research. For instance, participants are often asked to follow a modulated pace/frequency [86, 106, 108],which is known to significantly increase energy cost [115] or to walk/run along a predefined path [32, 104, 106, 107]. In studies where walking behavior was not constrained, data were collected over several days at very low sampling rates to form a picture of overall "behaviors" rather than basic gait parameters such as step length and frequency [26, 70, 76, 85, 110].

We conducted a study of unconstrained outdoor human walking that differed from previous studies in that we observed people walking for an extended period of time (1 h) and completely at their own volition [97]. We measured the position of the trunk and rotational rates of the trunk and head. The high accuracy required to capture trunk position outdoors was achieved by using a Carrier-Phase Differential GPS setup (C-DGPS). The C-DGPS utilizes a secondary static GPS unit (master station) to correct for errors in a mobile rover GPS (Novatel Propak, V3–L1). The rover was combined with an aviation grade, extremely light and compact antenna that was mounted onto a short pole fixed to a frame inside a backpack. Data were output at 5 Hz with a typical accuracy between 2 and 10 cm depending on environmental conditions (tree cover, reflections etc). For additional measures about movements of the trunk we used a 6-axis IMU (Crossbow Technology, IMU300), with measurement ranges of ± 19.6 m/s^2 and ± 100 °/s. The measuring unit was rigidly fixed to the bottom of the GPS antenna frame and logged data at 185 Hz. To measure the head we used a custom-built 3-axis IMU (ADXL202 and ADXRS150, logging at 1028 Hz) that was mounted on a head brace worn by the participants (total weight of less than 150 g). A strobe signal was used to align the data streams in post-processing. All devices plus data loggers and battery packs were fit in the backpack (just under 9 kg).

A task was designed that would induce the normal variability in walking behavior without imposing a stereotypical walking pattern. Fourteen participants walked through a residential area while searching for 30 predefined objects (e.g., street signs, statues) using a map of the area. The locations of the objects were indicated on the map by flags and participants were asked to note the time when they reached the location of an object. They were instructed to optimize the order in which they visited the targets such that they would visit the largest number of objects within one hour. Using recordings of the 3D position of the trunk, a wide range of walking parameters were computed including, step length (SL), step frequency (SF), and their ratio, also known as the walk ratio (WR). This ratio has been found to be invariant within a range of walking speeds [48, 90], and has been linked to optimal energy expenditure [56, 115]. Evidence of invariance in WR has been reported for walking at manipulated speeds along a 100 m straight athletic track [107] and a 400 m oval track [108], but never under free walking conditions. We also measured walking speed during straight and curved walking trajectories and starting and stopping behavior. Walking speed was calculated as the difference between consecutive positions of the trunk position in the horizontal (GPS) frame. Table 6.1 presents some individual and mean basic gait parameters computed from the GPS data. For a complete description of results please refer to [97].

Results demonstrated that when people walked on a straight path, the average walking speed was 1.53 m/s. This value is very similar to field survey data [41, 65]. Perhaps not surprisingly, walking speed decreased when people walked on a curved path. The magnitude of the decrease depended on both the radius and angle of the turn taken. For turn angle, walking speed decreased linearly with angle. Thus, it changed from 1.32 m/s at 45° angles to around 1 m/s at complete turnarounds (i.e., 180°). These values are in strong agreement with those observed in a controlled experiment conducted in a fully-tracked indoor lab space [98]. As for turn radius, walking speed was seemingly constant for turns with radii $\geq$10 m (1.49 m/s) and for turns with radii $\leq$5 m (1.1 m/s), while in between these radii values, walking speed changed in a fairly linear fashion.

Consistent with previous literature [90, 107] we found that WR was relatively invariant with respect to walking speed. After correcting for participant height (see [90]), we found that most of the adjusted values of WR were close to 0.4 m/steps/s. There were some outliers at slower walking speeds (i.e., below 1 m/s), which is again

Table 6.1 Individual and mean descriptive statistics of 1 h of unconstrained walking

	Demographics		Straight walking						Curve walking		No. Starts/Stops	Total distance (m)
	Age (year)	Height (m)	Time (%)*	WS (m/s)	SD WS (m/s)**	SL (m)	SF (Hz)	WR (m/step/s)	Time (%)*	No. turns		
Male	22	1.81	71.2	1.32	0.13	0.71	1.77	0.40	28.8	26	28	2439
	25	1.79	51.9	1.62	0.18	0.80	1.97	0.41	48.1	35	65	1885
	20	1.96	54.2	1.53	0.14	0.83	1.81	0.46	45.8	41	24	2743
	26	1.95	56.2	1.58	0.14	0.87	1.80	0.48	43.8	40	15	2558
	24	1.82	52.0	1.68	0.15	0.84	1.95	0.43	48.0	39	23	2207
	24	1.78	61.3	1.73	0.18	0.80	2.00	0.39	38.7	40	21	3143
	32	1.83	44.3	1.52	0.16	0.79	1.86	0.43	55.7	48	28	2677
	22	1.73	59.7	1.33	0.12	0.75	1.73	0.44	40.3	35	48	1680
Mean			56.4	1.54	0.15	0.80	1.86	0.43	43.7	38	31.5	2417
Female	21	1.65	64.8	1.41	0.14	0.73	1.86	0.39	35.1	27	34	1884
	27	1.65	43.5	1.49	0.13	0.73	1.99	0.37	56.5	59	31	2594
	26	1.68	58.9	1.50	0.13	0.71	2.00	0.35	41.0	36	22	2415
	22	1.64	40.4	1.62	0.15	0.69	2.20	0.31	59.6	68	34	2608
	20	1.70	42.3	1.63	0.17	0.78	2.00	0.39	57.7	48	27	2737
	31	1.63	47.3	1.40	0.15	0.70	1.94	0.36	52.7	38	12	2310
Mean			49.5	1.51	0.15	0.72	2.00	0.36	50.4	46.0	26.7	2425
Over all			53.4	1.53	0.15	0.77	1.92	0.40	46.6	41.4	29.4	2420

WS, walking speed; SL, step length; SF, step frequency; WR, height adjusted walk ratio. *Percentage of total walking time available to calculate descriptive statistics after excluding start/stop data. **Standard deviation of walking speed

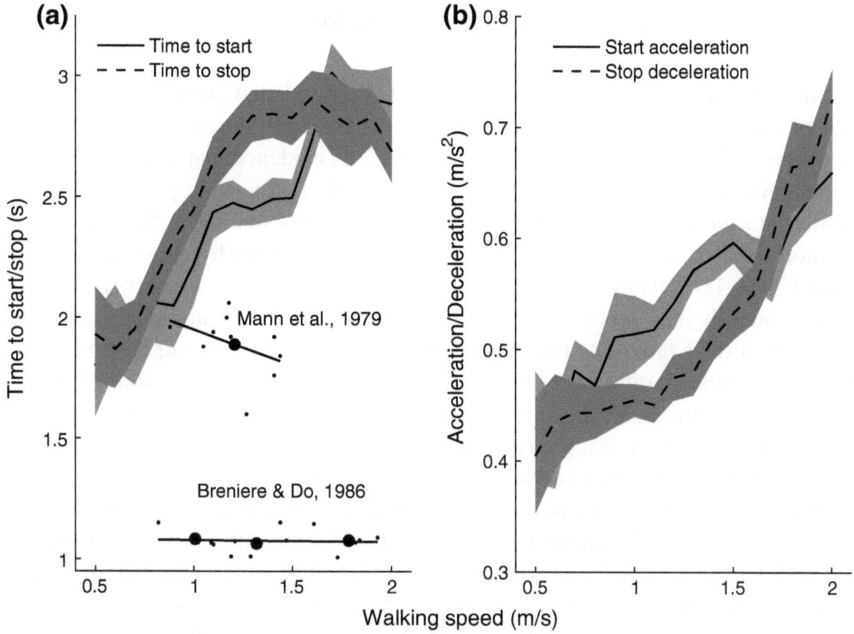

Fig. 6.1 Starting and stopping. The time (**a**) and accelerations (**b**) during starts (*solid line*) and stops (*dotted line*) as a function of steady walking speeds. *Shaded regions* indicate standard errors of the means. Also plotted (*panel a*) are the individual results (*small dots*), mean results (*large dots*) from two earlier studies. To illustrate the general trends, linear regressions are shown on all individual results (*black lines*)

consistent with earlier reports [90], and the WR at these slower walking speeds was also more variable. The relative invariance of WR in natural (and controlled) walking underlines its usefulness as a clinical diagnostic tool for detecting abnormal gait but also to the scientific study of human locomotion in general.

The time that it takes to reach a steady walking speed depends on the desired speed (see Fig. 6.1). It took an average of 2 and 3 s to reach walking speeds of 0.5 and 2 m/s, respectively. The relationship between the time it took to stop and walking speed was very much the same. The dependence on walking speed, however, contradicts findings by Breniere and Do [11] who found that the time it takes to reach the desired walking speed is independent of walking speed. Dependence on walking speed has been found by others [60, 71], although we observe that in natural walking humans take more time to start and stop than in laboratory settings. To illustrate these differences Fig. 6.1 also includes the data from Breniere and Do [11] and Mann et al. [71] together with our own results. One possible cause for this difference is the protocol used in laboratory experiments [96]. Specifically, whereas earlier studies typically use an external "go" signal, our participants were free to start and stop as they pleased.

6.2.2 Overground Versus Treadmill Walking

While treadmills allow for the observation of walking behavior over extended periods of time, it is still a matter of debate as to whether gait during treadmill walking is different than overground walking [3]. There is evidence that treadmill walking can significantly alter the temporal [3, 30, 100, 114], kinematic [3], and energetic [78] characteristics of walking. One apparently robust finding is that walking on a (motorized) treadmill increases step frequency (cadence) by approximately 6% [3, 30, 100, 114]. It has, therefore, been concluded by many researchers that motorized treadmills may produce misleading or erroneous results and that care should be taken in their interpretation. At the same time there are also studies that do not find any significant differences between overground and treadmill walking [75, 84]. Two possible sources for this discrepancy that we have addressed in our research are differences between walking surfaces and the availability of relevant visual feedback about self-motion during treadmill versus overground walking.

Treadmills are typically more compliant than the regular laboratory walking surfaces used in past studies, and it has been speculated that it is this difference in surface stiffness that affects locomotion patterns when directly comparing treadmill walking with overground walking (e.g., [30, 31]). Such speculations are warranted by other research showing significant effects of walking surface compliance on basic gait parameters such as step frequency and step length [72]. Interestingly, the one study that compared overground with treadmill walking using similar walking surfaces found no differences in gait parameters [84].

Another potential factor to consider is that participants typically have visual information available during walking. During natural, overground walking, dynamic visual information (i.e. optic flow), is consistent with the non-visual information specifying movement through space. However, during treadmill walking, a considerable sensory conflict is created between the proprioceptive information and the visual (and vestibular) information (see also Sect. 6.3.2) such that the former informs participants that they are moving, yet the latter informs them they are in fact stationary. Although it is not obvious how such a conflict might specifically alter gait parameters, there is evidence that walking parameters are affected by whether visual feedback is available or not. For instance, Sheik-Nainar and Kaber [91] evaluated different aspects of gait, such as speed, cadence, and joint angles when walking on a treadmill. They evaluated the effects of presenting participants with congruent and updated visuals (via a HMD projecting a simulated version of the lab space), compared to stationary visuals (real world lab space with reduced FOV to approximate HMD). These two conditions were compared to natural, overground walking. Results indicated that while both the treadmill conditions caused participants to walk slower and take smaller steps, when optic flow was consistent with the walking speed, gait characteristics more closely approximated that of overground walking. Further, Hallemans et al. [50] compared gait patterns in people with and without a visual impairment and compared the gait patterns of normally sighted participants under full vision and no vision conditions. Results demonstrated that participants with a visual impairment walked with a shorter

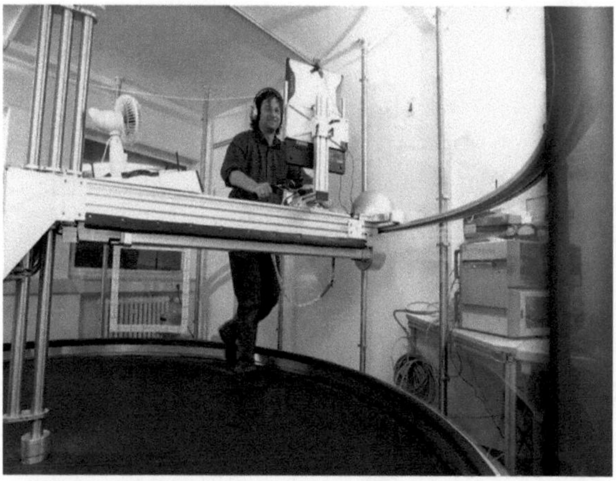

Fig. 6.2 The circular treadmill (CTM) at the Max Planck Institute for Biological Cybernetics. It consists of a large motorized wooden disc (Ø = 3.6 m) covered with a slip resistant rubber surface and a motorized handlebar. The disc and handlebar can be actuated independently from each other. The disc's maximum angular velocity is 73 °/s, and the handlebar can reach a maximum velocity of 150 °/s. Walking on the CTM is natural and intuitive and does not require any explicit training (see also [42]). For the overground versus treadmill study (Sect. 6.2.2) the setup was equipped with a TrackIR: Pro 4 (NaturalPoint) optical tracking device for tracking the position and orientation of the head. It was fixed on top of the depicted laptop monitor that was mounted in front of the participant. The device has a 46° field of view and provides 6 DOF tracking with mm and sub-degree precision for position and orientations, respectively. For the experiments described in Sects. 6.3.2 and 6.2.3, a custom-built pointing device was mounted on the handlebar within comfortable reaching distance of the right hand (at a radius of 0.93 m from the center of the disk). The pointing device consisted of a USB mechanical rotary encoder (Phidgets, Inc.) with a pointing rod attached and encased in plastic (see also [42]). Note that the CTM has since moved to the department of Cognitive Neurosciences at Bielefeld University. (Photograph courtesy of Axel Griesch)

step length than sighted individuals and that sighted participants who were blindfolded also showed similar changes in gait (see also [74]). Further, in the absence of vision, normally sighted participants walked slower and had lower step frequencies when blindfolded compared to when full vision was available, which was hypothesized to reflect a more cautious walking strategy when visual information was absent. However, it is not known whether walking is differentially affected by the presence and absence of congruent visual feedback.

Humans have a strong tendency to stabilize the head during walking (and various other locomotor tasks) in the sense that they minimize the dispersion of the angular displacement of the head [13]. Interestingly, visual feedback does not appear to be important for this stabilization [80]. However, the walking conditions under which this has been studied have been very limited. Participants were asked to walk at their own preferred speed or to step in place [80]. Very little is known about the generality of this lack of an effect of vision and whether there are differences between overground and treadmill walking.

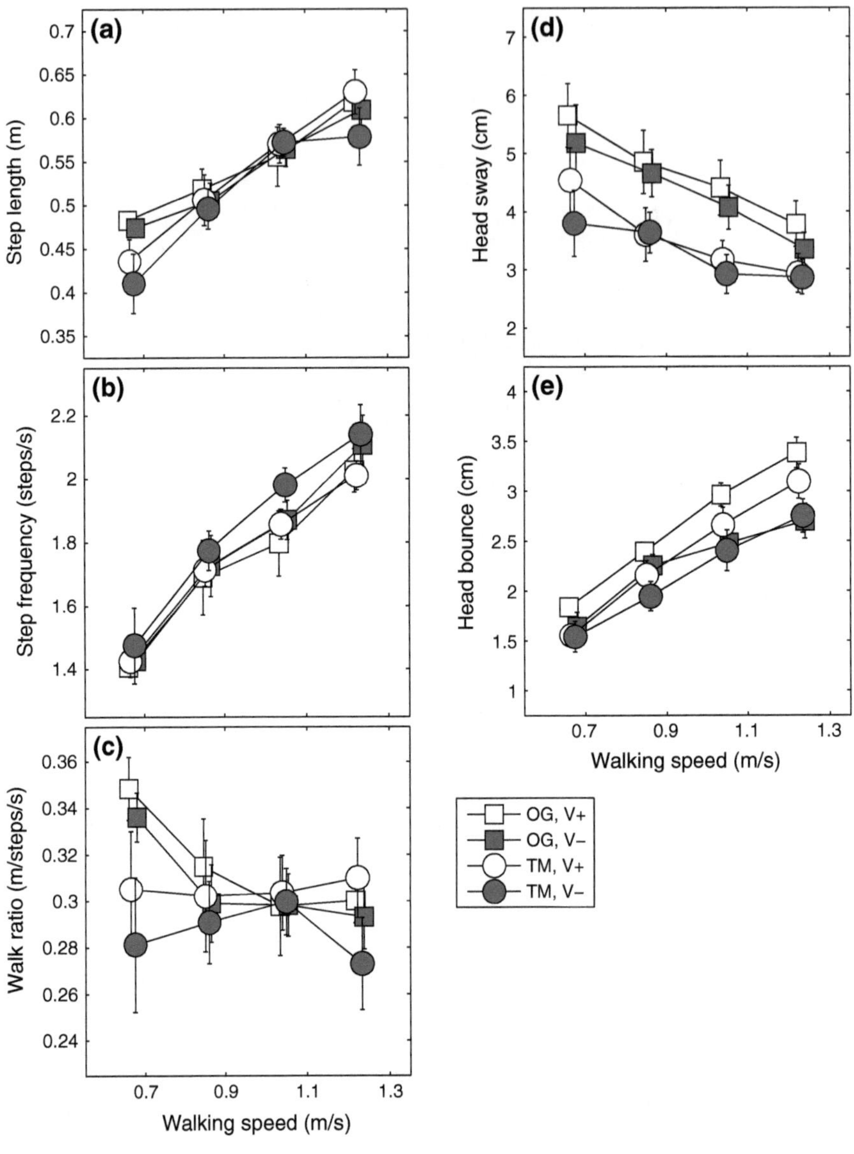

Fig. 6.3 Overground versus treadmill walking. Group means with SEM for **a** step length, **b** step frequency, **c** walk ratio, **d** head sway from left to right, and **e** vertical head bounce, as a function of walking speed (in m/s). Walking was either overground (OG) or stationary walking on the treadmill (TM), and there was either visual feedback (V+) or not (V−). Fourteen participants, between the ages of 19 and 33 (7 females), walked 3 times for 30 s at a constant velocity for each condition. Dependent measures were obtained from measurements of the position of the head, and step length and frequency were corrected for individual heights (see also Sect. 6.2.1)

We investigated the effects of walking surface and visual feedback on basic gait parameters and on the movement of the head in an integrated manner. This experiment was conducted using a circular treadmill (CTM) at the MPI (see Fig. 6.2 and caption for additional details). The effect of surface stiffness on gait characteristics was controlled for by having participants walk in place and walk through space on the same treadmill surface. Specifically, overground walking consisted of simply leading the participant around on the stationary disc using the motorized handlebar. Stationary ("treadmill") walking consisted of walking in place on the moving disc without moving through space. If the difference in surface is a major determinant in causing the previously reported differences between overground and treadmill walking, then we would expect this difference to disappear in this experiment. Visual feedback was also manipulated by having people walk while wearing a blindfold or not. Walking speeds were controlled by moving either the disc or the handlebar at one of four velocities (see caption of Fig. 6.3), for the stationary and walking through space conditions, respectively. The results demonstrated that there were indeed very few differences observed between the gait parameters measured during stationary walking versus overground walking. Step length (Fig. 6.3a) and walk ratio (Fig. 6.3c) were comparable across walking speeds. The exception was that for the slowest walking speed (0.7 m/s), the overground walking condition produced larger step lengths and walk ratios in comparison to stationary walking. This particular effect is consistent with previous findings that reflected higher walk ratios at slower overground walking speeds (e.g., [90]). This higher walk ratio at the slowest walking speed is likely due to an increase in step length given that step frequency was virtually identical across all conditions (see Fig. 6.3b). Results also demonstrated that during stationary walking there was a significant decrease in head sway (Fig. 6.3d) and head bounce (Fig. 6.3e) compared to overground walking. As for the effect of vision, the results demonstrated that, irrespective of the walking condition, step length and frequency were unaffected by the presence or absence of visual feedback. This is in contrast with above-described studies that did find significant decreases in both step length and frequency [50, 74].

In summary, with respect to basic gait parameters, there were hardly any differences between overground walking and stationary walking. Most notable was the complete absence of an effect on step frequency, which has typically been the most consistently observed difference in earlier studies. Our results are, however, consistent with several other earlier studies that also did not find a difference between overground and treadmill walking [75, 84] and lend support to the notion that previously reported differences may be (partially) due to the fact that walking surfaces were not controlled for. Another interesting finding is that stationary walking significantly reduced the lateral (sway) and vertical (bounce) head movements. It is currently unclear what the cause for this change is. However, it is thought that head stabilization behavior helps organize the inputs from the visual, vestibular, and even somatosensory systems [13]. It is possible that during treadmill walking head movements are reduced in order to establish a more stable reference frame because of the registered discrepancy between the proprioceptive sense that signals movement, and the vestibular and visual senses that signal a stationary position. As for visual feedback, the only statistically reliable effect of the visual manipulation was a reduction of the vertical movements of the head at the highest walking speeds during overground walking as compared to stationary walking. When visual feedback was not available, this produced some trends in the gait parameters (increases in step frequency and decreases in step length and walk ratio), although these were not statistically significant.

6.2.3 Potential Implications for CyberWalk

One specific finding that impacted the design specifications of the CyberWalk platform was that it took at least 2 s to accelerate the treadmill to the very slow speed of 0.5 m/s. As we will see in the following section, providing vestibular inputs by allowing movement through space is an important part of simulating natural locomotion. Thus, from this perspective it meant that the CyberWalk platform needed to ideally be big enough to accommodate such start up accelerations. The finding that stationary walking does not change the main walking parameters of step length and step frequency is encouraging as it means that the walking data on the treadmill should be representative of normal gait. This also affected the design of the platform, albeit in a more indirect fashion. We surmised that the platform should ideally have a surface that is as stiff as possible since the most typically studied walking surfaces are very stiff (e.g., sidewalks). Head movements, on the other hand, did change during stationary walking in that they were less pronounced than during overground walking. This might seem advantageous in light of the fact that on the CyberWalk, head mounted displays (HMDs) are the primary means of visually immersing the user in VR and therefore having less head bounce would reduce visual motion artifacts and potential tracking lags for rapid movements. However, it does raise the possibility that the normal head stabilization function during walking (e.g., [80]) may be different during treadmill walking, which may affect the role of the proprioceptive receptors in the neck and also the role of coincident vestibular inputs.

6.3 Multisensory Self-Motion Perception

A veridical sense of self-motion during walking is a crucial component for obtaining ecological validity in VR. Of particular interest to us is the multisensory nature of self-motion perception. Information about the extent, speed, and direction of egocentric motion is available through most of our sensory systems (e.g. visual, auditory, proprioceptive, vestibular), making self-motion perception during locomotion a particularly interesting problem with respect to multisensory processing. During self-motion perception there are important roles for the visual system (e.g. optic flow), the vestibular system (the inner ear organs including the otoliths and semicircular canals), the proprioceptive system (the muscles and joints), and efference copy signals representing the commands issued to generate our movements. There is also some suggestive evidence for a role of the auditory system (e.g., [99]) and somatosensory system (e.g., [33]). Much work has been done to understand how each of these sensory modalities contribute to self-motion individually, however, researchers have only recently begun to evaluate how they are combined to form a coherent percept of self-motion and the relative influences of each cue when more than one is available.

6.3.1 Multisensory Nature of Walking

Since no single sense is capable of operating accurately under all circumstances, the brain has evolved to exploit multiple sources of sensory information in order to ensure both a reliable perception of our environment (see [20]) and appropriate actions based on that perception [37]. A fundamental question in the cognitive neurosciences asks what mechanisms are used by the central nervous system to merge all of these sources of information to form a coherent and robust percept. It seems that it employs two strategies to achieve robust perception. The first strategy, sensory combination, describes interactions between sensory signals that are not redundant. That is, information is specified in different coordinate systems or units. The second strategy, sensory integration, reduces the variance of redundant sensory estimates, thereby increasing their reliability [37].

Human locomotion is particularly interesting from the perspective of sensory integration as it involves a highly dynamic system, meaning that the sensory inputs are continuously changing as a function of our movements. For instance, with each stride (i.e., from the heel strike of one foot to the next heel strike of the same foot) the head moves up and down twice in a near sinusoidal fashion [62, 106], thereby generating continuously changing accelerations that are registered by the vestibular system. Similarly, with each stride, the musculoskeletal system generates a set of dynamically changing motor signals, the consequences of which are registered by the proprioceptive system. Finally, the visual flow is likewise marked with periodic vertical and horizontal components. Thus, the various pertinent sensory inputs are in a systematic state of flux during walking. Moreover, findings that visual [54], somatosensory [116], and vestibular [6] signals exhibit phase-dependent influences on postural control during walking suggest the interesting possibility that the reliabilities of the sensory signals are also continuously changing and possibly in phase with the different stages of the gait cycle.

A particularly influential group of models of multisensory integration have considered the problem from the point of view of efficiency. These efforts are often referred to as the "Bayesian approach", which was originally applied to visual perception (e.g., [15, 17, 64]). It is acknowledged that neural processes are noisy [38] and consequently, so are sensory estimates. The goal is then for the brain to come up with the most reliable estimate, in which case the variance (i.e., noise) of the final estimate should be reduced as much as possible. If the assumption is made that the noise attributable to individual estimates is independent and Gaussian, then an estimate with the lowest variance is obtained using Maximum Likelihood Estimation (MLE) [35]. MLE models have three general characteristics. First, information from two or more sensory modalities is combined using a weighted average. Second, the corresponding weights are based on the relative reliabilities of the unisensory cues (i.e., the inverse of their variances); the cue with the lowest unimodal variance will be weighted highest when the cues are combined. Third, as a consequence of integration, the variance of the integrated estimate will be lower than those observed in either of the individual estimates. There is now mounting evidence that humans combine information from across the senses in such a "statistically optimal" manner (e.g., [37]). Most of this work has been aimed at modeling cue integration between the exteroceptive senses such as vision, haptics, and hearing [2, 4, 12, 35, 36], or within the visuomotor system (e.g., [63, 66]), but very few studies have considered whether the same predictions apply to multisensory self-motion perception.

The Bayesian perspective is now just starting to be considered in the field of human locomotion (e.g., [25]), and self-motion in particular [18, 19, 21, 23, 39, 42]. For instance, a study by Campos et al. [23] highlights the dynamic nature in which optic flow and body-based cues are integrated during walking in the real world. The study shows that the notion of optic flow as an all-inclusive solution to self-motion perception [46] is too simplistic. In fact, when body-based cues (e.g. proprioceptive and vestibular inputs) are available during natural walking they can dominate over visual inputs in dynamic spatial tasks that require the integration of information over space and time (see also [21] for supporting evidence in VR). Other studies have attempted to look at body-based cues in isolation and investigate how these individual sources interact with visual information. For instance, a number of studies have considered the integration of optic flow and vestibular information for different aspects of self-motion perception (e.g., [19, 39, 40, 51, 61]). Evidence from both humans [18, 39], see also [69]) and non-human primates [40, 49] shows that visual-vestibular integration is statistically optimal when making heading judgments. This is reflected by a reported reduction in variance during combined cue conditions, compared to the response patterns when either cue is available alone. Interestingly, when the visual signal lacks stereoscopic information, visual-vestibular integration may no longer be optimal for many observers [19]. To date, the work on visual-vestibular interactions has been the most advanced with respect to cue integration during self-motion in the sense that it has allowed for careful quantitative predictions. Studies on the combinations of other modalities during self-motion perception have also started to provide qualitative evidence that support the MLE. For instance, Sun et al. [102], looked at the relative contributions of optic flow information and proprioceptive information to human performance on relative path length estimation (see also [103]). They found evidence for a weighted averaging of the two sources, but also that the availability of proprioceptive information increased the accuracy of relative path length estimation based on visual cues. These results are supported by a VR study [21] which demonstrated a higher influence of body-based cues (proprioceptive and vestibular) when estimating walked distances and a higher influence of visual cues during passive movement. This VR study further showed that although both proprioceptive and vestibular cues contributed to travelled distance estimates, a higher weighting of vestibular inputs were observed. These results were effectively described using a basic linear weighting model.

6.3.2 Integration of Vestibular and Proprioceptive Information in Human Locomotion

Consider walking through an environment that is covered in fog or walking in the pitch dark. While these scenarios render visual information less reliable, evidence shows that humans are still very competent in various locomotion tasks even in the complete absence of vision (e.g., [22, 34, 67, 73, 83, 103, 109]). Past research often reports that when either walking without vision or when passively moved through space, body-based cues are often sufficient for estimating travelled distance [7, 21, 24, 51, 58, 67, 73, 92, 102, 103] and to some extent self-velocity [7, 22, 58, 92].

A series of studies have also looked specifically at the interactions between the two main sources of body based cues; the proprioceptive system and the vestibular system. Studies that have investigated the role of vestibular and/or proprioceptive information in self-motion perception have done so by systematically isolating or limiting each cue independently. Typical manipulations include having participants walk on a treadmill (mainly proprioceptive information), or passively transporting them through space in a vehicle (mainly vestibular information specifying translations through space). The logic is that walking in place (WIP) on a treadmill produces proprioceptive but no vestibular inputs associated with self-motion through space, while during passive movement (PM), there are vestibular inputs but no relevant proprioceptive information from the legs specifying movement through space. These conditions can then be compared to normal walking through space (WTS), which combines the proprioceptive and vestibular inputs of the unisensory WIP and PM condition. For instance, Mittelstaedt and Mittelstaedt [73] reported that participants could accurately estimate the length of a travelled path when walking in place (proprioception), or when being passively transported (vestibular). In their study, even though both cues appeared sufficient in isolation, when both were available at the same time (i.e., when walking through space) proprioceptive information was reported to dominate vestibular information. But what this study could not specify was by how much it dominates or, more generally, what the relative weights of the individual cues are.

There is, however, a fundamental problem that makes it very difficult to make assessments of cue weighting and studying the multisensory nature of self-motion in general. The problem is that there is a very tight coupling between vestibular and proprioceptive information during normal walking. The two signals are confounded in the sense that under normal circumstances there can be no proprioceptive activity (consistent with walking) without experiencing concurrent vestibular excitation.

In fact, this strong coupling has lead Frissen et al. [42] to argue for a "mandatory integration" hypothesis which holds that during walking the brain has adopted a strategy of always integrating the two signals. It also leads to substantial experimental difficulty when attempting to obtain independent measures from the individual senses (see also [24]). Consequently, during the often used, "proprioceptive only" walking in place condition, vestibular inputs are in fact concurrently present, yet specify a stationary position. This thus creates a potential sensory conflict when the aim is to obtain unbiased unisensory estimates. The reverse conflict occurs in the "vestibular only" PM condition, where the proprioceptive input specifies a stationary position. Although in this case, it should be noted, that there are numerous instances in which vestibular excitation is experienced without contingent proprioceptive information from the legs, including whenever we move our head, or when moving in a vehicle. In other words, in the case of passive movements, the coupling may not be as tight.

Despite the fact that it is difficult to obtain unisensory proprioceptive and vestibular estimates, it is possible to create conditions in which the conflict between the vestibular and proprioceptive cue are much reduced and, moreover, controllable. This will enable us to determine the relative weighting of the individual cues. One way is to use a rotating platform in combination with a handlebar that can be moved independently. An early example of this was a platform used by Pick et al. [79], which consisted of a small motorized turntable (radius 0.61 m) with a horizontal handle mounted on a motorized post extending vertically through the center. Using this setup Bruggeman et al. [14] introduced conflicts between proprioceptive and vestibular inputs while participants stepped around their earth-vertical body axis. Participants always stepped at a rate of 10 rotations per minute (rpm) (constituting the proprioceptive input), but because the platform rotated in the opposite direction, participants were moved through space at various different rates (constituting the vestibular input). They found that when the proprioceptive and vestibular inputs were of different magnitudes, the perceived velocity fell somewhere between the two presented unisensory velocities, thus suggesting that the brain uses a weighted average of vestibular and proprioceptive information as predicted by MLE (see also [5]). However, a limitation of this type of relatively small setup is that it only allows participants to perform rotations around the body axis. That is, it allows participants to step in place, which is a very constrained and rather unnatural mode of locomotion with biomechanics that are different from normal walking.

Such restrictions do not apply to the CTM (Fig. 6.2) which allows for full stride curvilinear walking. This unique setup also allows us to manipulate vestibular, proprioceptive (and visual) inputs independently during walking. In one of our recent studies we assessed multisensory integration during self-motion using a spatial updating paradigm that required participants to walk through space with and without conflicting proprioceptive and vestibular cues [42]. The main condition was the multisensory, "walking through space" condition during which both vestibular and proprioceptive systems indicated self-motion. This condition consisted of both congruent and incongruent trials. In the congruent trials, participants walked behind the handlebar while the treadmill disk remained stationary. Thus, the vestibular and proprioceptive inputs conveyed the same movement velocities; in other words, the proprioceptive-vestibular gain was 1.0. In the incongruent trials, systematic conflicts were introduced between the vestibular and proprioceptive inputs. This was achieved by having participants walk at one rate, while the disk was moved at a different rate. Specifically, proprioceptive gains of 0.7 and 1.4 were applied to two vestibular velocities (25 °/s and 40 °/s). To achieve a gain of 0.7, the disk moved in the same direction as the handlebar but at 30 % of its speed. To achieve a gain of 1.4, the disk moved at 40 % of the handlebar speed but in the opposite direction. We also tested two additional conditions. In the "walking in place" condition, participants walked in place on the treadmill but did not move through space. Like in previous studies, participants were instructed to use the proprioceptive information from their legs to update their egocentric position as if they were moving through space at the velocity specified by the CTM. In the "passive movement" condition, participants stood still while they were passively moved by the CTM. Spatial updating was measured using a continuous pointing task similar to that introduced by Campos et al. [22] and Siegle et al. [92], which expanded upon a paradigm originally developed by Loomis and colleagues [43, 67]. The task requires the participant to continuously point at a previously viewed target during self-motion in the absence of vision. A major advantage of this method is that it provides continuous information about perceived target-relative location and thus about self-velocity during the entire movement trajectory. The results were consistent with an MLE model in that participants updated their position using a weighted combination of the vestibular and proprioceptive cues, and that performance was less variable when both cues were available.

Unfortunately the results did not allow us to determine the relative weighting of the two cues (see [42]). We therefore conducted a new experiment which employed a standard psychophysical 2-interval forced choice (2-IFC) paradigm (see [45], for an introduction). Experimental details are provided in the caption of Fig. 6.4. In each trial participants walked two times and they indicated in which of the two they had walked faster. In one interval (the standard) participants walked under various conditions of conflicting vestibular and proprioceptive signals, while in a second interval (the comparison) they walked through space without cue conflict. By systematically changing the comparison (i.e., handlebar velocity) we can determine the point at which the standard and comparison were perceptually equivalent (i.e., the point of subject equality, or PSE).

Figure 6.4a shows the mean PSEs as a function of vestibular input. In the conditions with conflicting inputs, the PSEs lie between the two extreme cases (solid horizontal and diagonal line). Also, the PSEs are not on a straight line, indicating that the relative weighting depends on the vestibular input. This is illustrated in Fig. 6.4b where the vestibular weights are plotted for the different conflict conditions. The proprioceptive input is weighted higher in the two conditions where the vestibular input was smaller (20 or 30 °/s) than the proprioceptive input (40 °/s). However, when the vestibular input was larger (50 °/s) than the proprioceptive input, their respective weights were practically equal. This raises the question of whether, contrary to the instruction to judge their walking speed, participants were simply using their perceived motion through space (i.e., the vestibular input) to perform the task.

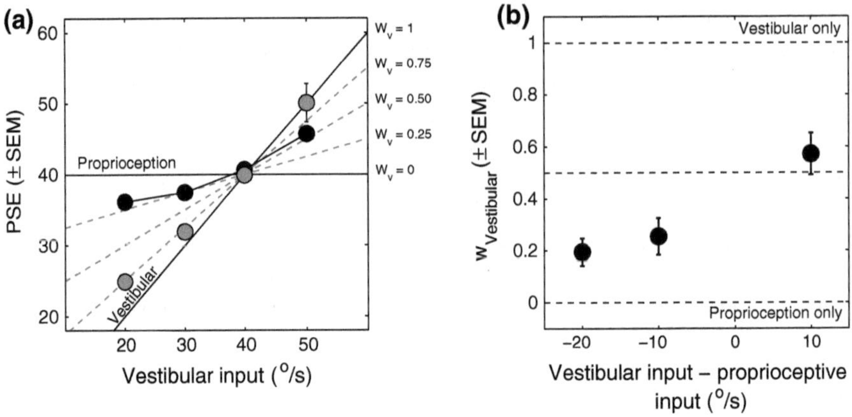

Fig. 6.4 Relative weighting of vestibular information. The relative weighting of the vestibular and proprioceptive inputs were investigated by fixing the proprioceptive input to a single value and varying the vestibular input (note that in [42] the proprioceptive inputs were varied). Eight participants were tested with a 2-IFC paradigm with four standards that all had the same walking speed but had different vestibular inputs, as explained below. The conflicts we tested differed in size and in direction. In the first condition there was no conflict and both the vestibular and proprioceptive inputs were the same (i.e., both at 40 °/s). In the second condition, the vestibular input was slower (20 or 30 °/s) than the proprioceptive input, and in the third condition, the vestibular input was larger (50 °/s) than the proprioceptive inputs. To illustrate how this last condition was achieved, the handlebar was moved at 50 °/s to establish the vestibular input. However since this by itself would also give a proprioceptive input of 50 °/s and not the desired 40 °/s, the difference was created by moving the disc at 10 °/s in the same direction as the handlebar. **a** The group means for the PSEs (and SEMs) for the main experiment (*black markers*) and the control experiment (*grey markers*, see text for details). The *dotted diagonal lines* illustrate hypothetical vestibular weighting schemes. **b** The estimated vestibular weights extracted from the results in panel (**a**). The horizontal *dotted lines* on the *top* and *bottom* of the *panel* represent hypothetical instances in which the perceived walking speed is entirely determined by the vestibular (*top*) or proprioceptive input (*bottom*)

This alternative interpretation is unlikely given the results of a control experiment in which two new participants were tested in the exact same experiment but with explicit instructions to judge how fast they were moving through space and to ignore how fast they were walking. The results are clearly different from those of the main experiment (Fig. 6.4a, grey markers). The PSEs are now close to the theoretical line for complete vestibular dominance. However, the PSEs are not exactly on the line but show an influence of the proprioceptive input, which is what we would expect under the mandatory integration hypothesis (i.e. even though participants were told to ignore their speed of proprioception, these proprioceptive cues still influenced their responses).

6.3.3 "Vection" from Walking

Under the mandatory integration hypothesis we expect that walking conditions, even with extreme conflicts between the proprioceptive and vestibular signals, will show evidence of a weighted averaging. Once again, walking in place creates a particularly interesting condition. Averaging a zero input (vestibular) with a non-zero input (proprioceptive) necessarily leads to a non-zero estimate. We therefore expect participants in this condition to experience illusory self-motion in the absence of actual movement through space (i.e., non-visual "vection"). There is indeed evidence that walking in place elicits nystagmus [9], and pseudo-coriolis effects [10], and self-motion aftereffects [8].

In one experiment we created five extreme sensory conflict conditions. The participants were moved through space at −10, −5, 0, 5, or 10 °/s while walking at a fixed speed of 40 °/s. Negative values indicate that the participant moved backwards through space. Thus, in two conditions the inputs were of the same sign (i.e., physical movement was in the same direction), but widely different in magnitude. In two other conditions, the sign was the opposite in direction such that participants stepped forward while being moved backwards through space. In the last condition they were walking in place. We used the same pointing task as in Frissen et al. [42] to measure perceived self-motion.

Figure 6.5a shows the perceived self-motion. An estimate of the proprioceptive weight was obtained from fitting the MLE model to the group means and was 0.07 with a corresponding vestibular weight of 0.93. The fit is, however, rather poor and, except for the −5 °/s condition, none of the pointing rates were significantly different from the test velocity, suggesting that participants used the vestibular input only. However, all participants at some point did experience illusory motion through space in the walking in place condition. Moreover, participants also confused the direction of motion on at least several trials. For instance, backward motion at 10 °/s was perceived as forward movement on 30 % of the trials. Therefore, simply averaging the signed mean pointing rate would give an incorrect impression of actually perceived motion. If we categorize the data according to whether the motion was perceived as backward or forward, this results in the two curves shown in Fig. 6.5b. For about 58 % of the trials this motion was perceived as forward (at ~7 °/s) and for about 42 % of the trials as backward (at ~6 °/s). Thus, walking in place clearly induces an illusion of self-motion. Interestingly, these new trends can still be described by a simple weighted averaging. The difference is that only the magnitudes of the inputs are used irrespective of direction. Thus, the magnitude of the trends in Fig. 6.5b are well described by, $\hat{S} = \sum_i w_i \, |S_i|$ where $\hat{S}$ is the multisensory estimate, $|S_i|$ the magnitude of the individual inputs, and w_i their relative weights. Estimates of the proprioceptive weights were obtained from fitting the adapted model to the group means. They were 0.12 and 0.07, for the motion that was perceived as forward and backward, respectively, which makes the corresponding vestibular weights 0.88 and 0.93.

What is most surprising about these results is that the odds of perceiving forward motion as opposed to backward motion were close to 1:1. This surprise comes

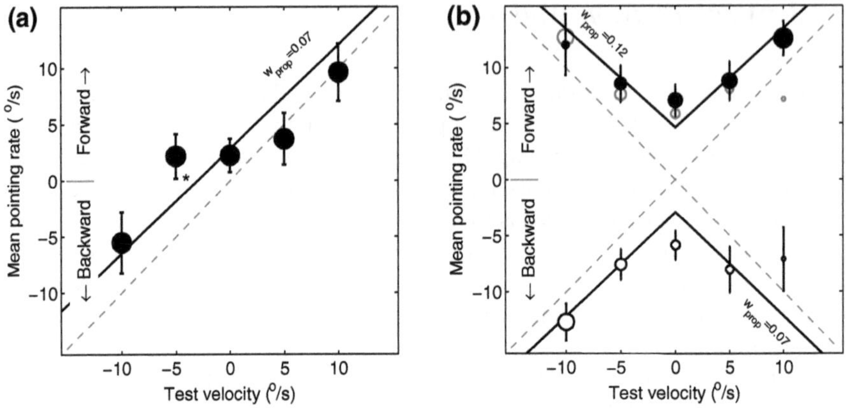

Fig. 6.5 Self-motion perception during walking with extreme conflicts between proprioceptive and vestibular signals. **a** The perceived self-motion for eleven participants after averaging across the six replications of each condition. The *solid black line* represents the fit of the MLE model to the group means. The *asterisk* indicates a significant difference between test velocity and mean pointing rate. **b** Data categorized according to whether motion was perceived as backward (*open black circles*) or forward (*filled black circles*). The sizes of the *circles* reflect the relative proportion of trials that contributed to the represented mean value. Clearly there were a substantial number of cases in which direction was confused. Pointing rates were virtual mirror images for the forward and backward perceived trials. To illustrate this, the *grey circles* show the backward perceived motion but with the sign inversed. The solid lines represent fits of the adapted MLE model to the group means. The annotations show the estimates for the proprioceptive weights that correspond to the fitted model

from the fact that the proprioceptive input is directionally unambiguous. Two subsequent experiments, in which we manipulated either the walking speed or the walking direction, clearly showed that there is an effect of the proprioceptive input on the distribution of the number of trials that are perceived as forward or backward motion. For instance, the proportion of trials perceived as forward was, as before, close to 50 % when mechanically walking forward in place, but dropped to around 25 % when mechanically walking backwards. In other words, stepping backwards also made the participant feel like they were moving backwards most of the time, but not always. The contribution of the proprioceptive input to the perceived direction is therefore only partial. It remains an open question as to what all of the determining factors are for perceived direction.

6.3.4 Potential Implications for CyberWalk

Taken together, these studies reveal the clear importance of vestibular inputs for self-motion perception during walking. The vestibular sense registers primarily accelerations and will gradually stop responding once a constant speed has been reached. However, this cessation of sensory stimulation does not mean that there is lack of motion information. After all, if no change in velocity occurs, this would indicate that self-motion had not ceased [92]. Nevertheless, the most salient moments are during the acceleration phase (i.e., start walking) and deceleration phase (stop walking). When simulating normal walking on a treadmill, it is therefore important to retain these inertial cues as accurately as possible. The CyberWalk effectively achieves this. Specifically, when the user starts to walk from a standstill, he/she initially walks on a stationary surface and accelerates through space as they would during normal, overground walking. Only once the user starts to reach a constant walking speed will the treadmill start to move. Gradually, the treadmill brings the user back to the center of the platform (ideally sub-threshold), by moving them backwards through space while they continue to walk. Similarly, when the user stops walking or changes walking direction, the treadmill only responds gradually, allowing the normal inertial input to the vestibular system to occur. For this scheme to work, the walking surface has to be large enough to accommodate several steps without large changes in treadmill speed. In preliminary studies this system has been shown to work very well for controlling treadmill speed on a large linear treadmill [94]. Through these studies, we determined that the minimum size of the walking surface needed to accommodate this control scheme is 6×6 m. However, financial and mechanical considerations limited the eventual size of the CyberWalk to 4×4 m.

6.4 Large Scale Navigation

One field in which the CyberWalk is expected to have a large impact is human navigation. Navigation requires estimates of perceived direction and position while moving through our environments. In order to achieve this we can use external devices such as maps, street signs, compasses or GPS systems, or we can use our internal representations of space that come from multiple cognitive and sensory sources. Much of what we know about human spatial navigation has come from studies involving spaces of relatively small scale (i.e. room size or smaller), while comparatively fewer human studies have considered large-scale navigation. In one recent extensive real world study by our group, we evaluated the extent to which humans are able to maintain a straight course through a large-scale environment consisting of unknown terrain without reliable directional references [93]. The scenarios were those in which observers were transported to the Tunisian Sahara desert or to the Bienwald forest in western Germany and were asked to walk in a completely straight trajectory. The area used for the forest experiment was selected because it was large enough to walk in a constant direction for several hours and has minimal changes in elevation. The thick tree cover also made it impossible to locate distant landmarks to aid direction estimation.

According to a belief often referred to in popular culture, humans tend to walk in circles in the types of desert or forest scenarios described above, yet there had been no previous empirical evidence to support this. The Souman et al. [93] study showed that people do indeed walk in circles while trying to maintain a straight course, but only when traversing in the absence of reliable external directional references.

This was particularly true when participants walked in a dense forest on a cloudy day, with the sun hidden behind the clouds. Most participants also repeatedly crossed their own path without any awareness of having done so. However, under conditions in which directional references such as landmarks or the solar azimuth were present, people were actually able to maintain a fairly straight path, even in an environment riddled with obstacles, such as a forest. A popular explanation for walking in circles is based on the assumption that people tend to be asymmetrical with respect to, for instance, leg length or leg strength. If this were true, it would be hypothesized that a particular individual would always turn in the same direction. However, this was not the case. In fact, inconsistency in turning and veering direction was very common across participants. Moreover, measured leg strength differences could not explain the turning behavior, nor could leg length.

Interestingly, the recorded walking trajectories show exactly the kind of behavior that would be expected if the subjective sense of straight ahead were to follow a correlated random walk. With each step, a random error is added to the subjective straight ahead, causing it to drift away from the true straight ahead. As long as the deviation stays close to zero, people walk in randomly meandering paths. When the deviation becomes large, it results in walking in circles. This implies that circles are not necessarily an indication of a systematic bias in the walking direction but can be caused by random fluctuations in the subjective straight ahead resulting from accumulating noise in the sensorimotor system, in particular the vestibular and/or motor system.

Another possible contribution to deviating from a straight path, not considered in Souman et al. [93] study is the instantaneous orientation of the head with respect to the trunk. It has been shown that eccentric eye orientation (e.g., [82]) and head orientations tend to be related to the direction of veer from a straight course. The most common finding is that people veer in the direction of eye/head orientation [113]. For instance, in a series of driving experiments, Readinger et al. [82] consistently found that deviations in a driver's gaze can lead to significant deviations from a straight course. Specifically, steering was biased in the direction of fixation. They tested a large range of eye positions, between $-45°$ and $+45°$. Interestingly, the largest effect was obtained with an eccentric eye position of as little as $10°$ and leveled off beyond that. Thus, even a small deviation of $5°$ created a significant bias. A very similar bias has been found during visually guided walking [28, 111]. Jahn et al. [59] asked participants to walk straight towards a previously seen target placed 10 m away while they were blindfolded. Their results demonstrated, contrary to all previous work, that with the head rotated to the left, participants' path deviated to the right, and vice versa. The effect of eye position showed the same pattern, but was not significant. The authors interpreted this as a compensation strategy for an apparent deviation in the direction of gaze due to the lack of the appropriate visual feedback.

Intrigued by the counterintuitive results of Jahn et al. [59] study we conducted a very similar experiment in an attempt to replicate these results. The results (see Fig. 6.6. and caption for details) suggest a bias in the direction of veering in the same direction as the head turn. The bias was asymmetric in that it was larger when the head was turned to the left than when the head was turned to the right. There was

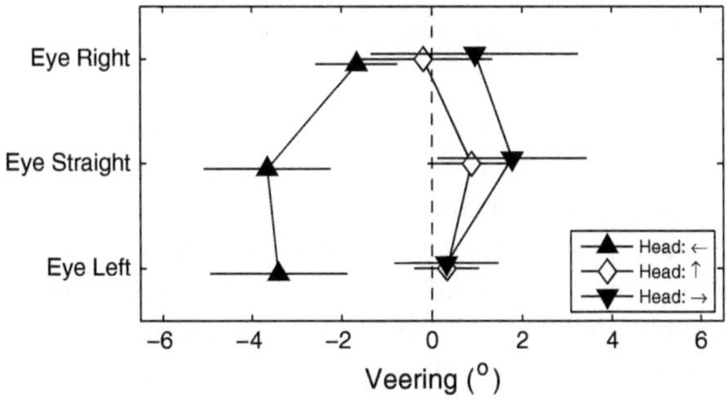

Fig. 6.6 The effect of head and eye orientation on veering. Thirteen participants walked in a large, fully tracked lab with different combinations of eye and head orientations. Each combination was tested 5 times for a total of 3 (Head: *Left, Straight, Right*) × 3 (Eyes, *Left, Straight, Right*) × 5 (repetitions) = 45 randomized trials. The head orientation was blocked and randomized across participants and within each block of head orientation, eye position was pseudo-randomized. For each trial the participant viewed the target straight ahead of them until they had a good mental image of its position and then walked to the target under the specified conditions of eye and head orientation. Except for when looking at the target, the participant was always blindfolded when walking. For safety an experimenter was always in the room with the participant and provided specific instructions prior to each trial. The participant's position was recorded using a Vicon tracking system. To specify eye position, a pair of safety goggles were customized with three red LEDs that were positioned on the outer surface such that looking at them would create an angular position of the eyes of approximately 45° to the *left* or to the *right*, or *straight* ahead, relative to the head. To control head orientation, on the other hand, no explicit reference was provided, but rather participants were instructed to turn their heads as far as possible (to the *left* or *right*) without causing any discomfort, and to hold their head there for the duration of a trial. Compliance was checked by the experimenter. To prevent any view of the environment, an *opaque black* veil was donned after orienting the eyes and head. However, the head's angle relative to the trunk was somewhat variable across trials and participants. The participant wore a wireless headset playing noise to mask any auditory feedback

also an apparent interaction between the head and eye orientation such that the bias tended to diminish when the eyes were turned away from straight ahead and was stronger when the head and the eyes were oriented in opposite directions. Statistical analyses, however, showed marginally significant effects of head orientation and its interaction with eye position. Whereas these results are qualitatively consistent with those of Cutting et al. [28] and Readinger et al. [82], they are opposed to those of Jahn et al. [59]. In fact, when we compare the average values, our and Jahn et al.'s results, are highly negatively correlated ($r = -0.84$). We can speculate that spontaneous head turns would have contributed to the effect of veering from a straight trajectory observed by [93], especially in the desert and forest experiments where participants were free to look around as they pleased.

6.4.1 Potential Implications for CyberWalk

The above-described large scale navigational studies demonstrate the need for a platform like the CyberWalk, more than to offer constraints on its design. Specifically, they demonstrate the real need for a laboratory setup that allows a walker to go in circles or to walk along meandering paths. Nevertheless, they show that more controlled environments are essential in studying human navigation. For instance, the forest experiment revealed that one apparently major factor in being able to stay on a straight trajectory was whether the sky was overcast or not. The CyberWalk achieves environmental control through the use of VR technologies which allow us to create large scale visual environments with high fidelity and control over environmental factors that are normally beyond control, such as the presence and position of the sun.

6.5 Putting it All Together: The CyberWalk Platform

The CyberWalk treadmill (Fig. 6.7) consists of 25 segmented belts each 5 m long and 0.5 m wide, which are mounted on two large chains in the shape of a torus. The entire setup is embedded in a raised floor. The belts constitute one direction of motion, while the chains form the perpendicular direction. The chains are capable of speeds up to 2 m/s, while the belts can run at 3 m/s. The chains are driven by four powerful motors placed at the corners of the platform and each belt segment has its own smaller motor. The drives are controlled such that they provide a constant speed independent of belt load. The walking surface is large enough to accommodate several steps without large changes in treadmill speed. This size allows for changes in treadmill speed which are low enough to maintain postural stability of the user, but makes it unavoidable that these accelerations will sometimes be noticeable to the user. To what extent this affects self-motion perception needs to be determined more closely, although Souman et al. [95] found that walking behavior and spatial updating on the CyberWalk treadmill approached that of overground walking.

The high-level control system determines how the treadmill responds to changes in walking speed and direction of the user in such a way that it allows the user to execute natural walking movements in any direction. It tries to keep the user as close to the center of the platform as possible, while at the same time taking into account perceptual thresholds for sensed acceleration and speed of the moving surface. The control law has been designed at the acceleration level to take into account the limitations of both the platform and the human user, while ensuring a smoothly changing velocity input to the platform (see [29]). The treadmill velocity is controlled using the head position of the user. The control scheme includes a dead-zone in the center of the treadmill where changes in the position of the user are not used when the user is standing still. This makes it much more comfortable for users to look around in the VE while standing still [95]. Users wear a safety harness connected to the ceiling to prevent them from falling and reaching the edge of the platform with their feet.

The setup is installed in a large hall (12 × 12 m walking area). The hall is equipped with a 16 camera Vicon MX13 optical tracking system (Vicon, Oxford, United Kingdom) that is used to track the position and orientation of the participant's head. To this end, participants wear a helmet with reflective markers. The tracking data are used to update the visualization presented through a head-mounted display (HMD) and to control the treadmill velocity. Presently the HMD used is an eMagin Z800 3DVisor (eMagin, Bellevue, USA) custom built into goggles, which prevents the participant from seeing anything else but the image on the displays. One advantage of this HMD is that it is lighter (<227 g) and less obtrusive than most other HMD systems, but also has a reduced field-of-view. If required, user responses can be collected via a wireless gamepad. When not in use, the treadmill can be covered with wooden boards with a thick rubber coating, creating one continuous, fully tracked walking area.

The omnidirectional capabilities of the platform form its largest contribution to the scientific study of human walking biomechanics. By definition, locomotion serves to transport us from one place to another. However, one of the major constraints on research has been space. For a typical research facility it is extremely expensive to maintain, and difficult to justify, a large instrumented, but otherwise empty room. Most locomotion laboratories are therefore rather small, especially in comparison to the scale of real walking. There is of course a relatively simple solution to the space limitation, and that is to put the participant on a treadmill so that she/he can walk forever. However, virtually all of these treadmills are relatively small and linear. Thus, the space limitation is only resolved for one dimension. In short, none of these restricted spaces enable truly normal walking behaviors like negotiating corners and walking along nonlinear trajectories. However, none of these spatial limitations apply to the CyberWalk platform. This then opens up a large range of possibilities for human locomotion research. One straightforward opportunity is the possibility of replicating the outdoor natural walking experiments described above (see Sect. 6.2.1). An issue with the natural walking study was the fact that turn angle and turn radius did not change independently from each other, another was the need for the 9 kilo backpack to hold all of the recording equipment. By utilizing a carefully designed virtual environment it becomes possible to control turn angles and radii. The backpack is no longer necessary since most of the measurements can be made directly through the optical tracking system, while other measurements (i.e., from the IMU) can be implemented such that there is no additional load on the walker. Such a study would effectively be an ideal marriage of the outdoor experiment [97] and the laboratory study on head-trunk interactions [98].

More generally, the platform's optical tracking system is capable of full body tracking which has enormous potential for extending studies of biomechanics and dynamics (e.g., [30]) during real, unconstrained walking. Understanding unconstrained walking is not only of scientific value but can also advance computer vision technologies for tracking and recognizing human locomotion behavior (e.g., [1]). The platform's tracking capability can be extended to support gaze tracking by including

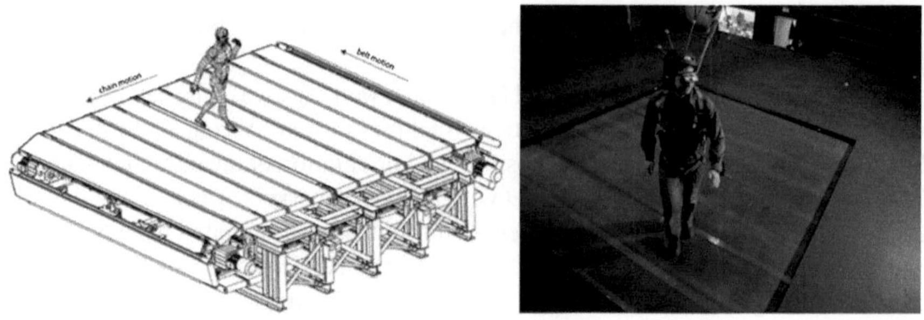

Fig. 6.7 The CyberWalk platform

a portable eye tracking device, which is of great value to the study of the coordination of the eye, head, and trunk while making turns [52, 53, 57, 81, 98]. Space has also been a major limitation to earlier research using tracking technologies. Thus, walkers have typically been tracked while walking short distances, making predefined turns (e.g., [27, 57, 81]), or walking in repetitive artificial patterns like circles [47], figure eights [52] or cloverleaf patterns (e.g., [53]). Sreenivasa et al. [98] had participants walk along trajectories that consisted of turns of various angles (between 45° and 135°, and 180° turns) interspersed with straight sections, in an attempt to simulate more closely the series of turns that occur in natural day-to-day walking. With the help of VE technologies it is also possible to strictly control the amount of visual information provided about upcoming turns. The effects of head/eye orientation on veering have only been studied when having participants walk for several meters. However, as the large scale navigation studies suggest, more complete evaluations are possible when assessing the effects of head/eye orientation on veering during walking trajectories that occur over longer periods of time, or across longer distances.

The CyberWalk platform also opens up a particularly large potential for human navigation research. For instance, recall the desert/forest experiments described in Sect. 6.4, for which it was necessary to travel to the Sahara desert. Without going through this level of effort and expense, conducting such experiments would be extremely difficult to test in the real world because of the need for a completely sparse environment through which an individual can walk for hours. However, such large scale experiments are now possible in the lab. VEs allow us to manipulate particular characteristics of the simulated world (e.g., position of the sun, or time of day) as a way of evaluating the exact causes of any observed veering behaviors, while still allowing for limitless walking capabilities in any direction. Other questions are now possible to address as well. Although, thanks to visual VE development programs, these large scale environments are relatively easy to create and manipulate, the platform is the first to enable truly unconstrained exploration of these environments. It thereby also creates much more ecologically valid, multisensory circumstances for studying questions about spatial cognition. The platform also creates unique opportunities for studying behavior in unfamiliar environments (e.g., [55]).

In conclusion, being able to physically walk through large VEs in an unrestricted manner opens up opportunities that go beyond the study of gait biomechanics, cognition, and spatial navigation in naturalistic environments [16, 105]. It also provides new possibilities for rehabilitation training [44], for edutainment (gaming, virtual museums), design (architecture, industrial prototyping) and various other applications. In summary, the CyberWalk treadmill has brought us a significant step forward towards natural walking in large VEs.

Acknowledgments The work reported in this chapter was funded by the European 6th Framework Programme, CyberWalk (FP6-511092). We would like to thank Jan Souman for his invaluable contributions to the work reported in this chapter.

References

1. Aggarwal JK, Cai Q (1999) Human motion analysis: a review. Comp Vis Image Underst 73(3):428–440
2. Alais D, Burr D (2004) The ventriloquist effect results from near-optimal bimodal integration. Curr Biol 14(3):257–262
3. Alton F, Baldey L, Caplan S, Morrissey MC (1998) A kinematic comparison of overground and treadmill walking. Clin Biomech 13(6):434–440
4. Battaglia PW, Jacobs RA, Aslin RN (2003) Bayesian integration of visual and auditory signals for spatial localization. J Opt Soc Am A 20(7):1391–1397
5. Becker W, Nasios G, Raab S, Jürgens R (2002) Fusion of vestibular and podokinesthetic information during self-turning towards instructed targets. Exp Brain Res 144(4):458–474
6. Bent LR, Inglis JT, McFadyen BJ (2004) When is vestibular information important during walking? J Neurophysiol 92(3):1269–1275
7. Berthoz A, Israël I, Georges-François P, Grasso R, Tsuzuku T (1995) Spatial memory of body linear displacement: what is being stored? Science 269(5220):95–98
8. Bles W (1981) Stepping around: circular vection and coriolis effects. In: Long J, Baddeley A (eds) Attention and performance IX. Lawrence Erlbaum, Hillsdale, NJ
9. Bles W, Kapteyn TS (1977) Circular vection and human posture: I. Does the proprioceptive system play a role? Aggressologie 18(6): 325–328.
10. Bles W, de Wit G (1978) La sensation de rotation et la marche circulaire (Sensation of rotation and circular walking). Aggressologie 19(A): 29–30
11. Breniere Y, Do MC (1986) When and how does steady state gait movement induced from upright posture begin? J Biomech 19(12):1035–1040
12. Bresciani J-P, Dammeier F, Ernst MO (2006) Vision and touch are automatically integrated for the perception of sequences of events. J Vis 6(5):554–564
13. Bril B, Ledebt A (1998) Head coordination as a means to assist sensory integration in learning to walk. Neurosci Biobehav Rev 22(4):555–563
14. Bruggeman H, Piuneu VS, Rieser JJ, Pick HL Jr (2009) Biomechanical versus inertial information: stable individual differences in perception of self-rotation. J Exp Psychol: Hum Percept Perform 35(5):1472–1480
15. Bülthoff HH, Mallot HA (1988) Integration of depth modules: stereo and shading. J Opt Soc Am 5(10):1749–1758
16. Bülthoff HH, van Veen HJ (2001) Vision and action in virtual environments: Modern psychophysics. In: Jenkin ML and Harris L (eds.), Spatial cognition research. Vision and attention, 233–252. Springer Verlag, New York.
17. Bülthoff HH, Yuille A (1991) Bayesian models for seeing shapes and depth. Comments on Theor Biol 2(4):283–314

18. Butler JS, Smith ST, Campos JL, Bülthoff HH (2010) Bayesian integration of visual and vestibular signals for heading. J Vis 10(11):Article 23
19. Butler JS, Campos JL, Bülthoff HH, Smith ST (2011) The role of stereo vision in visual-vestibular integration. Seeing Perceiving 24(5):453–470
20. Calvert GA, Spence C, Stein BE (2004) The handbook of multisensory processes. MIT Press, Boston
21. Campos JL, Butler JS, Bülthoff HH (2012) Multisensory integration in the estimation of walked distance. Exp Brain Res 218(4):551–565
22. Campos JL, Siegle J, Mohler BJ, Loomis JM, Bülthoff HH (2009) Imagined self-motion differs from perceived self-motion: Evidence from a novel continuous pointing method. PLoS ONE 4(11):e7793. doi:10.1371/journal.pone.0007793
23. Campos JL, Byrne P, Sun H-J (2010) The brain weights body-based cues higher than vision when estimating walked distances. Eur J Neurosci 31(10):1889–1898
24. Campos JL, Bülthoff HH (2011) Multisensory Integration during self-motion in virtual reality. In: Wallace M, Murray M (eds) Frontiers in the Neural Bases of Multisensory Processes. Taylor and Francis Group, London
25. Cheng K, Shettleworth SJ, Huttenlocher J, Rieser JJ (2007) Bayesian integration of spatial information. Psychol Bull 133(4):625–637
26. Cooper AR, Page AS, Wheeler BW, Griew P, Davis L, Hillsdon M, Jago R (2010) Mapping the walk to school using accelerometry combined with a global positioning system. Am J Prev Med 38(2):178–183
27. Courtine G, Schieppati M (2003) Human walking along a curved path. I. Body trajectory, segment orientation and the effect of vision. Eur J Neurosci 18(3):177–190
28. Cutting JE, Readinger WO, Wang RF (2002) Walking, looking to the side, and taking curved paths. Percept Psychophys 64(3):415–425
29. De Luca A, Mattone R, Robuffo Giordano P, Bülthoff HH (2009) Control design and experimental evaluation of the 2D CyberWalk platform. IEEE/RSJ International Conference on Intelligent Robots and Systems, St. Louis, MO, pp 5051–5058
30. Dingwell JB, Cusumano JP, Cavanagh PR, Sternad D (2001) Local dynamic stability versus kinematic variability of continuous overground and treadmill walking. J Biomech Eng 123(1):27–32
31. Dufek JS, Mercer JA, Griffin JR (2009) The effects of speed and surface compliance on shock attenuation characteristics for male and female runners. J Appl Biomech 25(3):219–228
32. Duncan MJ, Mummery WK, Dascombe BJ (2007) Utility of global positioning system to measure active transport in urban areas. Med Sci Sports Exerc 39(10):1851–1857
33. Eils E, Nolte S, Tewes M, Thorwesten L, Völker K, Rosenbaum D (2002) Modified pressure distribution patterns in walking following reduction of plantar sensation. J Biomech 35(10):1307–1313
34. Elliott D (1986) Continuous visual information may be important after all: A failure to replicate Thomson. J Exp Psychol: Human Percept Perform 12(3):388–391
35. Ernst MO (2006) A Bayesian view on multimodal cue integration. In: Knoblich G, Thornton IM, Grosjean M, Shiffrar M (eds) Perception of the human body from the inside out. Oxford University Press, New York, USA, pp 105–131
36. Ernst MO, Banks MS (2002) Humans integrate visual and haptic information in a statistically optimal fashion. Nature 415(6870):429–433
37. Ernst MO, Bülthoff HH (2004) Merging the senses into a robust percept. Trends Cogn Sci 8(4):162–169
38. Faisal A, Selen LPJ, Wolpert DM (2008) Noise in the nervous system. Nat Rev Neurosci 9(4):292–303
39. Fetsch CR, Turner AH, DeAngelis GC, Angelaki DE (2009) Dynamic reweighting of visual and vestibular cues during self-motion perception. J Neurosci 29(49):15601–15612
40. Fetsch CR, DeAngelis GC, Angelaki DE (2010) Visual-vestibular cue integration for heading perception: applications of optimal cue integration theory. Eur J Neurosci 31(10):1721–1729

41. Finnis KK, Walton D (2008) Field observations to determine the influence of population size, location, and individual factors on pedestrian walking speeds. Ergonomics 51(6):827–842
42. Frissen I, Campos JL, Souman JL, Ernst MO (2011) Integration of vestibular and proprioceptive signals for spatial updating. Exp Brain Res 212(2):163–176
43. Fukusima SS, Loomis JM, Da Silva JA (1997) Visual perception of egocentric distance as assessed by triangulation. J Exp Psychol: Human Percept Perform 23(1):86–100
44. Fung J, Malouin F, McFadyen BJ, Comeau F, Lamontagne A, Chapdelaine S, Beaudoin C, Laurendeau D, Hugheyh L, Richards CL (2004) Locomotor rehabilitation in a complex virtual environment. Proceedings of the 36th annual international conference of the IEEE-EMBS, San Francisco, Sept 1–5, pp 4859–4862
45. Gescheider GA (1997) Psychophysics: The fundamentals, 3rd edn. Lawrence Erlbaum, Mahwah, NJ
46. Gibson JJ (1950) Perception of the visual world. Houghton Mifflin, Boston
47. Grasso R, Glasauer S, Takei Y, Berthoz A (1996) The predictive brain: anticipatory control of head direction for the steering of control. NeuroReport 7(6):1170–1174
48. Grieve DW, Gear RJ (1966) The relationships between length of stride, step frequency, time of swing and speed of walking for children and adults. Ergonomics 9(5):379–399
49. Gu Y, Angelaki DE, DeAngelis GC (2008) Neural correlates of multisensory cue integration in macaque MSTd. Nat Neurosci 11(10):1201–1210
50. Hallemans A, Ortibus E, Meire F, Aerts P (2010) Low vision affects dynamic stability of gait. Gait Posture 32(4):547–551
51. Harris LR, Jenkin M, Zikovitz DC (2000) Visual and non-visual cues in the perception of linear self-motion. Exp Brain Res 135(1):12–21
52. Hicheur H, Vieilledent S, Berthoz A (2005a) Head motion in humans alternating between straight and curved walking path: Combination of stabilizing and anticipatory orienting mechanisms. Neurosci Lett 383(1–2):87–92
53. Hicheur H, Vieilledent S, Richardson MJE, Flash T, Berthoz A (2005b) Velocity and curvature in human locomotion along complex curved paths: a comparison with hand movements. Exp Brain Res 162(2):145–154
54. Hollands MA, Marple-Horvat DE (1996) Visually guided stepping under conditions of step cycle-related denial of visual information. Exp Brain Res 109(2):343–356
55. Hölscher C, Büchner SJ, Meilinger T, Strube G (2009) Adaptivity of wayfinding strategies in a multi-building ensemble: The effects of spatial structure, task requirements, and metric information. J Environ Psychol 29(2):208–219
56. Holt KG, Jeng SF, Ratcliffe R (1995) Energetic cost and stability during human walking at the preferred stride frequency. J Motor Behav 27(2):164–178
57. Imai T, Moore ST, Raphan T, Cohen B (2001) Interaction of the body, head, and eyes during walking and turning. Exp Brain Res 136(1):1–18
58. Israël I, Berthoz A (1989) Contributions of the otoliths to the calculation of linear displacement. J Neurophysiol 62(1):247–263
59. Jahn K, Kalla R, Karg S, Strupp M, Brandt T (2006) Eccentric eye and head positions in darkness induce deviation from the intended path. Exp Brain Res 174(1):152–157
60. Jian Y, Winter DA, Ishac MG, Gilchrist L (1993) Trajectory of the body COG and COP during initiation and termination of gait. Gait Posture 1(1):9–22
61. Jürgens R, Becker W (2006) Perception of angular displacement without landmarks: evidence for Bayesian fusion of vestibular, optokinetic, podokinesthetic and cognitive information. Exp Brain Res 174(3):528–543
62. Kerrigan DC, Viramontes BE, Corcoran PJ, LaRaia PJ (1995) Measured versus predicted vertical displacement of the sacrum during gait as a tool to measure biomechanical gait performance. Am J Phys Med Rehabil 74(1):3–8
63. Knill DC (2005) Reaching for visual cues to depth: The brain combines depth cues differently for motor control and perception. J Vis 5(2):103–115
64. Knill DC, Richards W (1996) Perception as Bayesian Inference. Cambridge University Press, Cambridge

65. Knoblauch RL, Pietrucha MT, Nitzburg M (1996) Field studies of pedestrian walking speed and start-up time. Transp Res Record 1538:27–38
66. Körding KP, Wolpert DM (2004) Bayesian integration in sensorimotor learning. Nature 427(6971):244–247
67. Loomis JM, Da Silva JA, Fujita N, Fukusima SS (1992) Visual space perception and visually directed action. J Exp Psychol: Hum Percept Perform 18(4):906–921
68. Loomis JM, Blascovich JJ, Beall AC (1999) Immersive virtual environment technology as a basic research tool in psychology. Behav Res Methods Instrum Comp 31(4):557–564
69. MacNeilage PR, Banks MS, Berger DR, Bülthoff HH (2007) A Bayesian model of the disambiguation of gravitoinertial force by visual cues. Exp Brain Res 179(2):263–290
70. Maddison R, Mhurchu CN (2009). Global positioning system: a new opportunity in physical activity measurement. Int J Behav Nutr Phys Act 6:Article 73. doi:10.1186/1479-5868-6-73
71. Mann RA, Hagy JL, White V, Liddell D (1979) The initiation of gait. J Bone Jt Surg 61(2):232–239
72. Menz HB, Lord SR, Fitzpatrick RC (2003) Acceleration patterns of the head and pelvis when walking on level and irregular surfaces. Gait Posture 18(1):35–46
73. Mittelstaedt ML, Mittelstaedt H (2001) Idiothetic navigation in humans: Estimation of path length. Exp Brain Res 139(3):318–332
74. Mohler BJ, Campos JL, Weyel M, Bülthoff HH (2007) Gait parameters while walking in a head-mounted display virtual environment and the real world. 13th Eurographics symposium on virtual environments and 10th immersive projection technology workshop (IPT-EGVE 2007), Aire-la-Ville, Switzerland, 85–88
75. Murray MP, Spurr GB, Sepic SB, Gardner GM, Mollinger LA (1985) Treadmill versus floor walking: kinematics, electromyogram, and heart rate. J Appl Physiol 59(1):87–91
76. Oliver M, Badland H, Mavoa S, Duncan MJ, Duncan S (2010) Combining GPS, GIS, and accelerometry: Methodological issues in the assessment of location and intensity of travel behaviors. J Phys Act Health 7(1):102–108
77. Patla AE (1997) Understanding the roles of vision in the control of human locomotion. Gait Posture 5(1):54–69
78. Pearce ME, Cunningham DA, Donner AP, Rechnitzer PA, Fullerton GM, Howard JH (1983) Energy cost of treadmill and floor walking at self-selected paces. Eur J Appl Physiol 52(1):115–119
79. Pick HL, Wagner D, Rieser JJ, Garing AE (1999) The recalibration of rotational locomotion. J Exp Psychol: Hum Percept Perform 25(5):1179–1188
80. Pozzo T, Berthoz A, Lefort L (1989) Head kinematic during various motor tasks in humans. Prog Brain Res 80:377–383
81. Prévost P, Ivanenko Y, Grasso R, Berthoz A (2002) Spatial invariance in anticipatory orienting behaviour during human navigation. Neurosci Lett 339(3):243–247
82. Readinger WO, Chatziastros A, Cunningham DW, Bülthoff HH, Cutting JE (2002) Gaze-Eccentricity Effects on Road Position and Steering. J Exp Psychol: Appl 8(4):247–258
83. Rieser JJ, Ashmead DH, Talor CR, Youngquist GA (1990) Visual perception and the guidance of locomotion without vision to previously seen targets. Perception 19(5):675–689
84. Riley PO, Paolini G, Della Croce U, Paylo KW, Kerrigan DC (2007) A kinematic and kinetic comparison of overground and treadmill walking in healthy subjects. Gait Posture 26(1):17–24
85. Rodriguez DA, Brown AL, Troped PJ (2005) Portable global positioning units to complement accelerometry-based physical activity monitors. Med Sci Sports Exerc 37(11 Suppl):S572–S581
86. Schutz Y, Herren R (2000) Assessment of speed of human locomotion using a differential satellite global positioning system. Med Sci Sports Exerc 32(2):642–646
87. Schwaiger M, Thümmel T, Ulbrich H (2007a) Cyberwalk: An advanced prototype of a belt array platform. Proceedings of IEEE International Workshop on Haptic Audio Visual Environments and their Applications, Ottawa, Canada
88. Schwaiger M, Thümmel T, Ulbrich H (2007b) Cyberwalk: Implementation of a ball bearing platform for humans. Proceedings of Conference on Human Computer Interaction, Beijing, China

89. Schwaiger M, Thümmel T, Ulbrich H (2007c) A 2D-motion platform: the cybercarpet. Proceedings of world haptics conference 2007, pp 415–420 Tsukuba, Japan
90. Sekiya N, Nagasaki H, Ito H, Furuna T (1996) The invariant relationship between step length and step rate during free walking. J Hum Mov Stud 30(6):241–257
91. Sheik-Nainar MA, Kaber DB (2007) The utility of a Virtual Reality locomotion interface for studying gait behavior. Hum Factors 49(4):696–709
92. Siegle J, Campos JL, Mohler BJ, Loomis JM, Bülthoff HH (2009) Measurement of instantaneous perceived self-motion using continuous pointing. Exp Brain Res 195(3):429–444
93. Souman JL, Frissen I, Sreenivasa MN, Ernst MO (2009) Walking straight into circles. Curr Biol 19(18):1538–1542
94. Souman JL, Robuffo Giordano P, Frissen I, De Luca A, Ernst MO (2010) Making virtual walking real: perceptual evaluation of a new treadmill control algorithm. ACM Trans Appl Percept 7(2):1–14 Article 11
95. Souman JL, Robuffo Giordano P, Schwaiger M, Frissen I, Thümmel T, Ulbrich H, De Luca A, Bülthoff HH, Ernst MO (2011) CyberWalk: enabling unconstrained omnidirectional walking through virtual environments. ACM Trans Appl Percept 8(4):Article 24.
96. Sparrow WA, Tirosh O (2005) Gait termination: a review of experimental methods and the effects of ageing and gait pathologies. Gait Posture 22(4):362–371
97. Sreenivasa M (2007) Statistics of natural walking (Master's thesis). Universität Duisberg-Essen, Duisberg, Germany
98. Sreenivasa M, Frissen I, Souman J, Ernst MO (2008) Walking along curved paths of different angles: the relationship between head and trunk turning. Exp Brain Res 191(3):313–320
99. Stoffregen TA, Pittenger JB (1995) Human echolocation as a basic form of perception and action. Ecol Psychol 7(3):181–216
100. Stolze H, Kuhtz-Buschbeck JP, Mondwurf C, Boczek-Funcke A, Jöhnk K, Deuschl G, Illert M (1997) Gait analysis during treadmill and overground locomotion in children and adults. Electroencephalogr clin Neurophysiol 105(6):490–497
101. Stopher P, FitzGerald C, Zhang J (2008) Search for a global positioning system device to measure person travel. Transp Res Part C 16(3):350–369
102. Sun H-J, Campos JL, Chan GSW (2004a) Multisensory integration in the estimation of relative path length. Exp Brain Res 154(2):246–254
103. Sun H-J, Campos JL, Young M, Chan GSW, Ellard C (2004b) The contributions of static visual cues, nonvisual cues, and optic flow in distance estimation. Perception 33(1):49–65
104. Tan H, Wilson AM, Lowe J (2008) Measurement of stride parameters using a wearable GPS and inertial measurement unit. J Biomech 41(7):1398–1406
105. Tarr MJ, Warren WH (2002) Virtual reality in behavioral neuroscience and beyond. Nat Neurosci 5:1089–1092
106. Terrier P, Ladetto Q, Merminod B, Schutz Y (2000) High-precision satellite positioning system as a new tool to study the biomechanics of human locomotion. J Biomech 33(12):1717–1722
107. Terrier P, Schutz Y (2003) Variability of gait patterns during unconstrained walking assessed by satellite positioning (GPS). Eur J Appl Physiol 90(5–6):554–561
108. Terrier P, Turner V, Schutz Y (2005) GPS analysis of human locomotion: Further evidence for long-range correlations in stride-to-stride fluctuations of gait parameters. Hum Mov Sci 24(1):97–115
109. Thomson JA (1983) Is continuous visual monitoring necessary in visually guided locomotion? J Exp Psychol: Hum Percept Perform 9(3):427–443
110. Troped PJ, Oliveira MS, Matthews CE, Cromley EK, Melly SJ, Craig BA (2008) Prediction of activity mode with global positioning system and accelerometer data. Med Sci Sports Exerc 40(5):972–978
111. Vallis LA, Patla AE (2004) Expected and unexpected head yaw movements result in different modifications of gait and whole body coordination strategies. Exp Brain Res 157(1):94–110
112. Van den Bergh M, Koller-Meier E, Van Gool L (2009) Real-time Body Pose Recognition using 2D or 3D Haarlets. Int J Comput Vis 83(1):72–84

113. Wann JP, Swapp DK (2000) Why you should look where you are going. Nat Neurosci 3(7):647–648
114. Warabi T, Kato M, Kiriyama K, Yoshida T, Kobayashi N (2005) Treadmill walking and overground walking of human subjects compared by recording sole-floor reaction force. Neurosci Res 53(3):343–348
115. Zarrugh MY, Radcliffe CW (1978) Predicting metabolic cost of level walking. Eur J Appl Physiol 38(3):215–223
116. Zehr EP, Stein RB (1999) What functions do reflexes serve during human locomotion. Prog Neurobiol 58(2):185–205

Part II
Technologies

Chapter 7
Displays and Interaction for Virtual Travel

Anthony Steed and Doug A. Bowman

Abstract Virtual travel can be accomplished in many ways. In this chapter we review displays and interaction devices that can be utilized for virtual travel techniques. The types of display range from desktop to fully immersive and the types of interaction devices range from hand-held devices through to motion tracking systems. We give examples of different classes of device that are commonly used, as well as some more novel devices. We then give a general overview of travel tasks and explain how they can be realized through interaction devices.

7.1 Introduction

Being able to move the viewpoint in a virtual environment (VE) is a critical facility: small movements allow the user to get new perceptual cues to understand 3D space; larger movements allow the user to access different parts of the VE so that the user can experience them at a closer range, or access them if they were previously not accessible. Thus, when we consider VE systems, we can identify that travel has a range of purposes, which might require a range of different devices and interaction techniques. If one were building a training operator for a user of a control console, the user might be seated, and would only need to move her head to see different parts of the console, or look in different directions. If one were building a training simulator for fire evacuation, it would be important to be able to simulate movement through the large-scale environment.

A. Steed (✉)
Department of Computer Science, University College London,
Gower St, London WC1E 6BT, UK
e-mail: A.steed@ucl.ac.uk

D. A. Bowman
Center for Human Computer Interaction, Virginia Tech, Blacksburg, USA

F. Steinicke et al. (eds.), *Human Walking in Virtual Environments*,
DOI: 10.1007/978-1-4419-8432-6_7, © Springer Science+Business Media New York 2013

In what has been called the "virtual reality model" of interaction [30, Chap. 20] or the highest level of "interaction fidelity" [20], the user would just carry out actions in the VE in the same manner as she would in the real world. The system would track the body of the user and recreate virtual images, sounds and other cues that mimicked cues and sensations that the user would get from the analogous real situation. The user would see the VE from a first-person point of view and would be able to effect natural interactions with her body. If the user wanted to pick up an object, she could reach out her hand and grasp the object and lift it. More importantly for this book, if she wanted to pick up an object that was out of reach, she could walk over and pick it up.

Of course, the virtual reality model is an ideal because the hardware and software we use can't simulate cues anywhere near as rich as the real world. Even if the participant isn't walking or moving, our technologies for simulating tactile and force cues are very limited in that they can provide a few points of contact or small area of stimulation, whereas the task could involve the whole body of the user. If the user walks or otherwise moves, then there is a much more pressing problem: virtual reality devices only allow actual movement within a small area. This is for various reasons: the displays might be static (e.g., small room), tethered (e.g., a wired head-mounted display) or otherwise limited through infrastructure (e.g., a tracking system that only functions within a bounded region). We will discuss such technologies in greater detail later in the chapter, but there is a more pertinent question: how can we simulate real walking, giving the impression of unconstrained motion, when physical motion is actually limited? What devices can give the impression of walking on different surface types, over long distances or on different inclines?

There are two fundamental problems: walking is implicitly a task that involves very complex simulation of the walking surface, and walking involves inducing momentum into the moving object: the walker's body. The first requirement might only be solved by what Sutherland called the *ultimate display* [33]. In his seminal paper he described that in this display the existence of matter would be controlled and that a speeding bullet could potentially be fatal. Thus the ultimate walking display would be a display that could simulate any surface by creating that surface. However even the ultimate display doesn't directly solve the second problem: creating momentum in an object. Researchers are only just starting to solve the problem of configurable surfaces (e.g., see Chaps. 9 & 17 in this volume), and the problem of momentum is recognized and some attempts have been made to simulate it by pushing on the body (e.g., see Chap. 6 in this volume).

Reproducing natural walking is thus one of the toughest challenges in human–computer interaction. We can try to imitate real walking, but we will be limited in the range we can support, or the naturalness of the interaction. The alternative is to provide interaction techniques that produce movement, or travel, through the VE using other metaphors and devices.

In this chapter we outline the broad range of displays and devices that are used for travel techniques in VEs. Other parts of the book focus on reproduction of natural walking through sophisticated devices. We will place these in context of supporting the general task in a broad range of VE systems.

Section 7.2 will cover display systems for VEs.
Section 7.3 will cover tracking and interaction devices.
Section 7.4 will describe the range of travel metaphors that are commonly used.

7.2 Display Systems

Display systems for VEs come in many sizes, form factors and capabilities. We distinguish a few types of display:

- Desktop displays
- Wall-sized displays
- Surround-screen displays
- Head-mounted displays
- Mobile augmented reality displays
- Hybrid situated displays

By a desktop display we mean a display such as a humble monitor or TV of the type that is found in most offices or homes. It may show a VE, or other media, but it does not cover a large proportion of the user's vision compared to other displays. The display might use stereo and/or head tracking, but the display doesn't surround the user, so it can only provide a small window into the VE. Obviously one can represent self-motion on such a display. Indeed, many modern video games focus on travel and exploration around a VE, but many of the perceptual cues received are unlike the ones received from actual self-motion. For example, the cues do not extend into peripheral vision.

Once a display becomes very large, such as a single wall of a room (Fig. 7.1), then it covers much more of the user's vision. One characteristic of such a display is that nearby items such as other humans can be depicted "life-size" (with realistic scale) and thus the representation of self-motion can exploit the fact that the user will be able to judge heights more accurately, since he can have an eye-level that can corresponds to his actual height. The displays might be flat (e.g., an actual room wall) or they might be large curved screens. A key characteristic is that the display system is not fully surrounding, so it is possible for the user to turn away from it. Thus any walking interface using such a display would normally require some facility to turn the viewpoint.

At this scale, it is much more common to provide stereoscopic viewing of the display. There are several technologies for this such as shutter glasses, polarising glasses, or color filters. Each works by presenting separate images to the left and right eyes. Most displays of this type support only a single user with head-tracked stereo, but multiple users can be supported [1]. The current state of the art is the C1X6 (Fig. 7.2), a prototype display that supports six users, each with a stereo view from her own viewpoint [15].

The next type of display is a surround-screen display (SSD), of which the most common type comprises multiple large flat display surfaces. These displays are often

Fig. 7.1 An example of a wall-sized display. The EVEREST display is 30 ft long by 8 ft tall and displays 35 million pixels. Image courtesy of the National Center for Computational Sciences, Oak Ridge National Laboratory

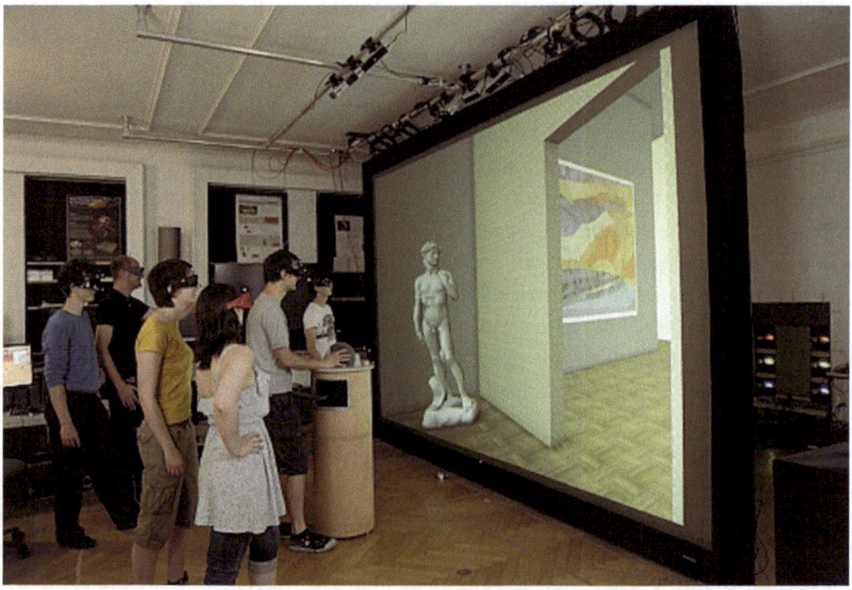

Fig. 7.2 The C1X6 supports six users, each seeing a first person stereo view [15]. Courtesy of Virtual Reality Systems Group, Bauhaus-Universität Weimar, Germany

Fig. 7.3 *Left* UCL's four-walled CAVE™-like display. *Right* looking down into the virtual pit

referred to as CAVE™-like displays (Fig. 7.3, left), named after the original presentation of such a display [6], and the current trademarked commercial version from Mechdyne Corporation. Again it is hard to draw a line between large wall displays and SSDs as the most common type of SSD is a CAVE™-like display with four sides (front, left, right, and floor). Such displays do not completely surround the user, but when standing in the center of such a system and facing forward, the user's vision is almost completely filled with the rendered display of the VE. Importantly, in the most common four-screen configuration the user has peripheral vision of the VE to his left and right, and also down. Thus the visual cues for motion are more powerful. The "virtual pit" scene that is often used as a demonstration of the effectiveness of virtual reality [38] is especially powerful in a CAVE™-like display with a floor because the drop is shown near the feet of the user (Fig. 7.3, right).

Six-sided CAVE™-like systems (Fig. 7.4) do exist, though there are only a handful in the world. The engineering of a fully surrounding system of this type is complex because the floor must be a back-projection screen while being safe for groups of users to stand upon.

Many variations of SSDs exist in all sorts of display configuration. Domed displays are common in certain forms of entertainment (e.g., planetariums), though these are typically used for audiences in the dozens rather than individuals. More exotic displays include the Allosphere (Fig. 7.5) at the University of California, Santa Barbara, a spherical display that supports moderately sized groups [12], and the Cybersphere (Fig. 7.6) [9], a spherical display that supports one person and also acts as a treadmill.

Head-mounted displays (HMDs) are the type of display that is most commonly associated with the phrase "virtual reality." While the basic technology has been available for over thirty years, recently there has been a resurgence in interest and new displays have come to the market. The advantage of HMDs is that they can create the impression that the VE fully surrounds the user: as he turns his head the visual and audio displays can be updated to reflect that motion. This requires some form of head tracking, whether built into the HMD or attached to it. There are many varieties

Fig. 7.4 HyPi-6 display at the Fraunhofer IAO, Stuttgart, Germany. Courtesy of Oliver Stefani

Fig. 7.5 Director JoAnn Kuchera-Morin on the bridge of the AlloSphere, photo taken from above the instrument. Photography by Paul Wellman. Courtesy of University of California, Santa Barbara

Fig. 7.6 Cybersphere. Courtesy of Professor Vinesh Raja, Director of Informatics and Virtual Reality, International Digital Lab, WMG, University of Warwick

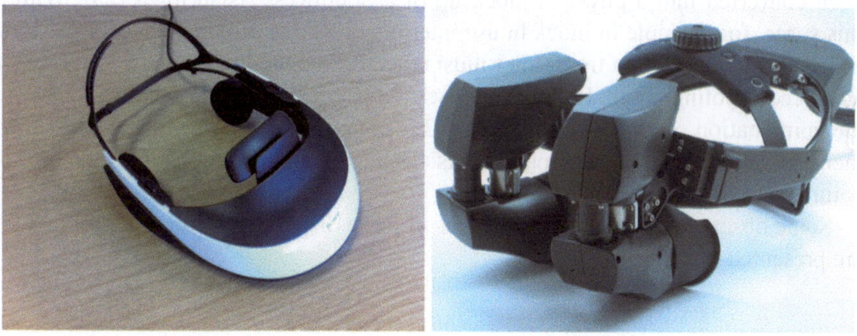

Fig. 7.7 *Left* Sony HMZ-T1. *Right* nVis SX111. Courtesy NVIS Inc

of HMD: not all provide separate images for the left and right eyes; they vary in resolution; and, most importantly, they vary in field of view (FOV). There are many older HMDs with very low resolutions and FOVs, but a modern HMD is the Sony HMZ-T1 (Fig. 7.7, left) released in late 2011, which used OLED panels, had twin screens at 1280 × 720 color pixels and a field of view of 45°. At the more specialist end of the market, the nVis SX111 (Fig. 7.7, right) has LCOS panels, 1280 × 1024 color pixels per eye, and a field of view of 111°.

A useful feature of HMDs is that since they are mounted on the user, the user can walk about. Thus a HMD can be placed in a relatively large space given a tracker that can track the user over that space. In practice this is limited by cabling, but wireless HMDs are now available.

Augmented reality (AR) displays are used to present virtual elements within a view of a real environment. That is, the VE is seen as an augmentation of or intervention into the real environment and both are to be understood by the user as a single consistent environment. We will focus on two sub-types: *mobile augmented reality* and *hybrid situated displays*. The key features in both are that the user walks about the "display" (where the real world is considered to be part of the display), and the display is typically larger than that provided by other technologies we have discussed. In particular, these two types of technology allow walking around much larger spaces.

Mobile augmented reality (Fig. 7.8) typically uses a see-through HMD or a hand-held display, which allows the user to see the real world with graphics overlaid on it. Early demonstrations of these systems included annotations that appeared fixed in place as the user walked [8]. In contrast, the *mixed-reality* project was an example of a project that attempted to create the impression that the physical environment had been extended with new buildings [36].

A hybrid situated display consists of a physical environment with a variety of situated displays within it. A good example is the Infantry Immersion Trainer (Fig. 7.9), a military training facility where soldiers walk around a warehouse space that has been converted into a physical mock-up of a Southwest Asian town [25]. Within this space, for example in mock house interiors, are large screen displays that show life-sized characters that the soldier must react to. The facility thus acts as a type of advanced shooting range. For our purposes the key feature is that the display, being the combination of physical props and VEs, comprises a fully immersive environment around which the participant walks. Theme park experiences are often of this nature as well.

Some observations on the affordances of these display types for walking in VEs are presented in Table 7.1.

7.3 Interaction Devices

In supporting walking interfaces, we can identify two important tasks:

1. Orientation of the viewpoint in the VE
2. Movement of the viewpoint in the VE

With real walking both are specified by the muscles and state of the body including legs, spine bend, neck orientation and eye rotation. As users move, each of these changes constantly in a complex manner, but we note that orientation of the viewpoints (i.e., the eyes) can be controlled separately from the gross direction of movement. We also note that while walking the eyes are in constant motion, not

Fig. 7.8 A later version of the touring machine system. Courtesy of Steve Feiner, Columbia University

just in the direction of travel, but up and down. All of these, and many other subtle effects, need to be recreated in VEs if a purely natural walking interface is desired.

In the virtual reality model discussed previously, the concept is that the system tracks the body and recreates the perceptual cues in a structure that is analogous to the real world. In practice this means that the system tracks the head of the user and uses this to orient a pair of virtual cameras, one for the left eye and one for the right eye. It would be ideal to track the eyes as well, both direction and focus, and then adapt to that, but that is beyond the current state of the art.

Since it is not possible to track the user in an unlimited area, interaction devices must be used to provide control input to effect the two tasks we identified above. While we will discuss the actual interaction tasks in the next section, it is worth pointing out the degrees of freedom that are required. The task of placing a camera in the world involves six degrees of freedom (DoF): three for orientation and three for translation. For two eyes, two cameras are required, but typically these are rigidly attached to the head, so a single six-DoF position/orientation needs to be calculated. In one extreme, say the desktop display type from Table 7.1, nothing about the user's

Fig. 7.9 Marines from 3rd Battalion, 1st Marines, confront avatars, or virtual humans, while clearing a room at the Office of Naval Research Infantry Immersion Trainer. U.S. Navy photo by John F. Williams from WikiMedia Commons

Table 7.1 Comparison of display technologies

Type	Viewing type	User own motion	Notes
Desktop	Through the window/Low FOV	None (typically) or very limited (<1 m)	Usually no head tracking, user typically seated
Wall-sized	Through the window/Medium FOV	None through to limited (<3 m)	Commonly a single person can be tracked. For a group, no tracking used
Surround-screen displays	Partially–fully surrounding/High FOV	Limited (<3 m) to moderately large (~10 m)	Commonly a single person can be tracked. For a group, no tracking used
Head-mounted displays	Fully surrounding/ low-high FOV	None through to wide area	Single user only
Mobile augmented reality	Fully surrounding/ FOV of virtual world limits	None through to wide area	Tracking quality and thus registration vary enormously
Hybrid situated displays	Fully surrounding/ FOV not applicable	None through to wide area	Tracking might be localized to certain areas of the display

actual head position is known, so the system must provide a metaphor for controlling the viewpoint using devices such as joysticks, mice, keyboards, etc. At the other extreme, such as mixed-reality displays, because the user can move through the full extent of the VE, head tracking is sufficient to give us the relevant viewpoint. In between there might be a hybrid: tracking the head over its limited range of motion, and providing a separate control mechanism for movement over longer distance.

We can thus identify several types of interaction device that might be used:

- Pose & movement sensors
- Position trackers
- Hand-operated devices
- Hybrid devices

As with the display types identified in the previous section, the distinctions between these types are quite subtle. Furthermore, any one device might incorporate elements of two or more of these types (e.g., a handheld device that contains a joystick and a position tracker). We also note that there are hundreds, perhaps thousands, of devices that are available for interaction in VEs. We'll thus pick some key examples that are either historically important or are very widely available and thus commonly used.

We start with pose & movement sensors (also called inertial sensors) because they are commonly available and cheap. This type of sensor is usually an integrated unit that can sense orientation and some types of movement: three DoF for an accelerometer that detects linear acceleration (including gravity, which can be used to estimate pitch and roll orientation), three DoF for a gyroscope that measures changes in rotation, and three DoF for a compass that can give an absolute heading. In 2012 a modern smartphone would typically include all of these sensors and a GPS unit to give ~5–10 m accurate global position information. This data is not itself sufficient to generate, for example, a precisely registered augmented reality (Fig. 7.10), as the sensors are not very high quality and do not give precise readings that are registered in local coordinates. It is not generally possible to build a local position tracking system from an accelerometer because the sensors drift over time. In addition the

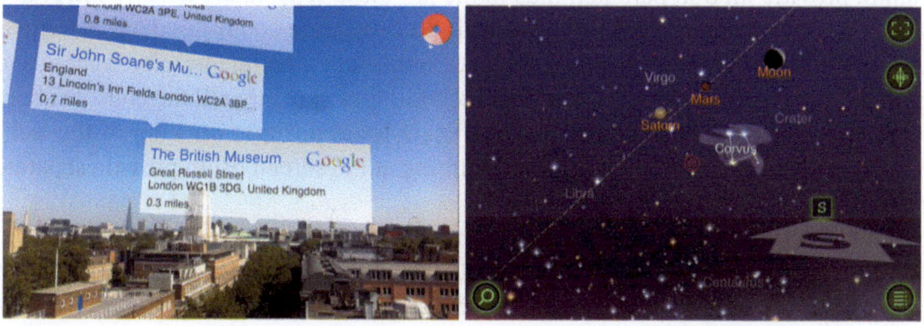

Fig. 7.10 *Left* acrossair browser, acrossair, showing museums south from University College London. *Right* Star Walk, Vito Technology, Inc

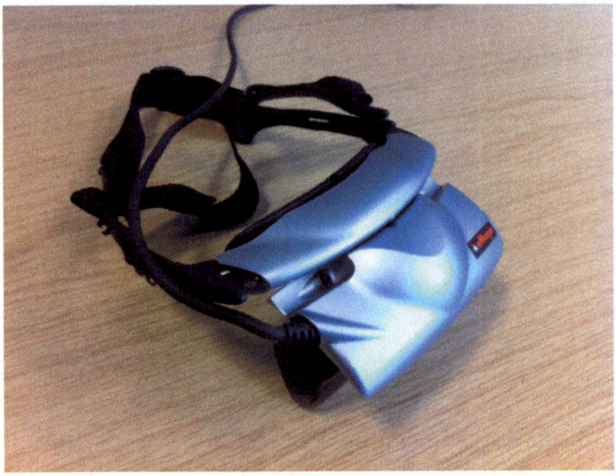

Fig. 7.11 The eMagin z800 helmet contains a three axis rotation sensor

accelerometers cannot tell the system the height above the ground or distance from any local feature without other calibration. Calibrating against GPS fixes is problematic because those fixes are inaccurate and the civilian implementations include a moving offset by design. Thus currently, outside of the lab, mobile augmented reality is limited to display modes that use gross position referencing, such as labels over whole buildings, rather than being able to accurate align an object to something the size of a window or door. Researchers that are prepared to integrate more equipment than can be found on current smartphones can build systems that maintain good registration, but the vision of being able to walk around and see an augmented real environment where the augmentations are seamlessly integrated is still a few years off.

The other common use for pose & movement sensors is for head-mounted displays and handheld devices. Some consumer HMDs include a two-DoF or three-DoF rotation tracker so that the user can rotate her head to directly control the orientation of the viewpoint (Fig. 7.11). These sensors do not track linear acceleration, so they cannot provide motion parallax, thus in the lab these sensors are usually integrated with a position tracking technology as well. Recent game controller devices such as the Nintendo Wii Remote and Sony Move controllers integrate accelerometers and gyroscopes. Again, these two controller technologies can only tell the console about local rotation of the device. This is insufficient to do position tracking of the technology, so it can't, on its own, support pointing at the screen. Both provide separate position tracking for this.

The basic component sensors thus form an important part of more sophisticated trackers. For example, the InterSense IS900, which is a current state of the art position tracker that is commonly used in virtual reality laboratories, includes an accelerometer as a component. Accelerometers in combination can create more sophisticated models of moving objects. The XSens system is a motion capture technology where the user wears several accelerometers. From the relative acceleration of the limbs a model of the skeleton of the user can be built. However, over time, the position of the user will drift if only the accelerometers are used to calculate his location.

Although we have already seen some crossover, we distinguish position trackers from the sensors already discussed because they return an absolute 3D position in a fixed coordinate system. That is, if the tracked device is moved and returned to the same position, the tracker reports the same position up to its own tolerance for precision and accuracy. This is unlike an accelerometer that will drift over time. Position trackers are a well-studied component of augmented reality and virtual reality. The technology changes relatively slowly, so previous surveys [21, 41, 44] give good overviews that are still very relevant. We thus detail the most common technologies in use and highlight their strengths and weaknesses.

For small spaces, the stalwart of the field has been the magnetic tracking technologies epitomized by the Polhemus Fasttrak. This can return six DoF of a sensor that is attached to or embedded in a control device. A standard unit might track 1–4 such sensors, and a common set up would be to attach one to a HMD and embed a second in a hand-held controller. Because the tracking is magnetic, there is no need for the sensor to have a line of sight to the base station, as with optical tracking (see below). Thus the trackers are commonly used in situations where occlusion is likely. However, the sensors typically do not work over spaces of more than 3 m by 3 m (the space is often less than this), and are affected by metal in the environment.

A cheaper, but more limited technology is the visual tracking systems that can track the head, hands, or other body parts in a small volume. In particular, the Wii Remote contains an IR sensor that can track a bar of LEDs and thus can estimate the relative position of the WiiMote from the bar. This is used to allow direct pointing at the screen. The Sony Move controller uses a camera near the display to track the 3D position of a large light source on the controller. Microsoft's Kinect uses a depth camera to track the skeletons of one or more people in front of the display. Alongside these three currently popular technologies, there are quite a few others that have the aim of giving a limited range of direct movement control in front of the display (e.g., NaturalPoint TRACKIR, Logitech Head Tracker). Most of these track only position, not orientation (although the Wii Remote and Sony Move controllers add pose and movement sensors to measure orientation indirectly), and all of them assume that the user is facing in the direction of the display.

In situations where larger tracking volumes, integrated position and orientation tracking, and greater flexibility of movement are needed, high-end optical tracking systems similar to those used in motion capture are frequently used (Fig. 7.12). Common systems include Vicon, OptiTrak, ARTrack and PhaseSpace. The first three systems use passive markers that are mounted on a device, an item of clothing or a full body suit. The passive markers are retro-reflective, and the camera has an infrared light source to illuminate the markers. The PhaseSpace system uses active markers similarly arranged. An important part of the technology is that the system tracks the positions of the individual points in 3D space, but does not itself track the orientation of the marker. Thus multiple markers on a rigid or near-rigid object are needed to determine the orientation of the object. Position and orientation tracking

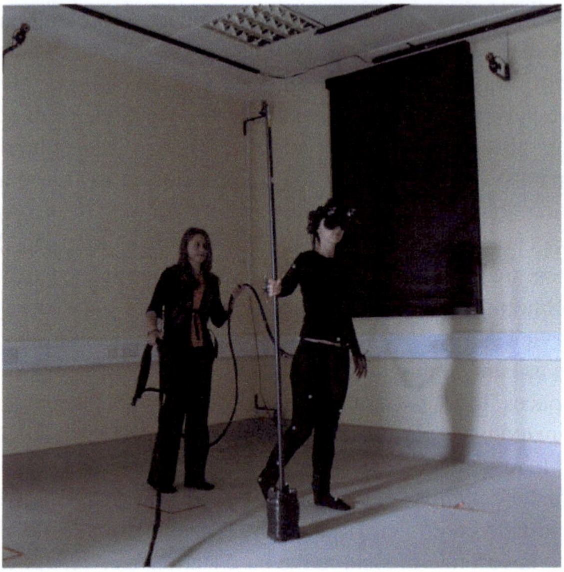

Fig. 7.12 A user wearing a motion capture suit within an OptiTrak installation. The user is taking part an experiment involving a simulation of a train. Courtesy of Angus Antley, University College London

of rigid bodies can be quite accurate and stable, so this approach is used for head tracking in order to ensure stability of the view. In some situations it is useful to have a full motion-capture setup where the body of the user is captured to monitor her limbs movements and potentially create a virtual representation of the user in real-time. In this case, a skeleton can be created and fit to the positions of the marker points. This tends to be slightly less accurate and stable.

A specific technology that is currently quite commonly used in similar situations is the InterSense IS900 (Fig. 7.13). This combines ultrasonic technology with inertial measurements. The IS900 is a common tracker in CAVE™-like installations or other larger spaces, or in situations otherwise not conducive to magnetic tracking.

Other common "cheap" position tracking technologies are marker-based and image-based. The seminal system is the ARToolkit system [14], where the position of a printed marker relative to a camera is derived. On its own, the tracking of a single marker is too noisy to, say, mount a camera on a head to track the head for a HMD display. The marker must be visible for a start. However with sets of markers and calibration of the locations of markers, a versatile position tracker can be built. This type of technology has been commercialized. An example is the InterSense IS1200 system. A novel system using virtual markers in a CAVE™-like environment was proposed by [13]. They proposed placing virtual markers behind the head of the user of the display. These could be used to accurately update the head position, which in turn could be used to move the markers out of the vision of user.

Fig. 7.13 *Left* a typical Intersense IS900 sensor bar installation. *Right* Intersense IS9000 MicroTrax Wand. Courtesy of Intersense, LLC

There are many ways of comparing tracking systems and we refer the interested reader to the previously mentioned reviews [21, 41, 44]. One specific point of comparison that is often overlooked but which is especially important for travel applications is how well the system scales. We can mention at the outset that certain designs do not scale because of the nature of the tracking technology. Magnetic tracking for example does not scale because multiple systems can interfere with each other.

The first wide area tracker of note was the UNC HiBall [42, 43] which is an optical tracker, but which, unlike the optical tracking systems that are now common, is "inside-out," in that the cameras are mounted on the user pointing towards the ceiling. The ceiling contains strips of LEDs that flash in sequence, and from recognizing points on the ceiling, the position of the camera can be calculated. This technology scales well because there is a single set of cameras, and the LED strips can be mounted over a wide area. Essentially, in the HiBall tracker, the environment is changed to include easily tracked points. In a similar manner, camera-based tracking systems based on markers, such as the InterSense IS1200, can also work over a large area. However the camera arrays are currently bulky, and these inside-out systems can suffer from problems of occlusions of the camera by the body.

Thus, although they are more expensive, large motion capture systems are often used to track large spaces. Motion capture systems such as Vicon are designed to scale up to large numbers of cameras: as the tracked volume increases more cameras are required to retain precision. An impressive system of this type is the VIRTSIM system from Motion Reality Inc. This can track a simple skeleton on 13 users simultaneously over an area of $50'$ by $100'$. It also drives 13 wireless HMDs to give users the impression they are training in a squad with 12 others.

The next class of device is by far the most diverse: hand-held devices. The space of physical and virtual devices that could be used to control travel is vast. It ranges from joysticks through props such as steering wheels, to versatile devices such as smartphones. Many variants have been tried and we will only mention a sampling.

Interested readers are referred to Bowman et al. [5], and the IEEE Symposium on 3D User Interfaces, where new devices are introduced every year.

We start with the simplest device: a button. The very simplest travel interfaces use one or more buttons on a device that the user holds. Indeed, many position-tracking systems include a hand-held device with buttons as standard. A button can be used to indicate that movement should be effected, and the direction of the movement might depend on the orientation of a tracker.

This is all one needs if the display fully surrounds the user (i.e., the field of regard is 360°), but otherwise some control is needed to rotate the viewpoint in the environment. One could achieve this with a button and a gesture to point, but simpler and possibly easier and more intuitive for the user is to provide a joystick (see Fig. 7.12, Right).

There is a broad range of devices that are common with desktop computers and can be used or customized to use in an immersive setup. A wireless handheld gyroscopic mouse is a common choice. As noted previously the gyroscope is only a rotational control, so it is not sufficient to give an accurately registered direction, but a position tracker can be attached.

Assemblies of controls can become quite complicated. Trackers and buttons can be embedded in props such as weapons or sports equipment. While such props have long been common in video games, Hinckley et al. proposed using them for other applications, including embedding a tracker in a doll's head to aid with a neurosurgical visualisation [11]. It may be that the task includes complex controls beyond travel, selection and manipulation of objects. In this case a controller might be custom built or a controller with a wide range of control patterns might be used. For example, the cubic mouse [10] is a tracked cube with three orthogonal rods passing through it.

For more complex user interfaces using multiple dynamic interfaces such as menus, an obvious and common choice is to utilize a mobile device such as a tablet or smartphone as a control device. This can display virtual buttons, virtual sliders, and other controls to facilitate travel. A common use of mobile devices is to act as a secondary data display, by providing maps, e.g., [23], text or numeric information, or alternate views of the virtual world. A metaphor that can be used is the concept of the magic lens or 3D magic lens [39] through which the 3D world can be seen in a different way. This then can act as an interface to tasks in the VE, including travel.

An example of the type of more complex travel interface that can be built with such devices is the World-in-Miniature (WIM) technique, which creates a hand-held miniature map of the world that users can use to move themselves around the VE [26].

Such devices don't need to be physical devices; they can be simulations of physical devices or other representations of virtual controls. For example, the *virtual tricorder* and *pen-and-tablet* approaches embed virtual controls in the 3D environment that are anchored to the tracked position of a hand-held device [4, 35, 46]. These controls can be used for a variety of tasks, including travel.

Our final category of device is hybrid devices. This is a catchall for novel and interesting technologies that can be used to control travel. For travel techniques, an obvious place to put the technology is on the floor. Consumer games utilize devices

Fig. 7.14 Joyman. Courtesy of Julien Pettré, INRIA-Rennes, France

such as dance mats and the Wii Fit Balance Board; these can be made into controllers for VEs very easily. To integrate a larger force sensing surface into a VE requires a larger engineering effort with multiple force sensors. An example of a successful system is the one at McGill University [17]. Because the floor is force sensitive (and in this case includes a haptic display), controls can be integrated into the floor so that the user can step on them. Other novel devices for travel include Joyman (Fig. 7.14) [27], which is based on the concept of a "human-scale joystick." The user stands on a platform that they tilt to control locomotion. A related system is the Virtual Motion Controller (VMC) (Fig. 7.15) [45] which measures the position of a user standing on a plate and then moves the user when they stand on its rim. A variety of foot-based interaction devices exist, such as the Interaction Slippers, which are tracked and sense contact between the feet [16]. Finally, systems such as GAITER (Fig. 7.16) can track foot and leg position and movement but can also support the user's doing actions such as kneeling and going prone [37].

Before moving on to discussing travel techniques themselves, it is worth analysing what these input devices give us in terms of control input. A useful tool is the analysis of Mackinlay et al. [18], which classified interaction devices based on what types of

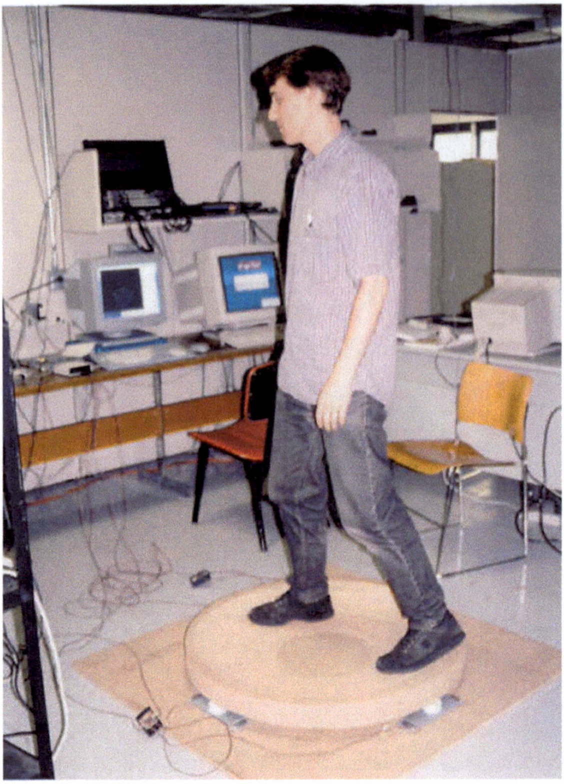

Fig. 7.15 Virtual Motion Controller. Courtesy of Thomas Furness, Human Interface Technology Laboratory, University of Washington

motion they sense. The type of diagram that they use is shown in Fig. 7.17. It classifies a single device as a union (indicated by connecting dashed lines) of individual sensors (the circles connected with solid lines). We can see that the device might sense linear absolute position (P), linear relative position (dP), linear absolute force (F), linear relative force (dF), rotary absolute position (R), rotary relative position (dR), rotary absolute force (T for tensor), or rotary relative force (dT). We also see how many degrees of freedom each device senses (X, Y, Z for linear degrees of freedom, and rX, rY, rZ for rotary degrees of freedom), and the resolution: from continuous (Inf) through to discrete (1). Thus a common mouse is a combination of a two-dimensional relative position (X, Y) sensor with continuous movement plus typically 2 discrete buttons (Z, but only two position values, indicating 0 or 1) plus a scroll wheel reporting a number of discrete positions of the wheel as relative rotation. The diagram thus depicts the mouse as three separate units (position, buttons, scroll wheel) connected by dotted lines. The two dimensions of the mouse are joined together by a solid line because they are reported by the same sensor in the mouse.

Fig. 7.16 GAITER. Courtesy of Jim Templeman, U.S. Naval Research Laboratory

We have plotted several other common input devices. The Polhemus Fastrak tracker is a widely used magnetic tracker which reports the six-DoF position and orientation of a device in a coordinate system centered on a transmission device. A Wii Remote with MotionPlus is a relatively complex device with four sets of sensors. The first is that the device has 11 buttons on it that (plus an on-off switch that isn't included in the diagram). The second is that the IR camera on the Wii Remote can sense a sensor bar placed on or under the display. This gives effectively four coordinates: three rotations of the Wii Remote relative to the sensor bar, plus a distance of the Wii Remote from the sensor bar. The third sensor is the accelerometer (actually the lowest set of nodes in this connected set), which reports the acceleration of the Wii Remote including gravity. The fourth set is the Wii MotionPlus, which adds gyroscope functionality and thus provides another set of relative rotations in a different coordinate system (the device's coordinate system rather than the screen

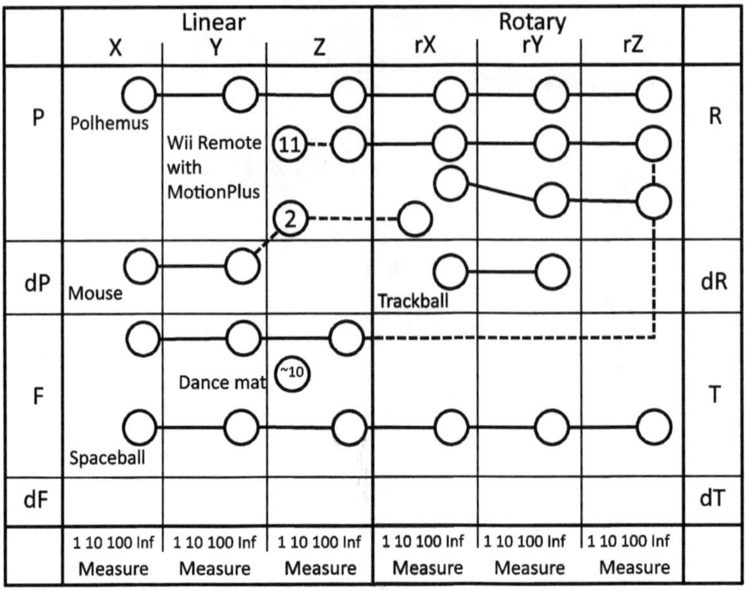

Fig. 7.17 Several input devices plotted using the analysis of Mackinlay et al. [18]

coordinate system). In covering other sections of the diagram we have included some other common devices. A trackball senses two degrees of rotation. A dance mat as might be used for popular dancing games supports approximately 10 buttons that respond to force applied to them, but do not themselves move. Finally the Spaceball is a type of controller that sense forces on it, but does not move. One current example is the HP Spaceball® 5000.

A note: we have not covered treadmills and similar locomotion devices in this chapter. From a control point of view, these provide one or two degrees of input because the user can walk in one or two directions. However the issues in using this for control are subtle. Some treadmill technologies are discussed in Chap. 6 of this volume.

7.4 Travel Techniques

7.4.1 Travel as a Control Task

In the virtual reality model of interaction, to travel the user would walk or use a vehicle, just as he would do in real life. Of course, as already discussed, this ideal is limited by the technology that is available. Most notably, VE systems don't support the user moving themselves over long distances and might even require that the user be stationary. Therefore some sort of virtual travel technique is necessary. Given what we know about sensors and displays, it is first important to identify what we know about the user, and thus what a travel technique actually moves.

We already noted that viewpoint control in the real world is a consequence of movements of various parts of the body. We also note that sometimes the user's head is tracked, and sometimes it is not. If the user stands inside a CAVE™ -like device, then they can physically move a little bit, but may utilize a joystick to move longer distances. Thus in many systems there are effectively two travel techniques: a short-range physical travel technique and a long-range virtual travel technique. This is unlike a standard desktop-style metaphor where the whole of the viewpoint motion control is under a single control metaphor (e.g., a joystick or mouse plus keyboard combination).

In order to understand how these two types of movement work together, we need to describe the various coordinate systems involved, and the relationships among them.

A logical separation which is reified in some implementations is to separate a *display center coordinate* system from a tracker coordinate system. To explain this we need to delve into one potential way of structuring coordinate systems within a display system. An example is shown in Fig. 7.18. This depicts an SSD-type system but the same coordinate systems would usually exist in other display types.

- Tracker Base is the fixed origin of the tracking system coordinate system. Typically this is centered on a physical base unit or some physical component of the tracking system that is static. The tracking unit reports positions relative to this.
- Head Tracker is the relative position of the tracking unit attached to the head. Note that it is not exactly the same as the position of the head, but it is commonly assumed that the tracking unit is in a fixed position relative to the user's head (e.g., on the glasses or on a cap).
- Hand Tracker is the relative position of the tracking unit attached to the hand. Again this is not exactly the same as the hand position as it is assumed there is a known offset between the two.

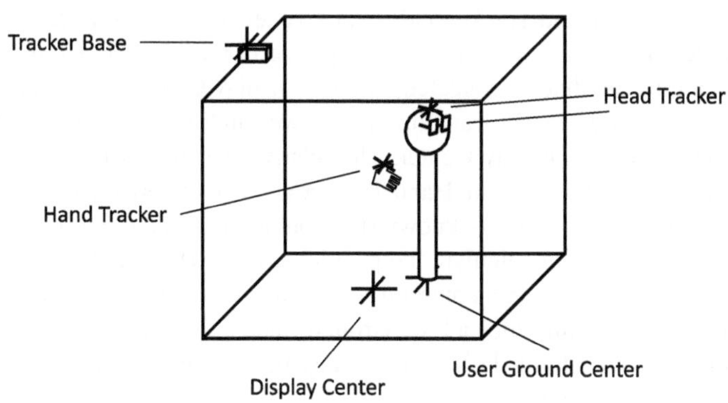

Fig. 7.18 Coordinate systems within a SSD-type display

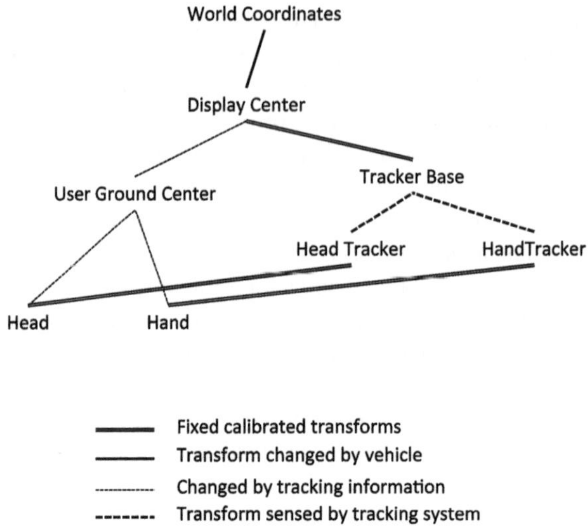

Fig. 7.19 Relationships between coordinate systems in a typical VE system

- Display Center is a fixed position in the display from which measurements of user position are reported. In Fig. 7.18 the display center is the center of the floor.
- User Ground Center is a calculated position relative to the Display Center where the user "is." Typically this is the centroid of his two feet or just the point directly below his head.

The relationships between these coordinate systems are represented in Fig. 7.19, which is an abstract scene graph. The lines represent parent-child relationships between coordinate systems. Each line also represents the transformation from one coordinate system to another. We see that the Display Center is a child of the World Coordinates. The position of the Tracker Base is known relative to the Display Center. This will usually be a one-time calibration at the installation of the tracking system. We can then see that Hand Tracker and Head Tracker coordinates are known relative to the Tracker Base. These are the values reported in real-time by the tracking system. We can thus see that by concatenating the Display Center to Tracker Base and Tracker Base to Hand Tracker transformations, we can find the location of the Hand Tracker relative to the Display Center. The relative position of the Head from the Head Tracker and Hand from the Hand Tracker are also calibrated and fixed. Given that all these transformations are known (the double lines in the figure and the large dashed line), we can then calculate the positions of the Head and Hand relative to the Display Center. However, a common convention, especially when implementing travel techniques is to introduce a User Ground Center coordinate system as defined above. Thus the three light dashed lines must be updated whenever the tracking system reports new positions.

Having explained the roles of coordinate systems, we can see that short-range physical movements are captured in the movements of the Head Tracker and Hand Tracker relative to the Display Center.

For virtual travel over longer distances, the position of the Display Center can be moved relative to the World coordinates. There are many options here. Fundamentally they involves changing the transformation of the Display Center relative to the World origin. This might involve translation, rotation and scale of the Display Center. However, it is very common for certain restrictions to be set. For example, there might be translation only: if the person is surrounded by the display system, they can simply physically turn to see any direction, whereas in a typical SSD with three walls there will need to be a rotation metaphor. If there is translation, then it might be 2D only (i.e., on a plane), or use some form of surface-following algorithm so that the user is always at the same height above the ground. Other questions must be answered as well. If there is translation, in which direction should it be? If there is rotation or scaling, around which point should the scale or rotation occur (e.g., the Display Center, the User Ground Center, or the Head)? And how does the user indicate how to start and stop traveling?

A task decomposition for travel techniques is provided by Bowman et al. [2]. Their decomposition, which is shown in Fig. 7.20, decomposes the travel task into different sub-tasks: start to move, indicate position, indicate orientation, stop moving. To start to move, the user might press and hold a button, and to stop moving she might release

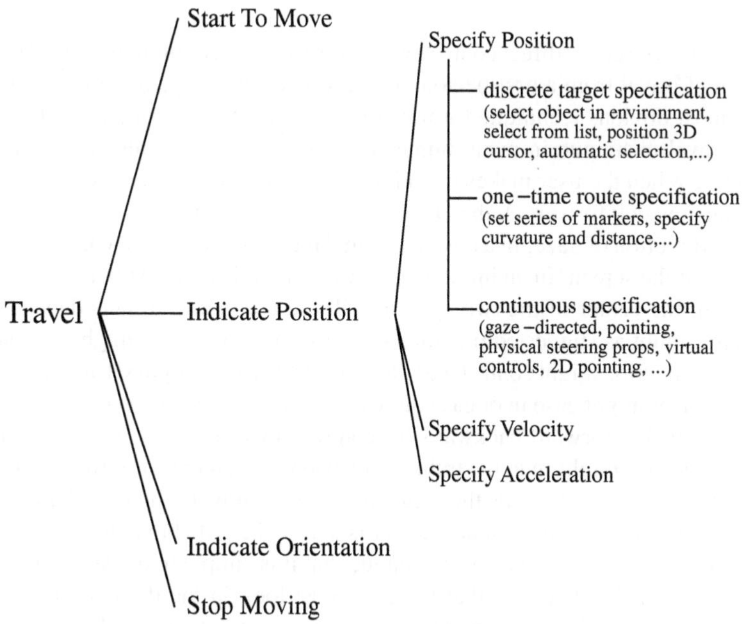

Fig. 7.20 Taxonomy of travel techniques focusing on level of user control [2]

a button. Alternatively, for example, the start and stop might be implicit in the user manipulating a joystick. In our terminology, the indicate position sub-task effects a change in translation of the Display Center, either directly, or by giving the dynamics of change over time (i.e., giving a velocity or acceleration). The task decomposition shows that this can be achieved in three ways: specify position, specify velocity and specify acceleration. Each of these can in turn be achieved in many ways. For example specifying a position can be achieved by selecting a target, giving a route, or by continuous specification. The last of these is the most common in real-time, interactive systems: the user can point or gaze towards the target and effect a control to start and continue travelling. There would be a similar breakdown for orientation: e.g., one might set a target orientation, one might turn using a direct angular control, or one might set a rotation speed.

In looking at the options here we can make a connection to Fig. 7.17, where the input devices had different dimensionality that might map conveniently to the different options here. Obviously a joystick is a common choice for specifying velocity, whereas a mouse might be better deployed for relative rotation or clicking to select targets in the world.

We examine some common control configurations in the following sections. The reader is also referred to Bowman et al. [5].

7.4.2 Direct Self Motion Control Techniques

In this section we cover direct control, where the user has continuous control over the direction of travel from a first-person view at every update cycle of the simulation.

The most obvious technique for travel is gaze-directed steering [22]. This is the default travel technique in many immersive systems, and it is also found in many 3D games. When the user makes the relevant control input (e.g., presses a button or moves the joystick forwards), the Display Center coordinate system moves forward along the direction of gaze. In a desktop virtual reality system, this would be through the center of the screen. In an immersive VE system, this is typically in the direction of a line in the center of the two eye lines. There are many variants depending on the control input as suggested previously. The velocity of travel might be constant, a joystick deflection might control the velocity of travel, or it might set an acceleration. The movement might ease in or ease out when the control is changed. If a joystick is used, typically it is forward/backward that controls travel along the direction of travel, but the other axis might map to strafing (sideways movement) or turning (rotation), or it might be ignored. Finally, the actual direction of travel might be clamped to be in only two dimensions or so that the user is at a set height above the ground.

Gaze-directed steering has the advantage that it is simple to explain to users, but it has the major disadvantage that the user must look in the direction they intend to travel. There are obvious variants: one could use any other coordinate system or relation between tracked points to set the travel direction. The most obvious ones are flying in the direction of pointing with a hand tracker [22], direction of gaze recorded by an eye-tracker, direction of the torso, the relative direction of hand from eye or the relative position of the two hands. All of these have been implemented in VE systems. Pointing with a hand tracker is a very commonly implemented technique because it decouples head orientation from travel so that users can look around while moving [2].

7.4.3 Indirect Self Motion Control Techniques

Indirect control techniques set the target of travel in an indirect or asynchronous manner. The most common technique is to indicate a target in the environment, and then to enable travel to that position over a period of time or instantly (when it is known as teleporting). Indicating the target might involve simply targeting an object by pointing directly (e.g., "go over there"), or on a map or miniature world. An example of the former is the ZoomBack technique [47] that uses a typical ray-casting metaphor to select an object in the environment, and then moves the user to a position directly in front of this object. Ray-casting has been used in other 3D interfaces for target-based travel as well, e.g., [3]. An example of the latter is the previously mentioned WIM technique [26], in which a small human figure represents the user's position and orientation in the miniature world. The user places the user representation in the WIM, and then a path is calculated that moves the camera to this location, taking into account any rotation that is necessary to reach the target orientation. In this technique the transition could be achieved by zooming in to the WIM itself, or it could be planned as a motion through the VE at the original scale.

7.4.4 Scene Motion Techniques

A common alternative to self-motion travel is to manipulate the scene. This is very common on desktop interfaces where the metaphor is that the camera is static and the object on the screen is moving. Less obvious is that this can be turned in to a travel technique, in that the object is considered stationary and the camera is moved around the object. The rotations of the camera for an immersive system would be violent, so this is not very commonly done, but the translation equivalent (pull and push objects) has been demonstrated. Most notable is the "grab the air" technique [19, 40]. In this technique the user can grab anywhere in the environment and when they move her hand back and forth, they move themselves through the environment. This motion can be scaled by using two hands. Similar techniques include Mine's Scaled-World Grab technique [24] and the LaserGrab technique of Zeleznik et al. [47]. A related set of techniques involve manipulation of objects using image plane techniques [28].

7.4.5 *Other Control Inputs*

The interaction techniques for travel that we have discussed in the section above cover a broad range of those that are used in practice. However, the area has seen a number of innovative techniques. Another way of controlling speed of rotation or velocity of travel involves measuring the distance between points on the body and using that as the rate. For example, the distance between the head and a measured or nominal foot position gives an estimate of lean, and this can be used to control velocity [7, 16, 32]. Alternatively the distance between hand and head can be used to control velocity in a point to fly technique [22].

An alternative to using a device to effect travel is to track a user movement that is similar to walking. Such techniques are called "walking in place" metaphors, where users move their feet to simulate walking without actually translating their bodies [31]. In the case of Slater et al., the user had to mimic walking, and a gesture recognition system detected that the user was performing this mime by monitoring his head movement. Walking in place metaphors have attracted a lot of interest because users have to physically exert themselves. A novel platform that allowed walking in place with extended leg movement was presented by Swapp et al. [34]. Walking in place techniques are covered in Chaps. 10 & 11.

Finally, we note that especially with four-walled CAVE™-like systems and HMDs with restricted tracker spaces, there has been a lot of interest in techniques that bias rotation to achieve the effect that the user doesn't look away from the main walls, or walks in the correct direction. These are covered elsewhere in the book (see Chap. 14), but we note the work by Razzaque et al. on redirected walking [29] which provides imperceptible rotation distortion, and explicit amplification of rotation [16].

7.5 Conclusion

We hope that in this short introduction to interaction devices and displays for virtual walking, we have conveyed some of the challenges of the field and the constant innovation that there has been over the past couple of decades. Travel is a very hard problem for virtual reality systems: it is inherently a "two-task" system because the user can move physically for short maneuvering tasks, but the user also needs a virtual travel technique to move over long distances. Other chapters in this volume indicate some of the work that is being done to alleviate the need for a virtual travel technique in some situations, but virtual travel techniques will likely be with us for some time yet. In describing the range of different display types, interaction devices and control methods, it should also be obvious that there are no one-size-fits-all solutions and that techniques need customization to fit application needs and system capabilities. While there are many options, there are also established best practices that can be uncovered by studying the literature. There will no doubt be further innovation, especially in gesture-based control that is enabled by recent advances in camera technology. We look forward to testing new techniques as they emerge.

References

1. Agrawala M, Beers AC, McDowall I, Fröhlich B, Bolas M, Hanrahan R (1997) The two-user Responsive Workbench: support for collaboration through individual views of a shared space. In: Proceedings of the 24th annual conference on computer graphics and interactive techniques (SIGGRAPH '97). ACM Press/Addison-Wesley, New York, pp 327–332
2. Bowman D, Davis E, Badre A, Hodges L (1999) Maintaining spatial orientation during travel in an immersive virtual environment. Presence: Teleoper Virt Environ 8(6):618–631
3. Bowman D, Johnson D, Hodges L (2001) Testbed evaluation of VE interaction techniques. Presence: Teleoper Virt Environ 10(1):75–95
4. Bowman D, Wineman J, Hodges L, Allison D (1998) Designing animal habitats within an immersive VE. IEEE Comp Graph Appl 18(5):9–13
5. Bowman D, Kruijff E, LaViola J, Poupyrev I (2004) 3D user interfaces: theory and practice. Addison Wesley Longman, Redwood City
6. Cruz-Neira C, Sandin DJ, DeFanti TA (1993) Surround-screen projection-based virtual reality: the design and implementation of the CAVE. In: Proceedings of the 20th annual conference on computer graphics and interactive techniques (SIGGRAPH '93). ACM, New York, pp 135–142
7. Fairchild K, Hai L, Loo J, Hern N, Serra L (1993) The heaven and earth virtual reality: designing applications for novice users. In: Proceedings of IEEE virtual reality annual international symposium, 1993. IEEE, pp 47–53
8. Feiner S, MacIntyre B, Höllerer T, Webster T (1997) A touring machine: prototyping 3D mobile augmented reality systems for exploring the urban environment. In: Proceedings of first international symposium on wearable computers. IEEE, pp 74–81
9. Fernandes KJ, Raja V, Eyre J (2003) Cybersphere: the fully immersive spherical projection system. Commun. ACM 46(9):141–146
10. Fröhlich B, Plate J (2000) The cubic mouse: a new device for three-dimensional input. In: Proceedings of the SIGCHI conference on human factors in computing systems (CHI '00). ACM, New York, pp 526–531
11. Hinckley K, Pausch R, Goble JC, Kassell NF (1994) Passive real-world interface props for neurosurgical visualization. In: Proceedings of ACM CHI'94 conference on human factors in computing systems. ACM New York, pp 452–458
12. Höllerer T, Kuchera-Morin J, Amatriain X, (2007) The allosphere: a large-scale immersive surround-view instrument. In: Proceedings of the 2007 workshop on emerging displays technologies: images and beyond: the future of displays and interacton (EDT '07). ACM, New York, Article 3
13. Hutson M, Reiners D (2011) JanusVF: accurate navigation using SCAAT and virtual fiducials. IEEE Trans Visual Comput Graph 17(1):3–13
14. Kato H, Billinghurst M (1999) Marker tracking and hmd calibration for a video-based augmented reality conferencing system. In: Proceedings of the 2nd IEEE and ACM international workshop on augmented reality (IWAR 99), pp 85–94
15. Kulik A, Kunert A, Beck S, Reichel R, Blach R, Zink A, Fröhlich B (2011) C1x6: a stereoscopic six-user display for co-located collaboration in shared virtual environments. In: Proceedings of the 2011 SIGGRAPH Asia conference (SA '11). ACM, New York, Article 188, 12 pp
16. LaViola J, Acevedo D, Keefe D, Zeleznik R (2001) Hands-free multi-scale navigation in virtual environments. In: Proceedings of the 2001 symposium on interactive 3D graphics (I3D '01). ACM, New York, pp 9–15
17. Law AW, Peck B, Visell Y, Kry P, Cooperstock J (2008) A multi-modal floor-space for experiencing material deformation underfoot in virtual reality. In: Proceedings of IEEE international workshop on haptic audio visual environments and games. IEEE, pp 126–131
18. Mackinlay J, Card SK, Robertson GG (1990) A semantic analysis of the design space of input devices. Human-Comput Inter 5(2):145–190
19. Mapes D, Moshell J (1995) A two-handed interface for object manipulation in virtual environments. Presence: Teleoper Virt Environ 4(4):403–416

20. McMahan R, Bowman D, Zielinski D, Brady R (2012) Evaluating display fidelity and interaction fidelity in a virtual reality game. IEEE Trans Visual Comput Graph (Proc IEEE Virt Real) 18(4):626–633
21. Meyer K, Applewhite H (1992) A survey of position trackers. Presence: Teleoper Virt Environ 1(2):173–200
22. Mine M (1995) Virtual environment interaction techniques. UNC Chapel Hill CS Dept
23. Miguel MM, Ogawa T, Kiyokawa K, Takemura H (2007) A PDA-based see-through interface within an immersive environment. In: Proceedings of The 17th international conference on artificial reality and telexistence (ICAT 2007), pp 113–118
24. Mine M, Brooks F, Sequin C (1997) Moving objects in space: exploiting proprioception in virtual environment interaction. In: Proceedings of the 24th annual conference on computer graphics and interactive techniques (SIGGRAPH '97). ACM Press/Addison-Wesley, New York, pp 19–26
25. Office of Naval Research (2012) Infantry immersion trainer. http://www.onr.navy.mil/Media-Center/Fact-Sheets/infantry-immersion-trainer.aspx, Retrieved July 2012
26. Pausch, R., Burnette, T., Brockway, D., & Weiblen, M. (1995). Navigation and Locomotion in Virtual Worlds via Flight into Hand-Held Miniature. In Proceedings of the 22nd annual conference on Computer graphics and interactive techniques (SIGGRAPH '95). ACM, New York, pp 399–400
27. Pettré J, Siret O, Marchal M, de la Rivire J-B, Lécuyer A (2011) Joyman: an immersive and entertaining interface for virtual locomotion. In SIGGRAPH Asia 2011 emerging technologies (SA '11). ACM, New York, Article 22
28. Pierce J, Forsberg A, Conway M, Hong S, Zeleznik R, Mine M (1997) Image plane interaction techniques in 3D immersive environments. In: Proceedings of the 1997 symposium on interactive 3D graphics (I3D '97). ACM, New York, 39-ff
29. Razzaque S, Swapp D, Slater M, Whitton M, Steed A (2002) Redirected walking in place. In: Proceedings of the workshop on virtual environments 2002 (EGVE '02), Eurographics Association, Aire-la-Ville, Switzerland, pp 123–130
30. Slater M, Steed A, Chrysanthou Y (2001) Computer graphics and virtual environments: from realism to real-time. Addison-Wesley, New York
31. Slater M, Usoh M, Steed A (1995) Taking steps: the influence of a walking technique on presence in virtual reality. ACM Trans Comput-Human Inter 2(3):201–219
32. Song D, Norman M (1993) Nonlinear interactive motion control techniques for virtual space navigation. In: Proceedings of IEEE virtual reality annual international symposium. IEEE, pp 111–117
33. Sutherland IE (1965) The ultimate display. Proc IFIP 65(2):506–508
34. Swapp D, Williams J, Steed A (2010) The implementation of a novel walking interface within an immersive display. In: Proceedings 2010. IEEE symposium on 3D user interfaces (3DUI). IEEE, pp 71–74
35. Szalavári Z, Gervautz M (1997) The personal interaction panel—a two-handed interface for augmented reality. Comput Graph Forum (Proceedings of EUROGRAPHICS'97. Budapest, Hungary) 16(3):335–346
36. Tamura H, Yamamoto H, Katayama A (2001) Mixed reality: future dreams seen at the border between real and virtual worlds. IEEE Comput Graph Appl 21(6):64–70
37. Templeman JN, Sibert LE, Page RC, Denbrook PS (2009) Designing user interfaces for training dismounted infantry. In: Nicholson D, Schmorrow D, Cohn J (eds) Handbook of virtual environments for training and education, vol 2. Greenwood Publishing Group, Westport, pp 219–239
38. Usoh M, Arthur K, Whitton M, Bastos R, Steed A, Slater M, Brooks F (1999) Walking > Walking-in-place > flying. In: Virtual environments. Proceedings of the 26th annual conference on computer graphics and interactive techniques (SIGGRAPH '99). ACM Press/Addison-Wesley, New York, pp 359–364
39. Viega J, Conway MJ, Williams G, Pausch R (1996) 3D magic lenses. In: Proceedings of the 9th annual ACM symposium on user interface software and technology (UIST '96). ACM, New York, pp 51–58

40. Ware C, Osborne S (1990) Exploration and virtual camera control in virtual three dimensional environments. In: Proceedings of the 1990 symposium on interactive 3D graphics (I3D '90). ACM, New York, pp 175–183
41. Welch G (2002) Tracking bibliography. http://www.cs.unc.edu/~tracker/ref/biblio/index.html. Retrieved July 2012
42. Welch G, Bishop G, Vicci L, Brumback S, Keller K, Colucci D (1999) The HiBall tracker: high-performance wide-area tracking for virtual and augmented environments. In: Proceedings of the ACM symposium on virtual reality software and technology (VRST '99). ACM, New York, pp 1–10
43. Welch G, Bishop G, Vicci L, Brumback S, Keller K, Colucci D (2001) High-performance wide-area optical tracking. Presence: Teleoper Virt Environ 10(1):1–21
44. Welch G, Foxlin E (2002) Motion tracking: no silver bullet, but a respectable arsenal. IEEE Comput Graph Appl 22(6):24–38
45. Wells M, Peterson B, Aten J (1996) The virtual motion controller: a sufficient-motion walking simulator. In: Proceedings of IEEE virtual reality annual international symposium. IEEE, pp 1–8
46. Wloka M, Greenfield E (1995) The virtual tricorder: a uniform interface for virtual reality. In: Proceedings of the 8th annual ACM symposium on User interface and software technology (UIST '95). ACM, New York, pp 39–40
47. Zeleznik R, LaViola J, Feliz D, Keefe D (2002) Pop through button devices for VE navigation and interaction. In: Proceedings of IEEE virtual reality 2002. IEEE, pp 127–134

Chapter 8
Sensing Human Walking: Algorithms and Techniques for Extracting and Modeling Locomotion

Franck Multon

Abstract This chapter reports the most popular methods used to evaluate the main properties of human walking. We will mainly focus on: global parameters (such as step length, frequency, gait asymmetry and regularity), kinematic parameters (such as joint angles depending on time), dynamic values (such as the ground reaction force and the joint torques) and muscle activity (such as muscle tension). A large set of sensors have been introduced in order to analyze human walking in biomechanics and other connected domains such as robotics, human motion sciences, computer animation... Among all these sensors, we will focus on: mono-point sensors (such as accelerometers), multi-point sensors (such as flock of sensors, opto-electronic systems and video analysis), and dynamic sensors (such as force plates or electromyographic sensors). For the most popular systems, we will describe the most popular methods and algorithms used to compute the parameters described above. All along the chapter we will explain how these algorithms could provide original methods for helping people to design natural navigation in VR.

8.1 Introduction

Measurement of human motion has a long history since J.E. Marey and E. Muybridge have proposed chronophotography to record animals in motion [20]. Motion was then expressed as a sequence of poses which is still widely used to describe human motion nowadays. Numerous sensors have been introduced to capture human motion and especially human walking. In this chapter we propose a summary of the most studied parameters in human walking and direct/indirect methods to obtain them.

Human walking is one of the most commonly used motion in everyday life and it has been studied from a long time. However it is still a very active field of research

F. Multon (✉)
University Rennes2, Campus La Harpe, Av. Charles Tillon, CS24414, F35044 Rennes, France
e-mail: franck.multon@univ-rennes2.fr

F. Steinicke et al. (eds.), *Human Walking in Virtual Environments*,
DOI: 10.1007/978-1-4419-8432-6_8,

in many scientific domains, including biomechanics, neurosciences, robotics and computer animation. One of the key points is to extract the most relevant parameters of human walking according to the specific requirements of a given application. In this section, we propose a summary of the most popular parameters and give some methods to retrieve them with various sensors.

Nowadays, there are many possible devices to measure human motion and, consequently, human walking. Most of them are generic but some devices have been specifically designed for human walking. In virtual reality, capturing the intentions of the user is necessary in order to navigate properly in the simulated environment or to animate an avatar. While motion capture devices were very expensive and difficult to use in the past, it is now possible to use cheap and easy-to-use systems such as the Wii-mote (product of Nintendo) or the Kinect (product of Microsoft) devices. Whatever the system, the key issue is to design algorithms to extract the relevant parameters of the user's gait that enable the system to react realistically.

The first part of this chapter is dedicated to global parameters such as walking speed, step length, frequency and global walking trajectory. The second part of the chapter focuses on kinematic and dynamic parameters such as joint angles, torques and muscle activity. We then conclude and give a summary of the studied parameters and their potential use in walking in VR.

8.2 Sensing and Interpreting Global Gait Parameters

Human being can be represented by more or less complex models. The simplest one consists in considering human being as a point which corresponds to his center of mass. Analyzing human walking with such a model consists in dealing with global gait parameters such as trajectory, velocity, step length, and frequency. It provides us with relevant information about the global performance of gait. As described in Chap. 3 those parameters enable us to associate the real-time performance of the user with multisensory feedbacks, such as adapting the movement of the virtual camera according to velocity when navigating in virtual environments [35]. In this section, we describe how these parameters are defined and measured.

8.2.1 Step Length and Frequency

When dealing with human walking, one can focus on global parameters such as step length S_L and frequency S_F. These two parameters are in relation with walking speed V:

$$V = S_L * S_F$$

As the model is restricted to the user's center of mass, walking speed can be approximated by integrating the signal delivered by an accelerometer placed on the pelvis (assuming that the center of mass is close to the Pelvis). It's thus possible to deduce the instantaneous velocity and to drive a virtual camera in the virtual world in navigation tasks [35]. However this approach has two main limitations. Firstly it assumes that acceleration is noise-free which is not true with common accelerometers. As a consequence velocity computed this way may become false after a moment. Secondly the initial velocity is required when computing $V(t)$:

$$V(t) = \left(\sum_{i=0}^{n} a(t_i) * \Delta t \right) + V(0)$$

where Δt is the sample time, $V(0)$ is the initial velocity, n is the number of samples (t_n corresponds to t) and a stands for accelerations provided by the accelerometer. Any error in estimating $V(0)$ could thus lead to an error in $V(t)$. One has to notice here that users generally have a limited space to move whereas the virtual environment could be very large. To solve this problem, one of the most famous solutions consists in using an instrumented treadmill. In that case, forward speed could be delivered by the treadmill while the other components could be given by the accelerometer.

An alternative consists in detecting footstrikes in the signal delivered by the accelerometer and to deduce step frequency S_F. If we assume that the step length is constant and relative to the user's size, it is thus possible to deduce speed. The resulting speed is less noisy than integrating acceleration and is not subject to deviations in time. However it does not provide accurate speed as the step length is not actually measured.

Other systems such as the GAITRite (see http://www.gaitrite.com) have been introduced to measure the step length and frequency when walking along limited distances. It consists of a cable which is attached to the user's ankles and which length is measured at a predetermined sampling frequency. It is widely used in medicine because of its simplicity (especially no calibration is required). In addition to S_L and S_F, the system returns the instantaneous distance between a fixed reference frame and the two ankles. It is thus possible to analyse the longitudinal trajectory of the ankles within the gait cycle. However, it is limited to straight line walking in a limited space (generally a few meters).

Recording step length and step width is more difficult in curved walks. As described in Chap. 3 these parameters are still difficult to define in a strict manner.

8.2.2 *Curvature and Non-linear Walking*

Retrieving the curvature of non-linear walking is still a complex problem. Indeed, the instantaneous global orientation of the body is difficult to define. As explained in Chap. 3 some authors focuses on the footprints, the orientation of the pelvis, the torso or the head. It leads to different results for determining the global trajectory of non-linear walking. Hence, positioning a unique sensor on the body to accurately analyze the instantaneous orientation of the body in curved walking is still debated.

The most popular approach consists in tracking the orientation of the pelvis. Hence placing a sensor on the pelvis, such as inertial, magnetic or gyroscope sensors (or a combination of these sensors), could provide a good compromise. Indeed, some authors have shown that the pelvis orientation can provide an early prediction of the running orientation of a rugby player even in deceptive motions where the subject tries to provide fake information to an opponent [4]. The pelvis' trajectory is thus strongly linked to the actual orientation of the user while other body parts may also perform other tasks while walking.

In the same way, detecting a change in walking direction is very difficult. Indeed, the trajectory of the center of mass and of a point placed on the pelvis could be viewed as a series of arcs of circles, even for straight-line walking. Some authors [24] have proposed to model the natural sinusoidal trajectory of the center of mass as a sequence of arcs of circles (see Fig. 8.1). This model assumes that the body could be viewed as an inverted pendulum which basis is the contact foot.

The authors defined a statistical relation between speed and curvature for straight-line walking. Thus turning could be determined when a new point in the speed-curvature space does not fit this relation.

This is a very important issue in immersive environments as we need to detect a subtle change in direction when the subject is walking in order to react in a proper manner when he wishes to change the walking direction. In some works in virtual reality, it leads to exaggerating some indices such as the head inclination in Walk-In-Place interfaces [34].

As shown by many authors [28] the head is viewed as a stable inertial platform which acceleration profile is unchanged even for various gait styles [11] and for uneven terrains [21]. To stabilize this inertial platform, the orientation of the head changes in direction 200 ms before the remaining of the body turns. In case of natural walking, the coordination between head and pelvis could enable us to early determine the orientation of the next step. However, in VR, this could be a problem if the screen is static and placed in front of the user. As the user will orient his head in the direction of the screen this natural behavior disappears. Because of constraints due to the visual

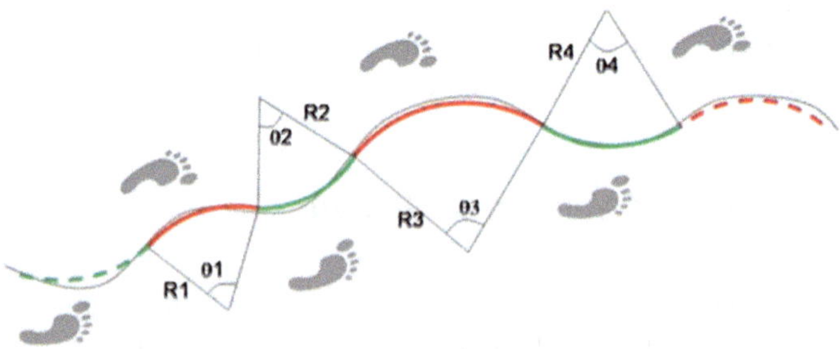

Fig. 8.1 The trajectory of the center of mass is modeled as a sequence of arcs of circles which parameters are the radius of gyration R_i and the velocity (represented here by the angle $\theta_i = (V_i * \text{duration})/R_i$ (where V_i is the average speed within the ith arc). Adapted from [24]

feedbacks and other interaction devices, it seems to be difficult to address this problem without introducing metaphors.

An alternative might be using a force plate under the feet which can bring relevant information of the user's gait. In biomechanics, force plates are used to measure the ground reaction force (GRF) below the feet, the moment of this force around the main ground axes and the location of the instantaneous center of pressure (COP). The ground reaction force is used to compute the acceleration of the center of mass if no other force than gravity and GRF occur. For the global mechanical system (restricted to its center of mass):

$$W + GRF = m\ddot{q}$$

where W stands for the body weight, m is the mass and q is the center of mass position. If we can estimate q and its derivative at some times (especially the initial value but it is sometimes possible to get these values for each foot-strikes event) it is thus possible to integrate the signal:

$$q(t) = \frac{1}{m}\int\int_{t_0}^{t} W + GRF(\tau)d\tau^2$$

where t_0 states for the beginning of the studied sequence. Practically, the initial center of mass velocity $\dot{q}(t_0)$ and position $q(t_0)$ are required to compute $q(t)$. As a consequence these two values should be either measured or imposed at the beginning of the motion (such as starting straight with a null velocity). With a simple Euler integration scheme, this equation becomes:

$$\dot{q}(n) = \dot{q}(0) + \sum_{i=1}^{n}\left(\frac{1}{m}\left(W + GRF(i)\right)\Delta t\right)$$

$$q(n) = q(0) + \sum_{i=1}^{n}\dot{q}(i)\Delta t$$

where Δt stands for the sampling frequency.

A force plate is generally calibrated at the beginning of the sequence so that the value is set to zero when nothing is placed over it. However, if the user jumps over the forceplate or goes out and in several times, the initial calibration could be inappropriate. From the numerical point of view, we obtain:

$$\dot{q}(n) = \begin{cases} \dot{q}(0) + \sum_{i=1}^{t_s-1} \frac{1}{m}\left(W_0 + GRF\,(i)\right) \times \Delta t \quad if\ t < t_s \\ \dot{q}(0) + \sum_{i=1}^{t_s-1} \frac{1}{m}\left(W_0 + GRF\,(i)\right) \times \Delta t + \sum_{i=t_s}^{n} \frac{1}{m}\left(W_{t_s} + GRF(i)\right) \times \Delta t \end{cases}$$

where W_0 is the initial body weight measured prior to an important strike, t_s is the frame number when the strike occurs, and W_{ts} is the body weight after foot trike. It is thus possible to deduce oscillations of the center of mass. It could be used in Walk-In-Place interfaces based on GRF measurements (such as using a Nintendo Wii Balance Board). Information about other axes provides interesting information for navigation, such as the velocity in the forward direction or lateral displacements involved in rotations. However, they are more difficult to use in virtual reality as they require the subject to move in all the directions which generally leads to going outside the measurement area in a very near future (generally the next step).

One has to notice here that the two feet must be placed over the forceplate surface to enable correct estimation of the above displacements. It is also possible to use footscan devices (from RSscan company http://www.rsscan.com) to measure the pressure below the feet. It consists in introducing a sole in the shoe which enables the user to move freely in the real environment. This type of device delivers the pressure in space and time which enables us to compute the resulting vertical component of GRF. The other components cannot be accurately deduced but the location of the center of pressure (COP) may help to guess how this force is oriented in the other main directions.

COP can also be deduced with a forceplate if this latter provides 3D GRF and the corresponding global momentum M over the ground. M and GRF are linked by:

$$\begin{cases} M_x = GRF_z \times y_{COP} - GRF_y \times z_0 \\ M_y = -GRF_z \times x_{COP} + GRF_x \times z_0 \end{cases}$$

where x_{COP} and y_{COP} stand for the local coordinates of the COP in the forceplate reference frame, and z_0 stands for the vertical coordinate of the forceplate surface. It becomes:

$$\begin{cases} x_{COP} = -\frac{(M_y - GRF_x \times z_0)}{GRF_z} \\ y_{COP} = \frac{(M_x + GRF_y \times z_0)}{GRF_z} \end{cases}$$

In quasi-static condition, it is possible to assume that the average value of COP is the projection of the center of mass on the ground. Hence if COP moves the pose of the user has changed, for example to lean in a given direction. This type of information has been used in biomechanics to determine if a subject was turning while walking [24, 34]. It can thus be used to indicate a main direction so that the user could be viewed as a kind of joystick. It is used in some Nintendo Wii games based on the Balance Board device (one has to notice that this device is not a force plate as it does not provide the system with 3D GRF and Momentums ; it provides the vertical component of the GRF and the location of the COP). The Joyman interface [19, 26] is an extension of the idea of using the user's body as a joystick.

8.2.3 Gait Asymmetry and Regularity

As stated above, walking is a quasi-cyclic and almost symmetrical motion. Measuring gait asymmetry and loss of regularity is widely used for diagnosis [17], especially in double-tasks protocols with cognitive loads (such as counting down while walking). There are two main methods to measure symmetry. The simplest one consists in computing a ratio between left and right values [30]:

$$\left[\frac{(R-L)}{0.5} \times (R+L)\right] \times 100$$

where R and L are respectively measurements performed on the right and the left respectively (such as step length or step frequency). The other approach consists in computing auto-correlation of the studied signal [2]. This signal is generally twice the frequency of the stride. Hence auto-correlation of this signal between one step and the following will provide information on asymmetry. Auto-correlation between several strides will provide information on regularity.

As described above irregularity and asymmetry mainly occur when the user has to perform a cognitive activity while walking. In immersive systems based on metaphors, the cognitive load of the user may be different compared to natural walking. Regularity and symmetry may then be affected even if it has not been demonstrated in VR yet.

8.3 Joint Angles, Torques and Muscle Activity

In the previous section human body was modelled as a point (his center of mass). We have seen that it could provide relevant information about the main gait pattern, such as speed, step length and frequency. In VR this type of information could be used to globally adapt the motion of the virtual camera or to capture basic gait parameters to drive the simulation. However more accurate information could be required in order to drive an avatar or capture more accurate information about the user's gait. To this end, it is necessary to have multi-point measurements and to capture more complex parameters, such as joint angles and position, but also to compute dynamic parameters such as joint torques or muscle activation patterns.

8.3.1 Measuring Joint Displacements

The most popular method to get joint position depending on time is to track visual markers placed over standardized anatomical landmarks. The International Society of Biomechanics (ISB) has standardized this markers' placement in order to enable people to compare their results to previously published ones [39, 40]. The key idea is to place markers on accurate anatomical position where skin markers do not significantly slide over the bones. The problem is then to retrieve the location of the internal joints according to the external position of the markers. In this chapter we describe the methods commonly used to process motion capture data in order to compute joint centers and angles.

Let us consider that the position of all the markers m_i is known (assuming that there is no missing information due to occlusions) as shown in Fig. 8.2. There are mostly three approaches to deduce the joint centers according to this marker placement. The first approach consists in applying regressions that express joint centers as a linear function of external markers' positions [12, 29]. For example, the right shoulder joint rShoJC could be expressed as:

$$\begin{cases} rShoJC_x = RSHO_x \\ rShoJC_y = RSHO_y + 0.43\cos(\frac{11\pi}{180})||CLAV - C7|| \\ rShoJC_z = RSHO_z - 0.43\cos(\frac{11\pi}{180})||CLAV - C7|| \end{cases}$$

where RSHO, CLAV and C7 are markers depicted in Fig. 8.2.

The main advantage of this type of method is simplicity. However this approach is not very accurate as it is based on average values while anthropometric data in humans can vary very significantly from one user to another.

The other approach, named functional approach, consists in searching for the joint centers that would generate the observable displacements [5, 7, 9]. For example, the right hip joint (rHip) is assumed to be a ball and socket. Its joint center should be the center of a sphere that covers the various positions of a point of the femur expressed in the pelvis reference frame. For example, any position of RKNE should satisfy the following constraint in the pelvis reference frame:

$$\left(RKNE_x - rHip_x\right)^2 + \left(RKNE_y - rHip_y\right)^2 + \left(RKNE_z - rHip_z\right)^2 - l^2 = 0$$

where l is the distance between the hip joint center and RKNE (length of the femur). Thus recovering the joint center consists in solving an optimization problem:

$$argmin_{rHip,l} \left[\left(RKNE_x(i) - rHip_x\right)^2 + \left(RKNE_y(i) - rHip_y\right)^2 + \left(RKNE_z(i) - rHip_z\right)^2 - l^2\right]^2$$

Of course, to find a good solution, RKNE should have large displacements in all the possible directions, in order to cover most of the sphere's surface. As a consequence, this method is generally applied to "range of motion" protocols where the user is moving each joint in all directions and with large displacements. However if the hip joint is actually not a ball and socket joint, the result could be inaccurate. Because of

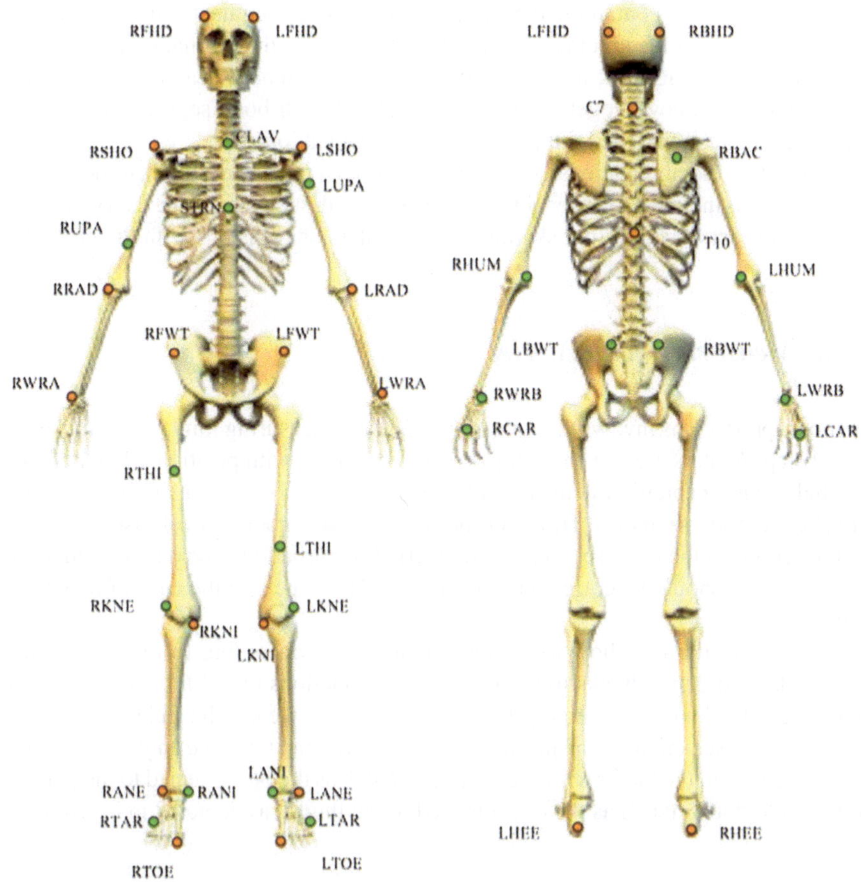

Fig. 8.2 Common marker placement used in Biomechanics [39, 40] to retrieve joint centers and to use anthropometric tables

skin displacements artefacts, this method will also provide the system with virtual bones that could change in length at each frame.

Another approach consists in generalizing the above idea to the whole skeleton. Let us consider that there exists a model of the skeleton based on rigid bodies and perfect joints (such as ball and socket and pivot joints). Knowing the external markers m_i^* in each local reference frame, the distance between m_i^* and m_i depends on the size of the body segments and the angles $\theta = \{\theta_1 \ldots n\}$ (where n is the number of degrees of freedom of the model). If there are enough markers on the body compared to the number of unknowns (i.e. number of degrees of freedom and of body segments), the problem can again be rewritten as a global optimization problem [15]:

$$argmin_{\theta,l} \frac{1}{2} \left(m - m^*\right)^T \left(m - m^*\right) \ with\ m^* = f(\theta, l)$$

where l is the vector containing the lengths of all the body segments, m is the number of markers and f is the kinematic function that computes the estimation of a marker placement according to the angles and the lengths of each body segment. As a result, joint angles are known together with the length of each body segment. Again the hypothesis here is that the joint are perfect mechanical joints.

In most of the applications in VR high accuracy is not needed when measuring joint positions in time. However animating avatars with such inaccurate data generally leads to artifacts, such as foot skating, flying avatars or collisions with the ground.

8.3.2 Measuring Joint Angles

In most applications involving motion capture data, measuring joint position is only a first step. Avatars are driven with joint angles and not with positions. Magnetic or inertial motion capture systems provide the user with global orientation of sensors attached to body segments. However, because of inaccuracies, local displacements of the sensor on the body segments and external perturbations, the data provided by these systems should be corrected (see [3, 22, 33] for the specific case of magnetic sensors).

Let us consider now how to compute joint angles according to local reference frames defined either thanks to the ISB recommendations [39, 40] or the H-ANIM norm (see http://www.h-anim.org for a description). If we use the Euler-like angles for a ball-and-socket joint, the problem consists in finding the three angles θ_x, θ_y and θ_z that transform the father reference frame F_i (such as the one attached to the pelvis) to the child one F_j (such as the one attached to the thigh), as depicted in Fig. 8.3.

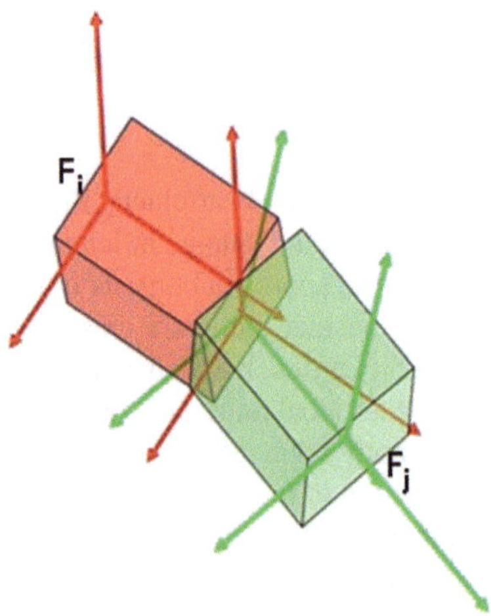

Fig. 8.3 Local reference frames associated with two adjacent body segments: F_i the parent segment (such as the pelvis) and F_j the child segment (such as the thigh)

Let Ω be the global reference frame. Let X_i, Y_i and Z_i (resp. X_j, Y_j and Z_j) be the three axis of reference F_i (resp. F_j). Let us define:

$$T_{i\to\Omega} = \begin{pmatrix} X_i & Y_i & Z_i \end{pmatrix}$$

$$T_{j\to\Omega} = \begin{pmatrix} X_j & Y_j & Z_j \end{pmatrix}$$

where $T_{i\to\Omega}$ (resp. $T_{j\to\Omega}$) stands for the transformation matrix from the local frame F_i (resp. F_j) to the world Ω. These matrixes can be computed according to the external markers and joint centers for each time:

$$T_{j\to\Omega} = T_{j\to i} \times T_{i\to\Omega}$$

where $T_{j\to i}$ stands for the transformation matrix from F_j to F_i, i.e. the transformation due to the action of the joint between F_i and F_j. $T_{j\to i}$ is thus given by:

$$T_{j\to i} = T_{j\to\Omega} \times {}^{T}T_{i\to\Omega} = \begin{pmatrix} t_{11} & t_{12} & t_{13} \\ t_{21} & t_{22} & t_{23} \\ t_{31} & t_{32} & t_{33} \end{pmatrix}$$

In the numerical model, this matrix is the product of three elementary symbolic matrixes associated with each degree of freedom and which depends on the chosen sequence of Euler angles. Let T_x, T_y and T_z be the elementary matrixes for a rotation along the X, Y and Z axes respectively. From the theoretical point of view, for a ZYX sequence (recommended by the International Society of Biomechanics), $T_{j\to i}$ could also be expressed as the product:

$$T^*_{j\to i} = T_x \times T_y \times T_z$$

The resulting matrix has the following shape:

$$T^*_{j\to i} = \begin{pmatrix} \cos(\theta_y)\cos(\theta_z) & -\cos(\theta_y) * \sin(\theta_z) & \sin(\theta_y) \\ \cdots & \cdots & -\cos(\theta_y)\sin(\theta_x) \\ \cdots & \cdots & \cos(\theta_x)\cos(\theta_y) \end{pmatrix}$$

As $T^*_{j\to i}$ should be equal to $T_{j\to i}$:

$$\theta_z = -\text{atan}\left(\frac{t_{12}}{t_{11}}\right)$$

$$\theta_x = -\text{atan}\left(\frac{t_{23}}{t_{33}}\right)$$

$$\theta_y = \text{atan}\left(\frac{t_{13} \times \cos(\theta_z)}{t_{11}}\right)$$

When using this type of approach, the results are very sensitive to the well-known Gimble-Lock problem. In the above equation, if $\theta_y = \pi/2$, then the denominator of the three equations is zero which leads to numerical problems. Other researchers use quaternions or exponential maps to avoid this problem. Whatever the selected formalism for the angles, noise or inaccuracies have a high impact on the result. In biomechanics more and more researchers propose to use a higher number of markers (named cloud of markers) and to apply the method described in the previous section: global optimization to fit the simulated markers m_i^* to the measured ones m_i [10].

Whatever the method, joint angles give a large quantity of information about the user's gait. These angles are directly used to animate avatars or to analyse if a user is avoiding a virtual obstacles, which is almost impossible if just considering parameters linked to the user's center of mass. Another application is to evaluate the quality of the user's gait when walking in virtual environments. For example, it has been used to demonstrate that treadmill walking in VR affects gait [31]. Moreover some authors have shown that joint angles were also adapted during nonlinear walking [25]. It could thus be used to detect turning in walk-in-place interfaces or to adapt the motion of an avatar in real-time when turning.

8.3.3 Estimating Joint Torques with Inverse Dynamics

Some authors have demonstrated that even if kinematic gait parameters were similar in ground and treadmill walking after some training, the joint torques and muscular activity remains different [14]. Taking this type of parameter into account could thus help to evaluate the performance of the user's gait in virtual environment, using or not a treadmill. Indeed, this type of parameter seems to enable us to highlight unperceivable modifications of gait even if joint angles look similar. However, there is no direct method for measuring joint forces and torques. Indirect methods named "inverse dynamics" have been introduced to compute these forces and torques according to kinematic data associated with a physical model of the skeleton. There are merely two main approaches to address this problem: isolated segments and use of controllers.

8.4 Isolated Segments

The first approach is based on the Newton formalism and consists in considering each body segment separately, assuming that the mechanical system is limited to body segment S_i [16, 38]. For each body segment S_i external forces could be separated into two families: external forces and torques (F_e^i, τ_e^i) exerted by the environment (such as gravity and contacts), and the internal ones due to muscles (F_m^i, τ_m^i). Thanks to kinematic data it is possible to compute the body segment center of mass acceleration. The Newton equation gives:

$$\begin{cases} m_i \gamma_i = F_e^i + F_m^i \\ \frac{dL_i}{dt} = \mathcal{M}_{F_e^i} + \mathcal{M}_{F_m^i} + \tau_e^i + \tau_m^i \end{cases}$$

where γ_i, m_i and L_i stand for the acceleration of segment S_i center of mass, its mass and its angular momentum respectively. Hence knowing the external forces, torques, the mass and the acceleration of the body segment it is possible to deduce (F_m^i, τ_m^i) if there is only one unknown for each equation. It means that there is only one force F_m^i and only one muscle torque τ_m^i. However for a given segment S_i with two adjacent segments S_{i-1} and S_{i+1}, F_m^i is the result of two forces: the forces exerted respectively by each neighboured segment at contact point.

To solve this problem, there are two main methods. The first one consists in solving the system from the extremities (without contact, such as the hands and the head) to the ground. For body segments placed at the extremities of the skeleton, there is only one unknown for F_m^i and τ_m^i (associated with the proximal joint attached to the body segment). F_m^i and τ_m^i could thus be rewritten $F_m^{j\to j}$ and $\tau_m^{j\to i}$ to express their relation to the joint attached to segments i and j. The problem can then easily be solved by inverting the above equation:

$$\begin{cases} F_m^{j\to i} = F_e^i - m_i \gamma_i \\ \tau_m^{j\to i} = \mathcal{M}_{F_e^i} + \mathcal{M}_{F_m^i} + \tau_e^i - \frac{dL_i}{dt} \end{cases}$$

When dealing with segment S_j, we can then reuse these results as known forces $F_m^{j\to i} = -F_m^{i\to j}$ and torques $\tau_m^{j\to i} = -\tau_m^{i\to j}$. The method is applied until the feet so that the GRF could be deduced at contact point with the ground. Comparison with measured GRF is a common method to estimate the errors due to this process. The second method, named bottom-up method, consists in starting with the segments which are in contact with the environment (and which external forces are known) and finishing with the extremities free of any contact. A mixed method can also be used.

8.5 Global System and Controllers

A second approach to solve inverse dynamics problems is to model the global system with the Lagrangian formalism (based on the principle of virtual works) in order to obtain the motion equations, including the torques at any joints τ_i:

$$\frac{d}{dt}\frac{\partial C}{\partial \dot{q}_i} - \frac{\partial C}{\partial q_i} - Q_i' = 0 \; with \; i = 1 \ldots n$$

where C is the difference between the potential and kinetic energies, and q_i is the ith state of the system. The applied forces and torques Q_i are expressed in the generalized coordinates as generalized forces,

Lagrangian multipliers are added to this equation in order to express that body segment displacement should correspond to motion capture data (see [8] for details). The model must be associated with a robust representation of contact forces with the ground to deliver realistic results. When simulating the system with a dynamic solver, Lagrangian multipliers will naturally compute forces and torques that are necessary to ensure that the resulting simulation is compatible with the imposed motion (generally motion capture data).

A famous solution consists in modelling the joint torques as proportional-derivative (PD) controllers which merely consists in associating damped springs to each joint:

$$\tau_i = k_p \left(\theta_i^d - \theta_i\right) - k_d \dot{\theta}_i$$

where θ_i^d stands for the desired joint angle for joint i, k_p and k_d are the proportionnal and derivative gains of the controller. If we consider that the desired joint angles correspond to those measured by the motion capture system, and if the gains are correctly tuned, it is thus possible to compute the joint torques that are required to perform the measured motion [42]. Torques obtained with the PD controller is applied to a physical model of the human body. This physical model can be obtained thanks to commercial software or opensource packages, such as OpenDynamicEngine (http://www.ode.org). This type of software provides us with a simulator of a physical model which inputs are internal and external forces applied to the system. In our case, external forces are obtained either by direct measurements with gauges or by using the above inverse dynamic method applied to the global whole-body system (human body is modelled by its center of mass). Internal forces and torques are computed using the PD controllers.

This approach is very difficult to tune, especially the values of the PD gains. However for well-known motions such as walking, many researchers have proposed semi-automatic methods to estimate these gains.

8.6 Conclusion About Inverse Dynamic Approaches

Joint torques and forces are mainly used in biomechanics and computer animation. In biomechanics it enables to distinguish different motor strategies that kinematic data fail to differentiate. In computer animation, it is mainly used to check if the square of joint torques of a given simulated motion is minimized, assuming that it corresponds to natural motions. When walking in VR, it could also help to evaluate gait efficiency as human walking is supposed to be associated with low energy consumption (see Chap. 3). For example, it has been used as a relevant criterion to distinguish overground and treadmill walking, as explained above [31]. Moreover animating the user's avatar walking on uneven terrain implies to adapt the joint trajectories performed by the user who is walking on a treadmill or a flat ground. Hence this motion adaptation is necessary to compensate differences between the constraints imposed in the virtual environment and those actually linked to the real movement of the user. In that case methods based on physical models enable to automatically adapt the motion to different kinematic and physical constraints [41].

However inverse dynamics is sensitive to noise or inaccuracies especially when computing joint accelerations. Another limitation is that it remains difficult to deal with closed-loop systems, such as dealing with the double support phase in walking where the two feet are in contact with the ground. In that case, it is very difficult to strictly separate the forces exerted below each foot.

8.6.1 Measuring or Estimating Muscle Activities

In some very specific applications, joint torques is not accurate enough to understand motion strategies. Indeed, joint torques provides us with the resulting action of a group of muscles whereas control strategies could have a direct link with the action of one isolated muscle. Slightly changing the axis of rotation of a motion may recruit different muscle groups even if the resulting joint torque looks the same.

The direct approach to measure muscle activity consists in sensing the electromyograms (EMG) of targeted muscles. EMG is a measurement of the electrical activity of skeletal muscles recorded with the placement of small electrodes over the skin (there exist more invasive electrodes but they are unusable for large movements). Thus EMG is limited to surface muscles. The signal returned by the EMG system is noisy and required heavy signal processing to estimate muscle tensions in Newtons. However, it gives an interesting point of view about muscle coordination if several muscles are measured concurrently. See [23] (among many others) for more details on EMG.

Some researchers have proposed to use indirect methods to retrieve the tension of all the muscles involved in the studied motion (including deeper muscles). Musculoskeletal models have been introduced in the early nineties [6] thanks to the increase of computation power of computers. The key idea is to model muscles thanks to action lines acting along an axis determined by to muscle insertions. Knowing the accurate location of each of these muscles and tendons insertion on bones, it is possible to retrieve these action lines, as shown in Fig. 8.4.

A muscle is supposed to work only by applying a positive tension (leading to contractions) and cannot push the bone. Hence for each muscle i, its tension T_i is positive. Each muscle is also limited to a maximum voluntary contraction (MVC) which is generally evaluated in isometric condition (i.e. no displacement of the bones but exertion of a force against an external load). If we consider the surface of the cross section area (perpendicular to the muscle fibers), this MVC is given by:

$$F_i^{max} = K(l) \times K_0 \times PCSA$$

where K_0 is a constant ranging from 15 to 33 N/cm^2 according to the authors, and K(l) is a value that depends on the muscle length l. Indeed there exists a relation

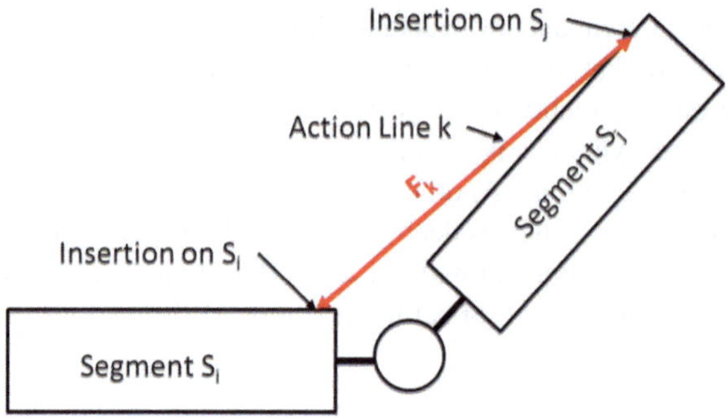

Fig. 8.4 Action line joining two adjacent segments S_i and S_j. The insertions of this action line in both segments are given by anthropometric tables in each local reference frame

between the maximum delivered force and the length of a muscle (named Force–Length relation). This relation is given in the biomechanical literature. Hence a muscle tension F_i for muscle i is constrained by:

$$0 \leq F_i \leq K(l) \times K_0 \times PCSA$$

l is then the distance between the two insertion points of the action line. Motion capture data enable to associate a local reference frame to each body segment, as shown previously for the computation of joint angles. It is thus possible to apply anthropometric tables that give the local coordinates of each line action insertion in each bone. These tables are generally embedded in dedicated software, such as OpenSim (see http://simtk.org/home/opensim for an example).

Knowing these insertion points for each action line, we can deduce the length of each action line and consequently its MVC in the current pose. If external torques τ_{ext} are also known, such as gravity and the momentum of all the contact forces, the torques applied by the muscles τ_{musc} in static condition are given by:

$$\tau_{ext} = -\tau_{musc}$$

In the remaining of this section we assume that the system is static for simplification but extension to dynamic situation simply involves adding to this formulae the terms linked to the derivative of the angular momentum of each body segment.

τ_{ext} could be expressed for each body segment i as:

$$\forall i \in [1, n], \sum_{j=1}^{m} M_i^j = -\tau_{ext}^i$$

where n is the number of body segments, m is the number of independent action lines, and M_i^j stands for the moment of action line j on the ith body segment. If the local reference frame O_i of segment i is placed on the proximal joint:

$$M_i^j = O_i P_i^j \times F_j$$

where P_i^j is the insertion point of muscle j on body segment i and F_j stands for the force applied by this muscle. Let us consider that the norm of F_j is denoted T_j. Using the superposition theorem we can express M_i^j as the product of Tj and I_i^j where I_i^j is the theoretical Momentum associated with a unit force vector Fj ($\| F_j \| = 1N$):

$$M_i^j = I_i^j * T_j$$

and

$$\forall i \in [1, n] \quad \sum_{j=1}^{m} I_i^j * T_j = -\tau_{ext}^i$$

which can be rewritten in a matrix form:

$$\beth . T_{musc} = -\tau_{ext}$$

where $\beth$ is the n × m matrix containing I_i^j for all i (body segment) and j (action line), T_{musc} is the m-vector containing the forces of all the action lines and τ_{ext} is the n-vector of the external torques and moment applied to all the body segments.

This linear system is not invertible because n is not equal to m. In general m is greater than n leading to redundancy of the actuators (the action lines). It consequently leads to an optimization problem with a space of solutions. In addition to this equality constraint, it is possible to add inequality constraints such as:

- A muscle force is positive and its maximum value is equal to K(l)*K_0*PCSA as shown previously,
- People naturally tend to minimize energy and torques ($\sum_{j=1}^{m} F_i^2$ in many motions [32] (or the normalized effort ($\sum_{j=1}^{m} \left(\frac{F_i}{F_i^{max}}\right)^2$) [18]),
- Taking some EMG signals into account in the solving process [1]

This domain is very active in biomechanics and many researchers try to improve the quality and to validate the results. One of the most important problems is to be able to take contraction of antagonist into account. Using EMG signals to check if the antagonist muscles are active is a very promising approach.

The approach presented in this section is named inverse dynamics but there exist two other approaches to solve this type of problem using direct dynamics within an optimization loop or directly exploiting EMG signals. Whatever the method, it is still difficult to validate the results. However, it is a promising contribution to analyse more accurately the actions performed by the user in a virtual environment, as it has been shown in ergonomics [27]. VR is more and more used to evaluate and train disable people as it provides a safe and reproducible environment. In this type of application EMG and muscle contraction feedbacks are very important.

Table 8.1 Summary of gait parameters and their potential use in walking in VR

Parameters	Methods	Examples of applications in VR		
		Sensing in VR	Multisensory feedbacks computation	Naturalness of the interaction
Global gait parameters	$V = S_L * S_F$ + direct measurements	Accelerometers on pelvis	Control of the virtual camera [35]	Treadmill versus overground walking in VR [14]
		Walk-in-place [34]	Synchronization with sound [13]	
		"Shake-your-head" [36]		
Curvature, trajectory	Head, center of mass, feet tracking		Could be used to improve camera trajectory and galvanic stimulation in nonlinear walking	Comparison between walking paths in real and in VR [35]
Balance status	Center of pressure or zero moment point (model)	Use of balance boards or pressure plates		
		Joyman [19, 26] interface based on balance status		
Asymmetry and regularity	Autocorrelation or $\left[\frac{(R-L)}{0.5} \times (R+L)\right] \times 100$ R (resp. L): right (resp. left) parameters	Asymmetry could be related to turning and could be used in walk-in-place interfaces	Could be used to adjust camera motion to the user's real-time performance	Criteria used to evaluate the cognitive load when walking [17]
Joint angles	Direct evaluation or kinematic model fitting	To improve walk-in-place interfaces when avoiding/stepping up obstacles	Real-time animation of avatars	Comparison to reference data & Individual trajectories
Joint torques and forces	Inverse dynamics		Vibration feedbacks according to the shape of the ground reaction force [37]	Comparison to reference values; used to evaluate treadmill walking in VR [14]
Muscle activation pattern	EMG + signal processing or musculoskeletal models	EMG to monitor the user's actions	Real-time feedback of the user's muscle coordination for rehabilitation in VR	Comparison to reference values; used to evaluate treadmill walking in VR [14]

As for the joint torques and forces, muscle tensions are good indicators of gait efficiency and can thus be used to determine subtle changes in gait patterns that kinematic data fail to identify. It could thus be used to evaluate if the user's gait is as efficient in VR compared to natural walking.

8.7 Conclusion

J. E. Marey and E. Muybridge were the first researchers who proposed objective measurements of animal and human locomotion [20]. Nowadays, numerous systems exist to analyse and measure human gait. The biomechanics community is still very active in this domain and collaboration with other domains will certainly lead to new systems, such as using depth-cameras (Kinect of Microsoft) or inertial sensors. It is difficult to predict what will be the future in this domain, but many researchers tend to propose non-invasive and light systems associated with more and more sophisticated numerical models in order to access to new parameters. Musculoskeletal models are clearly a step forward in this domain but many researches have to be carried-out in this domain for calibration and validation. The Table below summarizes the parameters introduced in this chapter and their potential use in walking in VR (Table 8.1):

- sensing the user's motion,
- delivering the most appropriate and accurate multisensory feedbacks,
- and evaluating the naturalness of the interaction for the user compared to reference values (as those reported in Chapter "Biomechanics of walking in real world").

Systems deliver more and more data (global parameters, joint positions, angles, torques and muscle forces) and a new problem occur: how to deal with all these parameters? In virtual reality researchers try to provide the user with more realistic feedbacks which could rely on using such big amount of knowledge. Getting more accurate data on the user's gait could help to select and compute the corresponding feedbacks. Let us consider the walk-in-place approach currently used in VR [36]. In this approach, signal processing is applied to the orientation of the head depending time. Extending this approach to more complex parameters would enable to deal with more complex situations and behaviours, such as getting up and down stairs, avoiding obstacles or taking specific gait style into account. As for future steps, one main challenge therefore remains to be able to compile all the available data and to compute the most appropriate feedbacks in real-time.

References

1. Amarantini D, Rao G, Berton E (2010) A two-step EMG-and-optimization process to estimate muscle force during dynamic movement. J Biomech 43(9):1827–1830
2. Auvinet B, Berrut G, Touzard C, Moutel L, Collet N, Chaleil D, Barrey E (2002) Reference data for normal subjects obtained with an accelerometric device. Gait Posture 16:124–134

3. Bodenheimer B, Rose C, Rosenthal S, Pella J (1997) The process of motion capture: dealing with the data. In: Thalmann D, van de Panne M (eds) Eurographics workshop on computer animation and, simulation. Springer-Verlag, Wien, pp 3–18
4. Brault S, Bideau B, Craig C, Kulpa R (2010) Balancing deceit and disguise: how to successfully fool the defender in a 1 versus 1 situation in rugby. Hum Mov Sci 29(3):412–425
5. Chang LY, Pollard NS (2007) Constrained least-squares optimization for robust estimation of center of rotation. J Biomech 40:1392–1400
6. Delp SL (1990) Surgery simulation: a computer graphics system to analyze and design musculoskeletal reconstructions of the lower limb. PhD thesis, Stanford University, Stanford, California
7. Ehrig RM, Taylor GN, Duda WR, Heller MO (2006) A survey of formal methods for determining the centre of rotation of ball joints. J Biomech 39:2798–2809
8. Faure F, Debunne G, Cani-Gascuel MP, Multon F (1997) Dynamic analysis of human walking (1997). In: Proceedings of eurographics workshop on computer animation and, simulation, pp 95–107
9. Gamage SSHU, Lasenby J (2002) New least squares solutions for estimating the average centre of rotation and the axis of rotation. J Biomech 35:87–93
10. Herda L, Fua P, Plaenkers R, Boulic R, Thalmann D (2001) Using skeleton-based tracking to increase the reliability of optical motion capture. Hum Mov Sci J 20(3):313–341
11. Kavanagh J, Barett R, Morrison R (2005) Age-related differences in head and trunk coordination during walking. Hum Mov Sci 24:574–587
12. Leardini A, Cappozzo A, Catani F, Toksvig-Larsen S, Petitto A, Sforza V, Cassanelli G, Giannini S (1999) Validation of a functional method for the estimation of hip joint centre location. J Biomech 32(1):99–103
13. Lécuyer A, Marchal M, Hamelin A, Wolinski D, Fontana F, Civolani M, Papetti S, Serafin S (2011) Shoes-your-style: changing sound of footsteps to create new walking experiences. In: Proceedings of workshop on sound and music computing for human-computer interaction, CHItaly, Alghero, 13–16 Sept 2011
14. Lee SJ, Hidler J (2007) Biomechanics of overground versus treadmill walking in healthy individuals. J Appl Physiol 104(3):747–755
15. Lu TW, O'connor JJ (1999) Bone position estimation from skin marker coordinates using global optimisation with joint constraints. J Biomech 32(2):129–134
16. Luh J, Walker M, Paul R (1980) On-line computational scheme for mechanical manipulators. Trans ASME J Dyn Syst Meas Contr 102(2):69–76
17. Maki BE (1997) Gait changes in older adults: predictors of falls or indicators of fear. J Am Geriatr Soc 45:313–320
18. Makshous M (1999) Improvements, validation and adaptation of a shoulder model. PhD thesis, Chalmers University of Technology, Göteborg
19. Marchal M, Pettré J, Lécuyer A (2011) Joyman: a human-scale joystick for navigating in virtual worlds. In: Proceedings of IEEE symposium on 3D user, interface 3DUI'11, pp 19–26
20. Marey EJ (1873) La machine animale, locomotion terrestre et aérienne. Germer Baillère, Paris
21. Menz H, Lord S, Fitzpatrick R (2003) Acceleration patterns of the head and pelvis when walking on level and irregular surfaces. Gait Posture 18(1):35–46
22. Molet T, Boulic R, Thalmann D (1999) Human motion capture driven by orientation measurements. Presence 8:187–203
23. Nigg BM, Herzog W (1999) Biomechanics of the musculo-skeletal system. Wiley, New York, p 349
24. Olivier AH, Crétual A (2007) Velocity/curvature relations along a single turn in human locomotion. Neurosci Lett 412(2):148–153
25. Orendurff M, Segal A, Berge J, Flick K, Spanier D, Klute G (2006) The kinematics and kinetics of turning: limb asymmetries associated with walking a circular path. Gait Posture 23(1):106–111
26. Pettré J, Marchal M, Siret O, de la Rivière J-B, Lécuyer A (2011) Joyman: an immersive and entertaining interface for virtual locomotion. ACM SIGGRAPH Asia emerging technologies. Hong-Kong, 12–15 Dec 2011

27. Pontonnier C, Dumont G (2010) From motion capture to muscle forces aiming at improving ergonomics of working stations. Virtual Phys Prototyp 5:113–122
28. Pozzo T, Berthoz A, Lefort L, Vitte E (1991) Head stabilization during various tasks in humans. II.Patients with bilateral vestibular deficits. Exp Brain Res 85(1):208–217
29. Reed M, Manary M, Schneider L (1999) Methods for measuring and representing automobile occupant posture. In: Proceedings of international congress and exposition, Detroit, Michigan, March 1–4
30. Sadeghi H (2003) Local or global asymmetry in gait of people without im-pairments. Gait Posture 17(3):197–204
31. Schellenbach M, Lövdén M, Verrel J, Krüger A, Lindenberger U (2010) Adult age differences in familiarization to treadmill walking within virtual environments. Gait Posture 31:295–299
32. Sereig A, Arvikar R (1989) Biomechanical analysis of the musculoskeletal structure for medicine and sports. J Anat 167:243–244
33. Silaghi MC, Plaenkers R, Boulic R, Fua P, Thalmann D (1998) Local and global skeleton fitting techniques for optical motion capture. In: Magnenat Thalmann N, Thalmann D (eds) Modeling and motion capture techniques for virtual environments, lecture notes in artificial intelligence, No1537, CAPTECH 1998. Springer, Geneva, pp 26–40
34. Taylor M, Dabnichki P, Strike S (2005) A three-dimensional biomechanical comparison between turning strategies during the stance phase of walking. Hum Mov Sci 24(4):558–573
35. Terziman L, Lécuyer A, Hillaire S, Wiener J (2009) Can camera motions improve the perception of traveled distance in virtual environments? In: IEEE international conference on virtual reality (IEEE VR), Lafayette, US
36. Terziman L, Marchal M, Emily M, Multon F, Arnaldi A, Lécuyer A (2010) Shake-your-head: revisiting walking-in-place for desktop virtual reality. In: Proceedings of the ACM symposium on virtual reality software and technology, Hong Kong, pp 27–34
37. Terziman L, Marchal M, Multon F, Arnaldi B, Lécuyer A (2012) The king-kong effects: improving sensation of walking in VR with visual and tactile vibrations at each step. In: Proceedings of the IEEE symposium on 3D user, interfaces 2012, Orange County, USA, pp19–26
38. Vaughan CL, Davis BL, O'Connor JC (1992) Dynamics of human gait. Human kinetics. Champaign, Illinois
39. Wu G, Siegler S, Allard P, Kirtley C, Leardini A, Rosenbaum D, Whittle M, D'Lima D, Cristofolini L, Witte H, Schmid O, Stokes I (2002) ISB recommendation on definitions of joint coordinate system of various joints for the reporting of human joint motion—part I: ankle, hip, and spine. J Biomech 35(4):543–548
40. Wu G, van der Helm FCT, Veeger HEJ, Makhsous M, Van Roy P, Anglin C, Nagels J, Karduna AR, McQuade K, Wang X, Werner FW, Buchholz B (2005) ISB recommendation on definitions of joint coordinate systems of various joints for the reporting of human joint motion—Part II: shoulder, elbow, wrist and hand. J Biomech 38:981–992
41. Yin k, Loken k, van de Panne M (2007) SIMBICON: simple biped locomotion control. ACM Trans Graph 26(3):Article 105
42. Zordan VB, Hodgins JK (1999) Tracking and modifying upper-body human motion data with dynamic simulation. In: Computer animation and simulation '99, eurographics, pp 13–22

Chapter 9
Locomotion Interfaces

Hiroo Iwata

Abstract A locomotion interface is a device that creates an artificial sensation of physical walking. It should ideally be equipped with three functions: (1) The creation of a sense of walking while the true position of its user is preserved, (2) Allowing the walker to change bearing direction, (3) The simulation of uneven walking surfaces. This chapter categorizes and describes four different methods for the design and implementation of such interfaces: Sliding shoes, Treadmills, Foot-pads, and Robotic tiles. It discusses related technical issues and potential applications.

9.1 Introduction

A locomotion interface is a device that creates a sense of walking in a virtual environment (VE). It provides for the experience of physical walking while a walker's body is maintained localized in the real world.

In many applications of VEs, such as immersive training or visual simulations, users can benefit from a good sensation of locomotion. It has often been suggested that the best locomotion mechanism for VEs would be walking [5]. It is well known that sense of distance or orientation while walking is much better than while riding on a vehicle. Proprioceptive and vestibular feedback during walking is particularly important for navigation [29–31].

Effects of proprioceptive and vestibular feedback have been tested in settings involving walking on real ground while immersed in VE. Loomis et al. [17] used a HMD in triangle completion walking tasks. Five conditions were employed for evaluating optic flow, vestibular, and proprioceptive stimulation as inputs to the path integration process. Two conditions involved walking (with and without vision), two

H. Iwata (✉)
Virtual Reality Laboratory, Department of Intelligent Interaction Technologies,
1-1-1 Tennodai, Tsukuba 305-8573, Japan
e-mail: iwata@kz.tsukuba.ac.jp

F. Steinicke et al. (eds.), *Human Walking in Virtual Environments*,
DOI: 10.1007/978-1-4419-8432-6_9, © Springer Science+Business Media New York 2013

involved wheelchair transport (with and without vision), and the fifth was a stationary (non-moving) condition with vision. The results indicated that the directional return toward the origin was much poorer when optic flow alone specified the outbound path. Chance et al. [3] set up a virtual maze, in which subjects encountered target objects along the way. Their task was to indicate the direction to these target objects from a terminal location in the maze. The scene of the virtual maze was provided by a HMD. Subjects controlled their motion through the mazes using one of three locomotion modes: Walk mode, Real Turn mode, and Visual Turn mode. The results showed that performance in the Walk mode was significantly better than that of Visual Turn mode. In another experiment, Bakker et al. [1] studied orientation performance in VE. They tested five stimulus conditions for turning: three with and two without visual stimuli, using one three different navigation metaphors to steer rotation. Their results showed that most accurate rotation was found when subjects turned using their legs.

All of these studies, in addition to more recent works such as from Ruddle and Lessels [23] and Suma et al. [25], have shown the positive effects of physical walking on navigation tasks.

To recreate the sensations related to physical walking in virtual environments there are theoretically three major technical issues to overcome:

- Creating a sense of walking while the true position is maintained: The device thus requires a mechanism that cancels the displacement of the walker in the real world.
- Allowing the walker to change direction: The device therefore requires a mechanism that cancels omni-directional displacement.
- Simulation of uneven surfaces: This is required since the terrain of the real world includes uneven surface, such as staircases. In general, the displacement of a walker is three-dimensional.

Pioneering work toward the development of locomotion interface devices began in 1988 [7] and many different prototypes have been fabricated until the present [13]. The four main existing approaches to the design of locomotion interfaces can be categorized as follows:

- Sliding shoes: The walker wears specialized shoes that generate relative motion between the foot and the floor.
- Treadmill: The walker stands on a belt conveyer that moves opposite to the direction of walking.
- Foot-pad: Two platforms are applied to the feet and move in accordance with the motion of the feet.
- Robotic tiles: Movable tiles provides a dynamic platform for walking. The tiles move opposite to the direction of walking.

The remainder of this chapter describes successively the hardware configurations of these four methods, and discusses related technical issues and potential usages.

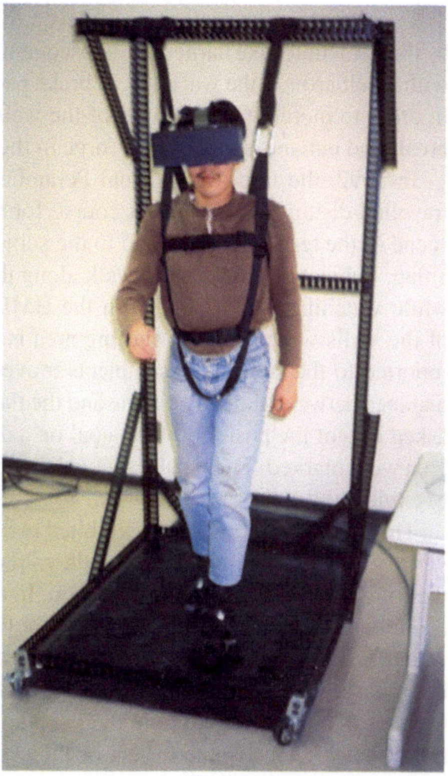

Fig. 9.1 The first prototype of virtual perambulator

9.2 Sliding Shoes

9.2.1 Virtual Perambulator

The first project aimed at developing a locomotion interface started in 1988 [7]. It was named "Virtual Perambulator", and had the primary object of allowing for changes in direction of the walker's feet. Figure 9.1 shows the overall view of the prototype. A user of the system wore a parachute like harness and omni-directional roller skates. The trunk of the walker is fixed to the framework of the system by the harness. An omni-directional sliding device is used for changing direction of the feet. To this end, a specialized roller skate was equipped with four casters, which enables two-dimensional motion. The walker could freely move his or her feet in any direction. Motions of the feet were measured via ultrasonic range detectors. From the results of measurements, images of the virtual space were displayed in the head-mounted display corresponding with the motion of the walker. The direction of locomotion in the virtual space was determined according to the direction of the walker's steps.

Users of this first prototype, however, felt uncomfortable due to pressure exerted by the parachute-like harness. In subsequent work, the harness was therefore replaced with a belt around the waist [8]. A brake pad was placed at the toe of the roller skate in order to increase the stability of the walker. While the walker steps forward, the break-pad enhances the friction force in the rear foot.

In 1992, the improved Virtual Perambulator was utilized to study the sense of travelled distance [8]. The test course for the experiment was a straight path. The scene of the test space provided to the subjects was a plain wall and floor as well as a flag. Subjects were asked to walk along the path from the starting area to the goal while watching the CG image in the HMD. The width of the path and the height of the walls were 3 m. The starting area is 5 m deep and the depth of this area was reported to the subjects. As subjects moved along the course, they memorized the distance between the starting line and the flag. After they finished walking, they were asked to plot the position of the goal on a data sheet in which the walls and starting area were marked. The results showed that Stevens' power law [26] could be applied to explain the estimated distances.

This second prototype still exhibited two major flaws: (1) The waist belt restricted vertical and turning motions of walkers' bodies, and (2) The weight and height of the roller skate affected natural motion. In order to overcome these problems, a new frame and sliding device was developed [9]. A hoop was set around the walker's waist in which he/she could physically walk and turn about. The diameter of the hoop was 60 cm. The walker could freely change the direction of walking. Novice users of the system could hold the hoop so that they can easily keep the balance of their bodies, while trained users of the system could push their waists against the hoop and learn to walk fast or can even run. Since no harness is used, walker's body had no restriction. A new sliding device based on rubber sandals instead of steel roller skates was also developed. A layer of low friction film was placed at the middle of the sole. Rubber in the toe region played the roll of a brake pad. Material of the floor surface was selected for compatibility with the low friction film and brake pad. The device was demonstrated at the Interactive Communities venue at SIGGRAPH'95. During the five-day conference, 235 people experienced the device. The behavior of the walkers was observed, and 94 % of them were found to succeed in rhythmical walking [9]. The main limitation of the Virtual Perambulator was found to be that the device itself is passive. Walkers had to slide their feet by themselves and they had to get accustomed to the sliding action.

9.2.2 Powered Shoes

Powered Shoes is a locomotion interface using motorized roller skates [15]. A compact and light-weighted drive mechanism is put underneath the sole. A large force is required to move the walker so that the motor for the locomotion interface could not, itself, be placed underneath the sole. A flexible shaft was instead used in order to separate the motor from the roller. Motors and batteries are the heaviest parts of the

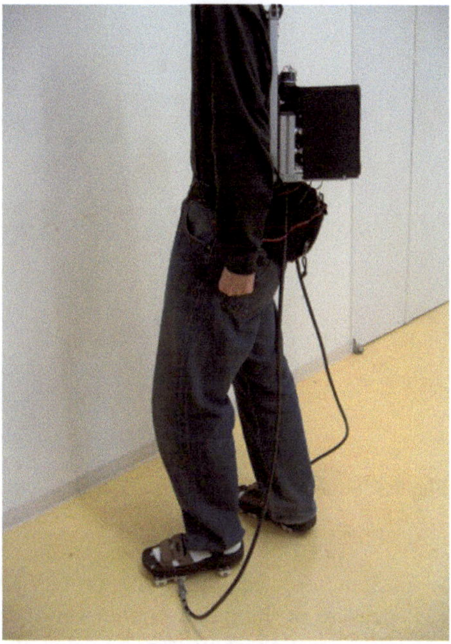

Fig. 9.2 Powered shoes

system, and were put in a backpack. The active roller skate was designed to fit in the sole. It has three rollers, two of which are connected by a timing belt and driven by a motor. The diameter of each roller is 16 mm and overall height of the mechanism is 20 mm. The overall weight of the mechanism is 700 g, roughly the same as a hiking shoe. The mechanism allows the user to walk at a speed of 600 mm/s. Figure 9.2 shows overall view of the Powered Shoes.

The major limitation of this system was that the direction of the traction force generated by the rollers is identical to the shoe, which only allows the walker in a straight line.

9.2.3 String Walker

The String Walker is a locomotion interface based on eight strings actuated by motor-pulley mechanisms mounted on a turntable [16]; see Fig. 9.3. The mechanism enables omni-directional walking. The device also allows the walker to perform various gait patterns, such as side-walking.

Four strings are connected to each shoe and are actuated by motor-pulley mechanisms. The maximum tension of each string is 25 Kgf. A touch sensor in the shoe detects stance phase and swing phase of walking. The signal is wirelessly transmitted to a host computer. No force is applied to the shoe while it is in the swing phase.

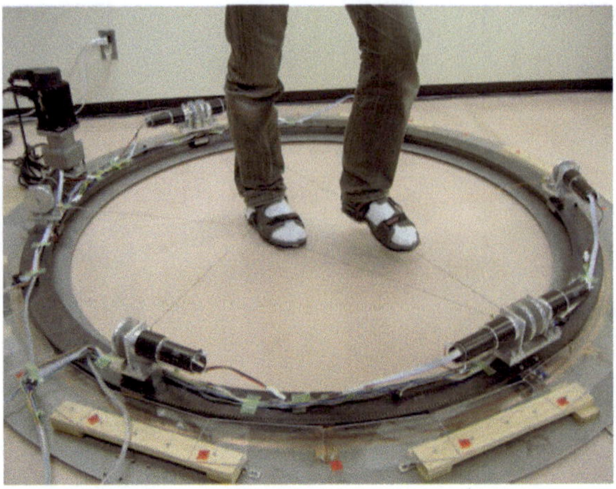

Fig. 9.3 String walker

Each motor is equipped with a rotary encoder and the motor-pulley mechanisms can measure position and orientation of the shoe. The strings pull the shoe in the opposite direction of walking, so that the displacement of the step is cancelled. The position of the walker is fixed in the real world by this computer-controlled mechanism. The motor-pulley mechanisms are themselves mounted on a motor-driven turntable that rotates according with the direction of the walker, thus enabling omni-directional walking.

A major limitation of the String Walker concerns the difficulty in controlling the tension in the strings. Durability of the strings is also a serious problem.

9.2.4 Evacuation Simulator Using the Virtual Perambulator

An "evacuation simulator" [32] based on the virtual perambulator was designed in a collaboration between the author's lab and the Ship Research Laboratory of the Ministry of transportation of Japan. The analysis of the evacuation of passengers during maritime accidents is very important for ship safety. However, it is impossible to carry out experiments with human subjects during an actual disaster. Consequently, these researchers introduced virtual reality tools for simulation of disaster in order to analyze evacuation of passengers. They built a virtual ship that models the generation of smoke and the inclination of the vessel. Experiments of evacuation were carried out for construction of mathematical model of passenger's behavior in disaster. Figure 9.4 shows the Virtual Perambulator integrated with the evacuation simulator.

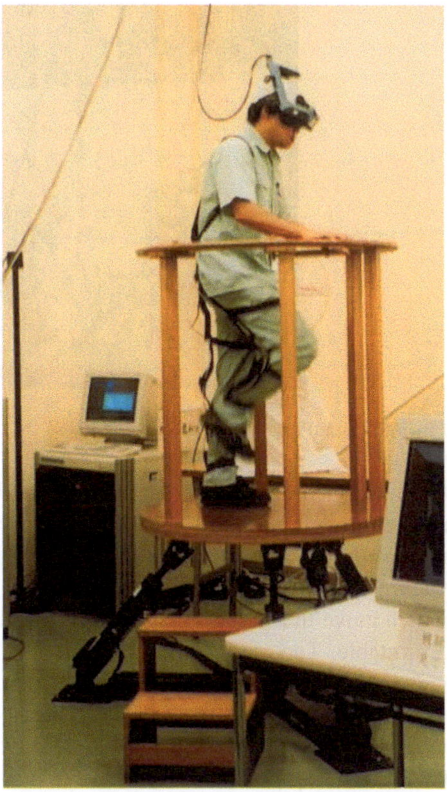

Fig. 9.4 Evacuation Simulator

9.3 Treadmills

9.3.1 Related Works in Treadmill-Based Locomotion Interface

A simple device for virtual walking is a treadmill, which is ordinarily used for physical fitness. An application of such a device to a virtual building simulator was developed at the University of North Carolina [2]. This treadmill had a steering bar similar to that of a bicycle. At ATR in Japan, Noma et al. developed a treadmill-based locomotion interface named ATLAS. Later, it was equipped with a series of linear actuators underneath the belt [19]. The device, named GSS, was able to simulate slopes of virtual terrain. The Treadport device (Fig. 9.5) developed at the University of Utah is a treadmill that is combined with a large manipulator connected to a walker [4]. The manipulator provides gravitational forces while the walker is traversing a slope.

The major problem of a treadmill-based locomotion interface is to allow the walker to change direction. One of the possible solutions for realizing omni-directional walking on a treadmill is to use small rollers to move the walker in perpendicular directions. The ODT, or Omni-directional Treadmill, employs two perpendicular

Fig. 9.5 TreadPort

treadmills, one inside of the other. Each belt is made from approximately 3400 separate rollers, woven together into a mechanical fabric. Motion of the lower belt is transmitted by the rollers to a walker. This mechanism enables omni-directional walking [6]. In another instance, the "Ball Array Treadmill", employs small balls, which move the walker in any direction. The balls are driven by a belt mounted on a turntable. The combination of balls, belt and turntable enables omni-directional walking [20].

An important limitation of these methods is that rollers or balls are hazardous when the walker falls down. The durability of the structure supporting the rollers or balls is also a problem.

Another approach for realizing omni-directional walking is to use a large sphere in which a walker stand. One of the examples is a product developed by Virtusphere Inc. The sphere rotates freely in any direction according to the walker's steps. The major problem of this method is that the inertia of the sphere is so large that the walker cannot stop while he/she is walking fast. Also, a curved walking area is not natural.

9.3.2 Torus Treadmill

An ideal solution for a treadmill-based locomotion interface is to create an omni-directional infinite, moving floor. In order to realize an infinite walking area, the geometric configuration of an active floor must be chosen. A closed surface driven by actuators has an ability to simulate an unlimited floor. The following requirements for implementation of the closed surface must be considered:

(1) The walker and actuators must be put outside the surface.
(2) The walking area must be a plane surface.
(3) The surface must be made of a material that stretches very little.

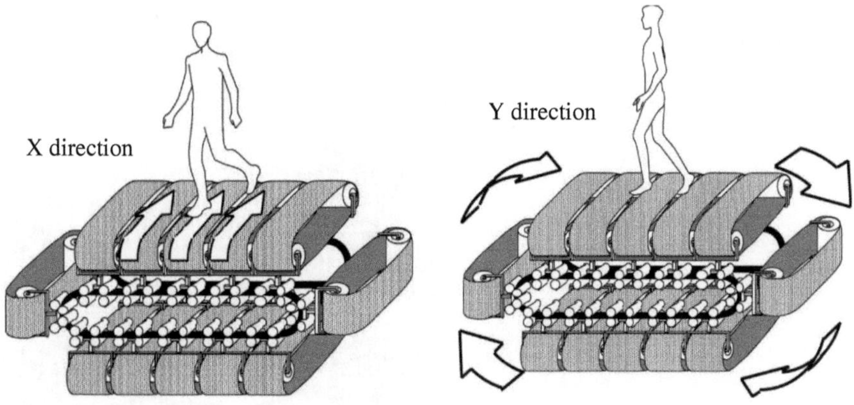

Fig. 9.6 Structure of Torus Treadmill

The shape of a closed surface, in general, is a surface with holes. If the number of holes is zero, the surface is a sphere. The sphere is the simplest infinite surface. However, the walking area of the sphere is not a plane surface. A very large diameter is required to make plane surface on a sphere, which restricts implementation of the locomotion interface.

A closed surface with one hole like a doughnut is called torus. A torus can be implemented by a group of belts. These belts make a plane surface for the user to walk on. A closed surface with more than one hole cannot make a plane walking surface. Thus, the torus is the only form suitable for a locomotion interface.

The "Torus Treadmill" device is implemented by a group of belts connected to each other. The Torus Treadmill is realized by these belts [10]. Figure 9.6 illustrates the basic structure of the Torus Treadmill. The first prototype of the Torus Treadmill employed twelve treadmills. These treadmills move the walker along an "X" direction. Twelve treadmills are connected side by side and driven in a perpendicular direction. This motion moves the walker along a "Y" direction.

Figure 9.7 shows overall view of the apparatus. Twelve treadmills are connected to four chains and mounted on four rails. The chain drives the walker along the Y direction. The rail supports the weight of the treadmills and the walker. An AC motor is employed to drive the chains. Each treadmill is equipped with an AC motor. In order to shorten the length of the treadmill, the motor is put underneath the belt.

A problem with this mechanical configuration is the gap between the belts in the walking area. In order to minimize the gap, a driver unit was put on each treadmill in alternating orientation. The gap achieved is only 2 mm wide in this design.

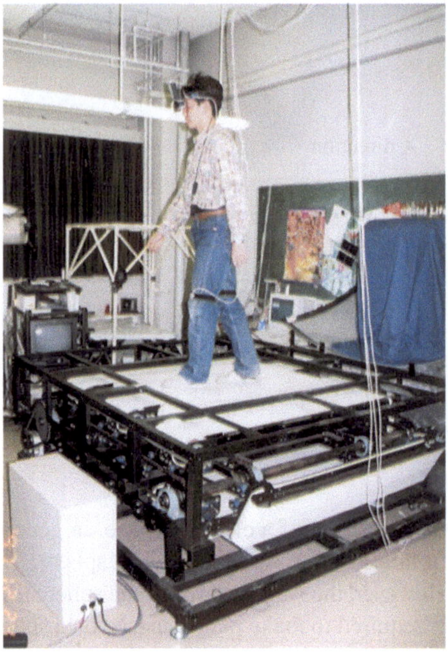

Fig. 9.7 Overall view of Torus Treadmill

9.3.3 Control Algorithm of the Torus Treadmill

A scene of the virtual space is generated corresponding with the results of motion tracking of the feet and head. The motion of the feet and head are measured by a Polhemus FASTRAK. The device measures 6 degree-of-freedom motion. Sampling rate of each point occurs at a rate of 20 Hz. A receiver is attached to each knee. Sensors cannot be put near the motion floor because a steel frame distorts magnetic field. The length and direction of a step is calculated by the data from those sensors. The user's view point in virtual space moves corresponding with the length and direction of the steps.

To keep the walker in the center of the walking area, the Torus Treadmill must be driven in correspondence with the walker. A control algorithm is required to achieve safe and natural walking. The walker is not connected to a harness or mechanical linkages, since such devices restricts the motion and inhibits natural walking. The control algorithm of the Torus Treadmill must be safe enough to allow removal of the harness from the walker. At the final stage of the Virtual Perambulator Project, the harness could be successfully removed using a hoop frame. The walker could freely walk and turn around in the hoop, which supports the walker's body while he/she slides the feet. The function of the hoop in the control algorithm of the Torus Treadmill was simulated by putting circular deadzone in the center of the walking area. If the walker steps out of the area, the floor moves in the opposite direction so that the walker is carried back into the deadzone (Fig. 9.8).

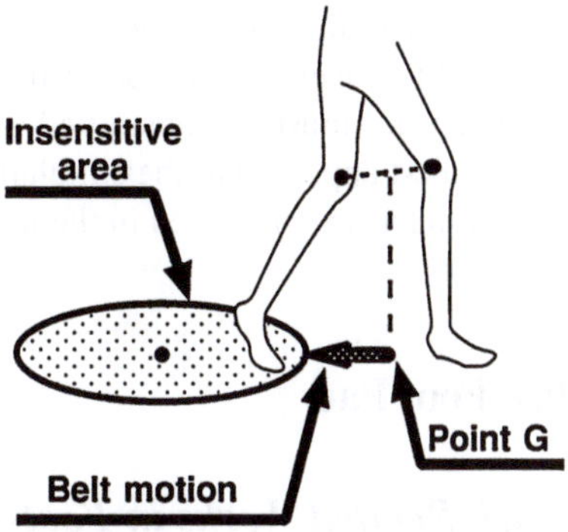

Fig. 9.8 Control algorithm for Torus Treadmill

9.3.4 Effects of Walking on the Torus Treadmill

The Torus Treadmill provides for natural turning motion. The walker on the Torus Treadmill can physically turn about on the active floor. Turning motion using the feet has major contribution to human spatial recognition performance. Vestibular and proprioceptive feedback is essential to the sense of orientation [11].

The same principle as the Torus Treadmill is used in the "CyberWalk Platform" that has a large walking area (Fig. 9.9). The device was constructed at Max Planck Institute and used for study on psychological characteristics [27, 28].

9.3.5 Limitation of Torus Treadmill

A major limitation of the Torus Treadmill is its inability to render uneven surfaces. Theoretically, such a device can be modified for simulation of uneven surface. If an array of linear actuators on each treadmill is installed, uneven floor can be realized by controlling the length of each linear actuator. However, this method is almost

Fig. 9.9 CyberWalk platform

impossible to implement, because a very large number of linear actuators would be required to cover the surface of the torus-shaped treadmills and control signal for each actuator must be transmitted wirelessly.

The best approach to the simulation of uneven surfaces is a foot-pad-based locomotion interface presented in the next section.

9.4 Foot Pad

9.4.1 Related Works in Foot-Pad-Based Locomotion Interface

A foot-pad-based locomotion interface uses a platform to apply displacements to each foot. The simplest way to realize this type of locomotion interface is via a pedaling device. In the battlefield simulator of NPSNET project, a unicycle-like pedaling device is used for locomotion in the virtual battlefield ([22]). A player of the system changes direction by twisting his/her waist. The OSIRIS, simulator for night-vision battle utilizes a stair stepper device ([18]). The device was as same as that used in athletic gyms. A player of the system changes direction by controlling a joystick or twisting at the waist.

In another work, two large manipulators driven by hydraulic actuators were developed at University of Utah and used to realize a locomotion interface. These manipulators are attached to feet of a walker. The device is named BiPort (http://www.sarcos.com). The manipulators can simulate the viscosity of a virtual ground surface. A similar device developed at the Cybernet Systems Corporation uses two 3 DOF motion platform for the feet [21]. These devices, however, have not been evaluated or applied to VE simulation.

Schmidt et al. developed a "Haptic Walker" solution which comprises two programmable foot platforms with permanent foot-machine contact [24]. It is applied to gait rehabilitation.

9.4.2 Gait Master

As was discussed in the first section, the third issue in locomotion interface is presentation of uneven surfaces. Locomotion interfaces are often applied for simulation of buildings or urban spaces. Those spaces usually include stairs. A walker should be provided with a sense of climbing up or going down those stairs. Some applications of locomotion interfaces, such as training simulators or entertainment facility, rough terrain should be presented.

The presentation of virtual staircases was tested at the early stage of the Virtual Perambulator project [9]. When the walker is climbing up a stair, the forward foot was pulled up and when the walker goes down stairs, the backward foot is pulled up. However, this method was not successful because it caused walker instability.

Later a 6 DOF motion platform was applied in a final version of the Virtual Perambulator. In it, a user walks in a hoop frame. The walker stood on the top plate of the motion platform. Pitch and heave motion of the platform were used. When the walker stepped forward to climb up a stair, the pitch angle and vertical position of the floor increased. After finishing climbing motion, the floor went back to the neutral position. When the walker steps forward to go down a stair, the pitch angle and vertical position of the floor decreases. This inclination of the floor is intended to present height differences between the feet. The heave motion is intended to simulate vertical acceleration. However, this method appeared to fail in simulation of stairs. The major reason was that the floor was flat.

A possible method of creation of height difference between the feet is application of two large manipulators. The BiPort is a typical example. A 4 DOF manipulator driven by hydraulic actuators is connected to each foot. The major problem with this method is that how the manipulators trace the turning motion of the walker. When the walker turns around, the two manipulators may interfere each other.

The "GaitMaster" is a locomotion interface that simulates an omni-directional uneven surface. The project started in 1999 [12]. The core elements of the device are two 6 DOF motion-bases mounted on a turntable. Figure 9.10 illustrates basic configuration of the GaitMaster. A walker stands on the top plate of the motion-base.

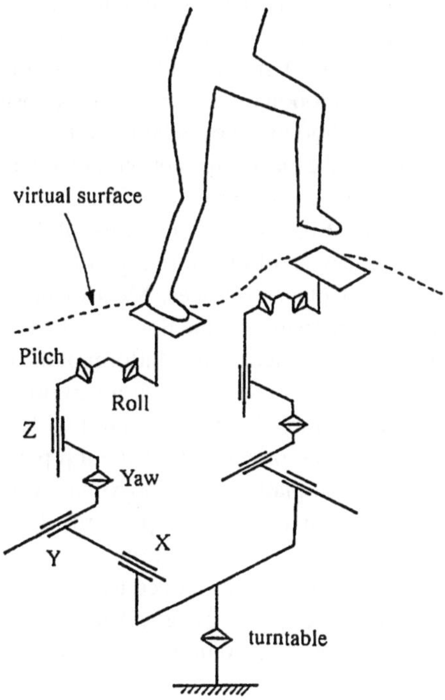

Fig. 9.10 Structure of GaitMater

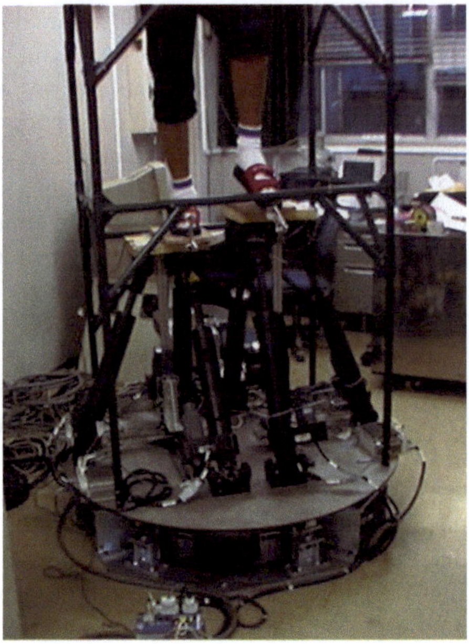

Fig. 9.11 Overall view of GaitMater

Each motion-base is controlled to trace the position of the foot. The turntable is controlled to trace the orientation of the walker. The motion of the turntable removes interference between the two motion-bases.

The X and Y motion of the motion-base traces horizontal position of the feet and cancel its motion by moving to the opposite direction. Rotation around the yaw axis traces the horizontal orientation of the feet. The Z motion traces vertical position of the feet and cancels this motion. The rotation around the roll and pitch axis simulates the inclination of a virtual surface.

Figure 9.11 shows overall view of the prototype GaitMaster. In order to simplify the mechanism of the motion-platform, the surface of the virtual space was defined as sets of planar surfaces. Most of buildings or urbane spaces can be simulated without inclination of the floor. Thus, the roll and pitch axes of the motion-platforms could be neglected. Each platform of the prototype GaitMaster is composed of three linear actuators top of which a yaw joint is mounted. A 6 DOF Stewart platform was disassembled and made into two XYZ stages. Three linear guides are applied to support the orientation of the top plate of the motion-platform. The payload of each motion-platform is approximately 150 Kg. A rotational joint around yaw axis is mounted on each motion platform. The joint is equipped with a spring that moves the feet to the neutral direction.

A turntable was developed using a large direct-drive (DD) motor. The maximum angular velocity is 500 deg/s. A three degree-of-freedom goniometer was connected to each foot. The goniometer measures back-and-forth and up-and-down motion as well as yaw angle. The control algorithm mentioned in the former section was implemented and successfully applied for the presentation of virtual staircases.

9.4.3 Control Algorithm of the GaitMaster

The control algorithm is required to keep the position of the walker at the neutral position of the GaitMaster. In order to keep the position constant, the motion-platforms have to cancel the motion of the feet. The principle of the cancellation is:

(1) Suppose the right foot is at the forward position and left foot is at the backward position while walking.
(2) When the walker steps forward with the left foot, the weight of the walker is placed on the right foot.
(3) The motion-platform of the right foot goes backward in accordance with the displacement of the left foot, so that the central position of the walker is maintained.
(4) The motion-platform of the left foot follows the position of the left foot. When the walker finishes stepping forward, the motion-platform supports the left foot.

If the walker climbs up or goes down stairs, a similar procedure can be applied. The vertical motion of the feet is canceled using the same principal. The vertical displacement of the forward foot is canceled in accordance with the motion of the backward foot, so that the central position of the walker is maintained at the neutral height.

The turntable rotates so that the two motion-platforms can trace the rotational motion of the walker. If the walker changes direction of walking, the turntable rotates to trace the orientation of the walker. The orientation of the turntable is determined according to direction of the feet. The turntable rotates so that its orientation is at the middle of the feet. The walker can physically turn around on the GaitMaster using this control algorithm of the turntable.

A usability test of the prototype GaitMaster was conducted by examining the behavior of novice users. Data were collected at the Emerging Technologies venue at the SIGGRAPH'2000 (July 23–28, 2000, New Orleans). Participants of our demonstration experienced virtual flat terrain as well as staircases.

The major interest of this pilot study related to tracking performance of the motion platforms. Safety straps were put at the level of the foot pad in case the motion platform failed in tracing the trajectory of the foot.

Through the course of empirical observations, tracking performance was found insufficient. The position sensor using strings has 0.3 s time delay. This delay implied an offset between the foot and pad. Walkers often unwittingly stepped off from the pad. However, the safety straps worked well enough that their feet didn't fall down from the pad. The toe strap played the main roll in safety. The heel strap was therefore removed on the 3rd day of the conference. None of the participants experienced any dangerous situation through the 6-day conference.

Velocity and acceleration of the foot while natural walking are very high. It is much difficult for the motion platform to trace the foot. This characteristic limits practical use of the GaitMaster.

9.4.4 GaitMater for Walking Rehabilitation

One major field of application for locomotion interfaces such as the GaitMaster is rehabilitation. If the foot of a patient is connected to the motion platform, it can assist walking. Interestingly, a high performance of the motion platform to follow the foot trajectory is not required for this application.

A simplified version of the GaitMaster was therefore designed for rehabilitation purpose [33, 34]. It moves each foot via a motion platform with two degrees of freedom (back and forth and up and down), which allows for repeated walking cycles, makes it possible to attach and detach easily, and permits to moderately restrain the body (Fig. 9.12). Given the range of movement of human joints, the device is so designed as to only move the feet, leaving to the user's voluntary control movements of the joints in the legs, the hips, and other parts of the body. By doing so, a compatibility between, a high amount of exercise and a relatively moderate resulting restraint could be achieved.

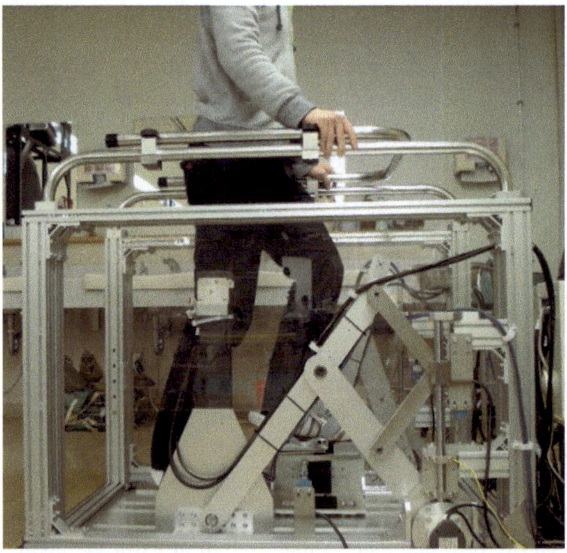

Fig. 9.12 Simplified GaitMater for walking rehabilitation

9.5 Robotic Tiles

9.5.1 The CirculaFloor

The "CirculaFloor" project was launched to profit from approaches taken by both the Torus Treadmill and GaitMaster devices [14]. Locomotion interfaces often require bulky hardware, since they have to carry the user's whole body. The Torus Treadmill is a typical example of large hardware. Also, the hardware is not easy to reconfigure to improve its performance or to add new functions. Considering these issues, the goals of the CirculaFloor project were:

(1) To develop compact hardware for the creation of an infinite surface for walking. The major disadvantage of existing locomotion interfaces being their difficult installation.
(2) To develop scalable hardware architecture for future improvement of the system.

In order to achieve these goals, a new configuration for a locomotion interface using a set of omni-directional movable tiles, or robotic tiles, was proposed. Each tile is equipped with a holonomic mechanism that achieves omni-directional motion. An infinite surface is simulated by the circulation of the robot tiles (Fig. 9.13).

The motion of the feet is measured with laser scanning sensors. The sensor is a non-contact laser measurement system that scans its surroundings two-dimensionally like laser radar. The sensor radially measures distance between the scanner and an object that reflects the laser. This system requires no auxiliary passive components such as reflectors or position markers, so that the walker never requires putting any obstructive sensors or markers.

The robot tile moves opposite to the measured direction of the walker so that the motion of the step is canceled. The position of the walker is fixed in the real world by this computer-controlled motion of the robot tiles. The circulation of the robot tiles has the ability to cancel the displacement of the walker in an arbitrary direction. Thus, the walker can freely change direction while walking. This advantage is as same as the Torus Treadmill.

The combination of robot tiles provides a sufficient area for walking, and thus precision tracing of the foot position as is needed in the GaitMaster is not required. It has the ability to create an uneven surface by mounting an up-and–down mechanism on each tile. Figure 9.14 shows robot tiles with lifting actuators presenting virtual staircases.

9.5.2 User Study of the Robot Tile Approach

The first prototype system of the CirculaFloor was demonstrated at SIGGRAPH 2004 venue (Los Angeles). The behavior of 325 participants was videotaped using a

Fig. 9.13 Circulation of robot tiles

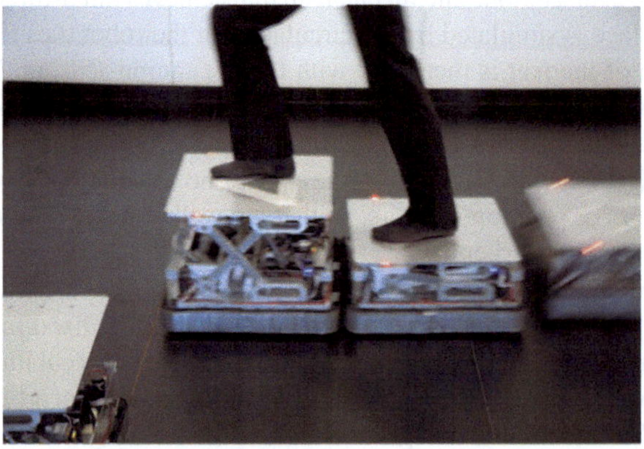

Fig. 9.14 Robot tiles present virtual staircases

DV video recorder. Before walking, subjects were given instructions that they should walk in small steps and could step on the edge between tiles.

As an index of stability of walking, maximum horizontal movements of the subjects were evaluated. The maximum horizontal amplitude of the center of the subject's lumbar on the monitor and the widths of the subject's lumbar on the monitor were measured. The maximum horizontal movement equals to the horizontal amplitudes divided by the width.

The results suggested that 78 % of the subjects were below 10 % horizontal movements. For comparison, the horizontal movement in the real world was measured.

As a result, most subjects marked under 10 % movements. It is assumed that 78 % of the subjects on the CirculaFloor can walk stably.

On the contrary, the remaining 22 % of the subjects could not walk stably. The reasons of unstable walk were categorized into following 5 classes: (1) tile was moving when landing forefoot; (2) gait style was suddenly changed. E.g. a change from wide based gait to narrow based gait was observed; (3) the subjects kicked back tiles by their hind leg; (4) the subjects strolled along the edge of tiles; (5) The subjects walked in an otherwise awkward fashion. These factors depended primarily on the gait style of the subjects. The movable tiles were controlled to maintain a desired velocity trajectory. However, when a subject applied a force to the tile that was too large, the tile could move in an unexpected direction. Improvements to the stability of the tile system are still required. Also, the subjects' shoes affecting their gait pattern. It is difficult for subjects who wear high-heel to walk stably on the device. The height of the movable tiles is 90 mm. Such subjects were fearful of even small displacements of the tile.

A major limitation of the robot-tile-based locomotion interface seems however to be its low walking speed. Very high circulation speed is required to follow natural walking speed. The maximum running speed of the robot tile is limited due to the upper limit of traction force of the holonomic mechanism (Fig. 9.15).

Still, the circulating tiles seem to remain a very attractive and popular interactive device. A more recent version of the robotic tiles was demonstrated at Tokyo Fiber '09 art exhibition (Triennale, Milan). This updated version encloses new conductive fibers mounted on each tile, making it possible to finely detect positions of the feet. It was, observed to be a successful demonstration.

Fig. 9.15 Robot tiles exhibited at Tokyo Fiber 20009 (Photo:)

9.6 Conclusion

This chapter introduced and described the various kinds of existing locomotion interfaces: Sliding shoes, Treadmills, Foot-pads, and Robotic tiles. Main advantages and limitations of each method were discussed. The preferred configurations for serious applications would probably remain treadmill-based locomotion interfaces. They can indeed create highest walking speeds in safe conditions. If a virtual environment application requires uneven surfaces, foot-pad-based locomotion interface should however be used. Specific mechanisms must be designed to better match specific applications. Safety issues should be carefully considered in applications of locomotion interfaces. A proper design of handrails can often solve this issue. Examples of application-driven devices were also presented such as for Gait rehabilitation and Evacuation simulator. But locomotion interfaces are still in an immature state of development. Significant trial-and-error will therefore be needed in order to realize further advances in this promising research area.

References

1. Bakker NH, Werkhoven PJ, Passenier PO (1998) Aiding orientation in virtual environments with proprioceptive feedback. In: Proceedings of the IEEE 1998 virtual reality annual international symposium, pp 28–33
2. Brooks FP Jr (1986) A dynamic graphics system for simulating virtual buildings. In: Proceedings of the 1986 workshop on interactive 3D graphics, Chapel Hill, NC, ACM, New York, Oct 1986, pp 9–21
3. Chance SS, Gaunet F, Beall AC, Loomis JM (1998) Locomotion mode affects the updating of objects encountered during travel: the contribution of vestibular and proprioceptive inputs to path integration. Presence 7(2):168–178
4. Christensen R, Hollerbach JM, Xu Y, Meek S (1998) Inertial force feedback for a locomotion interface. In: Proceedings of the ASME dynamic systems and control division, DSC, vol 64, pp 119–126
5. Darken R, Allard T, Achille L (1998) Spatial orientation and wayfinding in large-scale virtual space: an introduction. Presence 7(2):101–107
6. Darken R, Cockayne W, Carmein D (1997) The omni-directional treadmill: a locomotion device for virtual worlds. In: Proceedings of UIST'97, pp 213–222
7. Iwata H (1990) Artificial reality for walking about large scale virtual space. Hum Interface News Rep 5(1):49–52 (in Japanese)
8. Iwata H, Matsuda K (1992) Haptic walkthrough simulator: its design and application to studies on cognitive map. In: Proceedings of ICAT'92, pp 185–192
9. Iwata H, Fujii T (1996) Virtual perambulator: a novel interface device for locomotion in virtual environment. In: Proceedings of the IEEE 1996 virtual reality annual international symposium, pp 60–65
10. Iwata H (1999) Walking about virtual space on an infinite floor. In: Proceedings of the IEEE virtual reality'99, pp 236–293
11. Iwata H, Yoshida Y (1999) Path reproduction tests using a torus treadmill. Presence 8(6):587–597
12. Iwata H, Yano H, Nakaizumi F (2001) Gait Master: a versatile locomotion interface for uneven virtual terrain. In: Proceedings of IEEE virtual reality 2001 conference, pp 131–137

13. Iwata H (2004) Touching and walking; issues in haptic interface. Keynote lecture of EuroHaptics
14. Iwata H, Yano H, Fukushima H, Noma H (2005) CirculaFloor: a locomotion interface using circulation of movable tiles. In: Proceedings of VR2005, pp 223–230
15. Iwata H, Yano H, Tomioka H (2006) Powered shoes. In: SIGGRAPH 2006 conference DVD
16. Iwata H, Yano H, Tomiyoshi M (2006) String walker. SIGGRAPH 2007 conference DVD
17. Loomis JM, Beall AC, Klatzky RL, Golledge RG, Phibeck JW (1995) Evaluating the sensory inputs to path integration. Paper presented at the psychonomic society meeting, Los Angels, CA, 10–12 Nov 1995
18. Lorenzo M et al (1995) OSIRIS. In: SIGGRAPH'95 visual proceedings, p 129
19. Noma H, Sugihara T, Miyasato T (2000) Development of ground surface simulator for tel-E-merge system. In: Proceedings of IEEE virtual reality, pp 217–224
20. Nagamori A, Wakabayashi K, Ito M (2005) The ball array treadmill: a locomotion interface for virtual world. In: Proceedings of the IEEE VR2005
21. Poston R et al (1997) A whole body kinematic display for virtual reality applications. In: Proceedings of the IEEE international conference on robotics and automation, pp 3006–3011
22. Prat DR et al (1994) Insertion of an articulated human into a networked virtual environment. In: Proceedings of the 1994 AI, simulation, and planning in high autonomy systems conference, pp 7–9
23. Ruddle RA, Lessels S (2009) The benefits of using a walking interface to navigate virtual environments. ACM Trans Comput Hum Interact 16(1):1–18
24. Schmidt H et al (2005) HapticWalker—a novel haptic foot device. ACM Trans Appl Percept 2(2):166–180
25. Suma EA, Finkelstein SL, Reid M, Babu V (2010) Evaluation of the cognitive effects of travel technique in complex real and virtual environments. IEEE Trans Vis Comput Graph 16(4):690–702
26. Stevens SS (1957) On the psychological law. Psychol Rev 64(3):153–181
27. Souman JL, Robuffo Giordano P, Schwaiger M, Frissen I, Thümmel T, Ulbrich H, De Luca A, Bülthoff HH, Ernst M (2011) CyberWalk: enabling unconstrained omnidirectional walking through virtual environments. Trans Appl Percept 8:1-22
28. Souman JL, Robuffo Giordano P, Frissen I, Luca AD, Ernst MO (2010) Making virtual walking real: perceptual evaluation of a new treadmill control algorithm. Trans Appl Percept 7(2:11) 1–14
29. Usoh M, Arthur K, Whitton MC, Bastos R, Steed A, Slater M, Brooks FP (1999) Walking > walking-in-place > flying, in virtual environments. In: Proceedings of the SIGGRAPH'99, pp 359–364
30. Ware C, Slipp L (1991) Using velocity control to navigate 3D graphical environments: a comparison of three interfaces. In: Proceedings of the human factor society 35th annual meeting, pp 300–304
31. Witmer BG, Bailey JH, Knerr BW, Parsons KC (1996) Virtual spaces and real world places: transfer of route knowledge. Int J Hum Comput Stud 45:413–428
32. Yamao T, Ishida S, Ota S, Kaneko F (1996) Formal safety assesment—research project on quantification of risk on lives, MSC67/INF.9 IMO information paper
33. Yano H, Masuda T, Nakajima Y, Tanaka N, Tamefusa S, Saitou H, Iwata H (2008) Development of a gait rehabilitation system with a spherical immersive projection display. J Robot Mechatron 12(6):836–845
34. Yano H, Tamefusa S, Tanaka N, Saitou H, Iwata H (2010) Gait rehabilitation system for stair climbing and descending. Proc Haptics 2010:393–400

Chapter 10
Implementing Walking in Virtual Environments

Gerd Bruder and Frank Steinicke

Abstract In the previous chapter, locomotion devices have been described, which prevent displacements in the real world while a user is walking. In this chapter we explain different strategies, which allow users to actually move through the real-world, while these physical displacements are mapped to motions of the camera in the virtual environment (VE) in order to support unlimited omnidirectional walking. Transferring a user's head movements from a physical workspace to a virtual scene is an essential component of any immersive VE. This chapter describes the pipeline of transformations from tracked real-world coordinates to coordinates of the VE. The chapter starts with an overview of different approaches for virtual walking, and gives an introduction to tracking volumes, coordinate systems and transformations required to set up a workspace for implementing virtual walking. The chapter continues with the traditional isometric mapping found in most immersive VEs, with special emphasis on combining walking in a restricted interaction volume via reference coordinates with virtual traveling metaphors (e.g., *flying*). Advanced mappings are then introduced with user-centric coordinates, which provide a basis to guide users on different paths in the physical workspace than what they experience in the virtual world.

10.1 Introduction

Using sophisticated hard- and software technology, immersive virtual environments (VEs) provide users with a multisensory medium for exploring and interacting with computer-generated three-dimensional environments. In particular, ego-centric

G. Bruder · F. Steinicke
Department of Human-Computer-Media, University of Würzburg, Campus Nord,
Oswald-Külpe-Weg 82, D-97074 Würzburg, Germany
e-mail: gerd.bruder@uni-wuerzburg.de

F. Steinicke
e-mail: frank.steinicke@uni-wuerzburg.de

F. Steinicke et al. (eds.), *Human Walking in Virtual Environments*,
DOI: 10.1007/978-1-4419-8432-6_10,

perspectives and natural interaction metaphors can provide users with a compelling experience similar to interactions in the real world, which cannot be simulated using any other existing technology. In this context, the most natural technique for exploring a virtual world is *real walking*, which provides a greater sense of presence than other virtual traveling techniques [4, 24], such as flying or walking in-place [36], and naturally stimulates human spatial wayfinding and cognitive map building [27]. As described in Part 1, walking is a form of natural locomotion, which encompasses repetitive motions of legs or body for active self-propulsion [9], such that users in immersive VEs receive proprioceptive, kinesthetic and efferent copy signals from their physical movements, supporting the perception of self-motion in the virtual world.

In order to provide users with an unimpaired sense of place and plausibility during self-motions [29], virtual reality (VR) applications have to maintain simultaneous awareness of coordinate systems and transformations in both the real and virtual world. In this chapter, we describe the basic transformations that can be used to implement real walking user interfaces in VR laboratory workspaces. In particular, we show how the sense of moving in computer graphics environments can be stimulated with sequences of frame to frame changes of the position and orientation of a user in a VR workspace. If the changes from one frame to the next are large, we talk of *teleportation*, whereas if the changes are considerable small, the feedback from the virtual world causes a sensory flow (e.g., optic flow [15] or acoustic flow [30]), which engenders the sense of continuous motion.

We distinguish between two main characteristics of real walking user interfaces:

- *Isometric* transformations describe mappings that preserve motion distances and angles when movements of a tracked user in the physical workspace are mapped to changes of a virtual representation.
- *Nonisometric* transformations, in contrast, describe different mapping approaches to introduce a discrepancy between user movements and virtual feedback.

It is generally assumed that human spatial perception and cognition in virtual worlds is optimally supported in isometric user interfaces, since sensory motion feedback from the user's physical movements (e.g., proprioceptive and vestibular motion cues) match feedback from the virtual world (e.g., optic and acoustic flow). However, isometric user interfaces have a severe practical problem: With such mappings the size of the physical workspace limits the size of the virtual scene that a user can explore by natural walking. We show how such limitations can be alleviated with multimodal interfaces that combine walking over short distances with traveling over long distances. We introduce nonisometric mapping strategies that provide a different solution to the problem of unrestricted omnidirectional walking by guiding users on a different path in the real world than experienced in the virtual scene. Nonisometric mappings for walking user interfaces are encompassed under the term *redirected walking* [23].

The remainder of this chapter is structured as follows. Section 10.2 gives a short introduction to workspaces and coordinate systems in VR laboratories. In Sect. 10.3

we present the basic math and algorithms necessary to implement isometric virtual walking, and then show with reference coordinates how limitations of virtual interaction space can be alleviated with traveling techniques. In Sect. 10.4 we describe nonisometric transformations for redirected walking, give an overview of the basic algorithms with user-centric coordinates, and go into detail on linear and angular scaling transformations, as well as curvature mappings. We present a simple algorithm that allows practitioners to implement unrestricted redirected walking in VR workspaces. Section 10.5 concludes the chapter.

10.2 Virtual Reality Workspaces

In order to support real walking, user movements in a VR laboratory have to be tracked and mapped to motions in a three-dimensional virtual scene. In particular, movements of the user's head position in the physical workspace have to be measured and transferred to motions of camera objects in the virtual space in order to provide ego-centric visual feedback to the user's eyes from the virtual world.[1]

Physical workspaces in VR laboratories incorporate tracking systems to measure the position and/or orientation of objects located in the tracking space. Such tracking systems can differ in underlying technology, accuracy and precision of tracking data, as well as how the user is instrumented. In particular, some VR laboratories incorporate separate tracking systems for position and orientation measurements, such as optical marker tracking systems that measure the head position and inertial orientation sensors that measure the head orientation. The coordinate systems in which tracking systems provide position and orientation data are not standardized, such that usually the tracking coordinates have to be transformed into the coordinate system used for the virtual scene [8]. In the following, for convenience, we assume that virtual and physical coordinate systems are calibrated and represented in right-handed OpenGL coordinates [28]. Therefore, the y-axis is oriented in inverse gravitation direction, whereas the x- and z-axis are orthogonal to the y-axis and each other, thus defining the ground plane. These coordinates can easily be derived from arbitrary tracking coordinates by reassigning the x-, y- and z-axes, or multiplying the z-coordinate with -1 for changing the handedness.

Figure 10.1 illustrates such a coordinate system in a tracked workspace in a VR laboratory. Position and orientation of tracked objects can be described as a transformation from a specified origin of the tracking volume to the object's local coordinate system. Tracking systems often provide position data as a translation vector $(x_r, y_r, z_r) \in \mathbb{R}^3$, and orientation data as yaw, pitch and roll angles $(\tilde{y}_r, \tilde{p}_r, \tilde{r}_r) \in [0, 360)^3$, describing three subsequently applied rotational transfor-

[1] While most immersive VEs implement head tracking for visual feedback, some laboratories also implement tracking of other body parts to provide virtual body feedback or interaction methods.

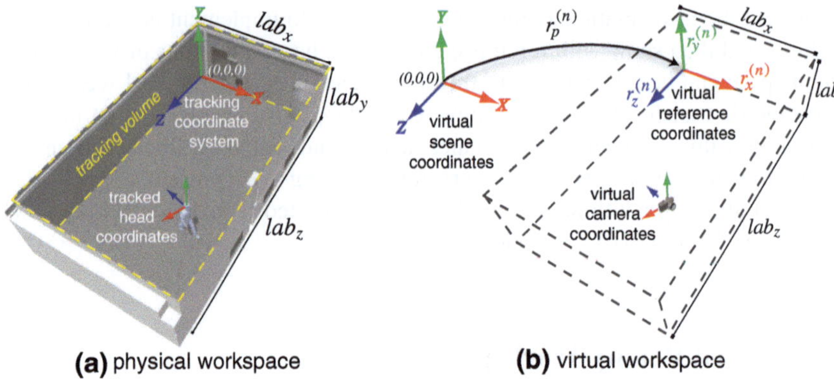

Fig. 10.1 Illustration of (**a**) three-dimensional tracking coordinates in a VR laboratory with a user's tracked local head coordinate system as defined by positional and orientational tracking data, and (**b**) virtual scene coordinates with interaction volume defined by reference coordinates

mations.[2] Although the axes and order of yaw, pitch and roll transformations are not standardized among tracking systems, practitioners tend to represent orientations first with yaw rotations around the y-axis, followed by pitch rotations around the x-axis, and roll rotations around the z-axis [35]. For instance, assuming the user's head position and orientation is tracked with a local coordinate system that is defined with the z-axis for the (inverse) *look*-direction, the y-axis in *up*-direction, and the x-axis in *strafe*-direction (see Fig. 10.1a), then yaw, pitch and roll rotations correspond to turning the head to the left or right, up or down, or around the view axis, respectively. In the following, we assume tracked head orientations to be provided using this representation.

In order to provide a user with visual feedback by rendering the three-dimensional scene onto one or more VR display surfaces, we have to consider a virtual analog of the user's head in the VE.[3] We assume virtual camera coordinates are represented with the triple of orthogonal axes as used for physical head tracking coordinates [8], with transformations from the origin of the virtual scene to the camera coordinates defined by a translation vector $(x_v, y_v, z_v) \in \mathbb{R}^3$, and yaw, pitch and roll angles $(\tilde{y}_v, \tilde{p}_v, \tilde{r}_v) \in [0, 360)^3$. Figure 10.1b illustrates local camera coordinates in virtual scene coordinates.

[2] Some tracking systems use quaternions as their native reporting format, which provides an alternative representation of the transformations, and can be converted from and to the angular notation used in this chapter [21].

[3] Depending on the display system (e.g., head-mounted displays or immersive projection technologies) the actual positions or orientations of computer graphics camera objects are usually specified relative to these head coordinates, such as transformations from the head center to the eye displays [25].

10.3 Isometric Virtual Walking

In this section we present the basic math and algorithms to implement isometric real walking transformations, i.e., mappings that preserve distances and angles of a user's movements.

10.3.1 One-to-One Mappings

Assuming the real and virtual workspaces are defined using the coordinate systems introduced in Sect. 10.2, basic one-to-one mappings can be implemented by using the tracked position and orientation of a user's head in the laboratory to define the position and orientation of a corresponding virtual camera object for each rendering frame. In particular, a tracked change of one unit (e.g., meter or degree) in the physical workspace is mapped to a translation or rotation of one unit in the virtual scene. Examples of such mappings are often found when displaying a virtual replica of a virtual reality laboratory to users in head-mounted display environments [12, 33], or in architectural passive haptics environments, in which real and virtual objects are registered to provide users with haptic feedback when touching virtual objects [10]. In such environments, one-to-one mappings can be implemented using the following simple pseudo code:

Algorithm 1 One-to-one mapping from tracked head to camera coordinates

for all rendering frames $n \in \mathbb{N}_0$ **do**
  // Get current head tracking state:
  $(x_r^{(n)}, y_r^{(n)}, z_r^{(n)}) \leftarrow$ tracked head position (in $\mathbb{R}^3$)
  $(\tilde{y}_r^{(n)}, \tilde{p}_r^{(n)}, \tilde{r}_r^{(n)}) \leftarrow$ tracked head orientation (in $[0, 360)^3$)

  // Set virtual camera state:
  $(x_v^{(n)}, y_v^{(n)}, z_v^{(n)}) \leftarrow (x_r^{(n)}, y_r^{(n)}, z_r^{(n)})$ *// position*
  $(\tilde{y}_v^{(n)}, \tilde{p}_v^{(n)}, \tilde{r}_v^{(n)}) \leftarrow (\tilde{y}_r^{(n)}, \tilde{p}_r^{(n)}, \tilde{r}_r^{(n)})$ *// orientation*
end for

In the pseudo code, $(x_r^{(n)}, y_r^{(n)}, z_r^{(n)}) \in \mathbb{R}^3$ denotes the current three-dimensional position, and $(\tilde{y}_r^{(n)}, \tilde{p}_r^{(n)}, \tilde{r}_r^{(n)}) \in [0, 360)^3$ the current yaw, pitch and roll orientation of the user's head in the physical workspace as provided by the tracking system for rendering frames $n \in \mathbb{N}_0$, as well as $(x_v^{(n)}, y_v^{(n)}, z_v^{(n)}) \in \mathbb{R}^3$ the computed new position and $(\tilde{y}_v^{(n)}, \tilde{p}_v^{(n)}, \tilde{r}_v^{(n)}) \in [0, 360)^3$ the new orientation of the camera object that is used as the basis for rendering the current frame to be displayed to the user.

10.3.2 Reference Coordinates

A simple extension of isometric mappings is to introduce virtual *reference coordinates*. Since one-to-one mappings only allow a user to explore a volume in the virtual scene that is exactly as large as the interaction volume in the laboratory, it becomes important to map the user's movements to specific regions of interest in the virtual scene. This can be accomplished by introducing an intermediate reference coordinate system when transferring position and orientation data from tracking coordinates to virtual scene coordinates. Introducing such virtual reference coordinates, the corresponding pseudo code in Sect. 10.3.1 changes to:

Algorithm 2 Isometric mapping with reference coordinates

for all rendering frames $n \in \mathbb{N}_0$ **do**

// Get current head tracking state:

$(x_r^{(n)}, y_r^{(n)}, z_r^{(n)}) \leftarrow$ tracked head position (in $\mathbb{R}^3$)

$(\tilde{y}_r^{(n)}, \tilde{p}_r^{(n)}, \tilde{r}_r^{(n)}) \leftarrow$ tracked head orientation (in $[0, 360)^3$)

// Set virtual camera state:

$$(x_v^{(n)}, y_v^{(n)}, z_v^{(n)}, 1)^{\mathrm{T}} \leftarrow \begin{pmatrix} r_{x_x}^{(n)} & r_{y_x}^{(n)} & r_{z_x}^{(n)} & r_{p_x}^{(n)} \\ r_{x_y}^{(n)} & r_{y_y}^{(n)} & r_{z_y}^{(n)} & r_{p_y}^{(n)} \\ r_{x_z}^{(n)} & r_{y_z}^{(n)} & r_{z_z}^{(n)} & r_{p_z}^{(n)} \\ 0 & 0 & 0 & 1 \end{pmatrix} \cdot (x_r^{(n)}, y_r^{(n)}, z_r^{(n)}, 1)^{\mathrm{T}} \quad \textit{// position}$$

$(\tilde{y}_v^{(n)}, \tilde{p}_v^{(n)}, \tilde{r}_v^{(n)}) \leftarrow (\tilde{y}_r^{(n)}, \tilde{p}_r^{(n)}, \tilde{r}_r^{(n)})$ *// orientation*

end for

In the pseudo code, the 4×4 transformation matrix for homogenous coordinates defines a reference position $r_p^{(n)} = (r_{p_x}^{(n)}, r_{p_y}^{(n)}, r_{p_z}^{(n)}) \in \mathbb{R}^3$ in the virtual scene, as well as coordinate axes with the direction vectors $r_x^{(n)} = (r_{x_x}^{(n)}, r_{x_y}^{(n)}, r_{x_z}^{(n)}) \in \mathbb{R}^3$, $r_y^{(n)} = (r_{y_x}^{(n)}, r_{y_y}^{(n)}, r_{y_z}^{(n)}) \in \mathbb{R}^3$, and $r_z^{(n)} = (r_{z_x}^{(n)}, r_{z_y}^{(n)}, r_{z_z}^{(n)}) \in \mathbb{R}^3$ along the transformed x-, y- and z-axes of the reference coordinates. The virtual yaw, pitch and roll transformations are applied to the reference coordinate axes. Figure 10.1 illustrates reference coordinates that are used to move the virtual interaction volume (limited by the size of the laboratory workspace $(lab_x, lab_y, lab_z) \in \mathbb{R}^3$) to regions of interest.

To account for changing regions of interest in the virtual scene, reference coordinates can be changed at run time. In particular, teleportation of the user's virtual viewpoint can be implemented by abrupt changes of the reference coordinates, whereas continuous traveling can be implemented by iterative changes in reference positions and orientations [3, 18, 24].

10.3.3 Virtual Traveling

Instead of just implementing natural walking, many immersive VEs make use of a hybrid walking-and-flying metaphor, in which the user's head is tracked in a limited interaction space, whereas the user can change the reference position or orientation in the virtual environment by using a hand-held controller. This interaction technique can easily be grasped by users when introduced as a *flying carpet* [40], i.e., the user can naturally walk over a limited carpet region, while the carpet itself can be flown through the virtual world. In contrast to real walking, which is identified by natural locomotion in the physical workspace, flying and steering techniques are denoted as virtual *traveling* [3]. A traditional implementation of virtual traveling is view-directed flying [3, 18, 24], which refers to user-initiated changes of reference coordinates relative to the user's virtual view, i.e., the coordinates of the virtual camera object. With view-directed flying in immersive virtual environments usually only the reference position is changed, whereas the orientation of reference coordinates is not affected, such that the virtual interaction volume remains level to the real world [18].

A basic virtual flying controller can be implemented using the following simple approach. For each rendering frame $n \in \mathbb{N}$ we compute the current view-direction $(v_{v_x}^{(n)}, v_{v_y}^{(n)}, v_{v_z}^{(n)}) \in \mathbb{R}^3$, the strafe-direction $(s_{v_x}^{(n)}, s_{v_y}^{(n)}, s_{v_z}^{(n)}) \in \mathbb{R}^3$, and the up-direction $(u_{v_x}^{(n)}, u_{v_y}^{(n)}, u_{v_z}^{(n)}) \in \mathbb{R}^3$ of the camera object in the VE (see Sect. 10.2). Providing commodity input hardware to the user, such as a keyboard, the user can initiate changes in reference coordinates by pressing different keys. For instance, if we detect that the user has pressed the up- or down-key on a keyboard, we compute the reference position for the next rendering frame as

$$(r_{p_x}^{(n)}, r_{p_y}^{(n)}, r_{p_z}^{(n)}) = (r_{p_x}^{(n-1)}, r_{p_y}^{(n-1)}, r_{p_z}^{(n-1)}) + (v_{v_x}^{(n)}, v_{v_y}^{(n)}, v_{v_z}^{(n)}) \cdot g_v^{(n)},$$

with $g_v^{(n)} \in \mathbb{R}$ defining a speed factor for virtual traveling in view direction, e.g., with $g_v^{(n)} > 0$ for forward motions if the user pressed the up-key, and $g_v^{(n)} < 0$ for backward motions if the user pressed the down-key on the keyboard. In particular, this means that the user can turn the head towards a target in the virtual scene, and fly towards the virtual target by pressing the up-key. Using corresponding keys and speed factors, we can allow the user to change the reference position in the virtual scene not only in view-direction, but also in strafe-direction $(s_{v_x}^{(n)}, s_{v_y}^{(n)}, s_{v_z}^{(n)})$, and up-direction $(u_{v_x}^{(n)}, u_{v_y}^{(n)}, u_{v_z}^{(n)})$. The speed factors may be as simple as a constant or a more sophisticated function based on sensor inputs.

10.4 Nonisometric Virtual Walking

Isometric mappings enable users to explore a virtual region by real walking. However, since with isometric mappings the virtual interaction space is limited, virtual traveling techniques have to be used to cover larger distances in the VE, which impair the user's sense of being able to explore a VE like the real world, and can significantly degrade spatial perception and task performance [27, 36]. To alleviate such problems, researchers proposed nonisometric mappings, which have the potential to enable unrestricted omnidirectional walking. In order to describe such mappings, we introduce relative user-centric coordinates.

10.4.1 User-Centric Coordinates

Instead of mapping position and orientation data from the tracking volume in the laboratory for each rendering frame to their respective absolute position and orientation in a fixed virtual interaction volume, redirection techniques are based on relative mappings, in which each change in position or orientation from one rendering frame to the next is addressed separately [32]. Using user-centric relative coordinates requires more transformations and may introduce numerical error propagation, but allows more sophisticated mapping strategies.

From Absolute Position and Orientation to Relative Changes

For each rendering frame the change in position and orientation of the user's head is measured in coordinates of the tracking volume. Changes can be computed as the difference of the current tracking data at rendering frame $n \in \mathbb{N}$ from the previous state at rendering frame $n-1$, defined by tuples consisting of the previous position $(x_r^{(n-1)}, y_r^{(n-1)}, z_r^{(n-1)}) \in \mathbb{R}^3$ and orientation $(\tilde{y}_r^{(n-1)}, \tilde{p}_r^{(n-1)}, \tilde{r}_r^{(n-1)}) \in [0, 360)^3$, and the current position $(x_r^{(n)}, y_r^{(n)}, z_r^{(n)}) \in \mathbb{R}^3$ and orientation $(\tilde{y}_r^{(n)}, \tilde{p}_r^{(n)}, \tilde{r}_r^{(n)}) \in [0, 360)^3$ in the real-world tracking volume. The three-dimensional head position change $(\Delta x_r^{(n)}, \Delta y_r^{(n)}, \Delta z_r^{(n)}) \in \mathbb{R}^3$, as well as the changes in yaw, pitch and roll head orientation angles $(\Delta \tilde{y}_r^{(n)}, \Delta \tilde{p}_r^{(n)}, \Delta \tilde{r}_r^{(n)}) \in [-180, 180)^3$ result as:

$$\begin{cases} \Delta x_r^{(n)} = x_r^{(n)} - x_r^{(n-1)}, \\ \Delta y_r^{(n)} = y_r^{(n)} - y_r^{(n-1)}, \\ \Delta z_r^{(n)} = z_r^{(n)} - z_r^{(n-1)}, \end{cases}$$

$$\begin{cases} \Delta \tilde{y}_r^{(n)} = atan2(\ sin(\tilde{y}_r^{(n)} - \tilde{y}_r^{(n-1)}),\ cos(\tilde{y}_r^{(n)} - \tilde{y}_r^{(n-1)})\ ), \\ \Delta \tilde{p}_r^{(n)} = atan2(\ sin(\tilde{p}_r^{(n)} - \tilde{p}_r^{(n-1)}),\ cos(\tilde{p}_r^{(n)} - \tilde{p}_r^{(n-1)})\ ), \\ \Delta \tilde{r}_r^{(n)} = atan2(\ sin(\tilde{r}_r^{(n)} - \tilde{r}_r^{(n-1)}),\ cos(\tilde{r}_r^{(n)} - \tilde{r}_r^{(n-1)})\ ). \end{cases}$$

It should be noted that computing the angular difference from one frame to the next is not trivial. In this computation we assume that the user's head rotation between the previous and current frame did not exceed 180° in each dimension, which is a reasonable assumption in real-time simulations. Therefore, we compute the smaller of the two angles in each direction that can lead from the previous orientation angle to the current angle (i.e., rotating clockwise or counterclockwise). The computed $\Delta\tilde{y}_r^{(n)}$, $\Delta\tilde{p}_r^{(n)}$ and $\Delta\tilde{r}_r^{(n)}$ angles are in the interval $[-180, 180)$.

Mapping Relative Changes

Linear and angular changes of the user's head pose in tracking coordinates have to be mapped to the virtual environment for each rendering frame, i.e., the virtual position and orientation result from accumulation of relative differences measured in the physical tracking volume. We can describe a one-to-one relative mapping for all linear and angular movements from the tracking volume to virtual coordinates in pseudo code as follows:

Algorithm 3 Relative mapping from tracked head to camera coordinates

for all rendering frames $n \in \mathbb{N}$ **do**
  // Get current head tracking state:
  $(x_r^{(n)}, y_r^{(n)}, z_r^{(n)}) \leftarrow$ tracked head position (in $\mathbb{R}^3$)
  $(\tilde{y}_r^{(n)}, \tilde{p}_r^{(n)}, \tilde{r}_r^{(n)}) \leftarrow$ tracked head orientation (in $[0, 360)^3$)

  // Compute relative changes:
  $(\Delta x_r^{(n)}, \Delta y_r^{(n)}, \Delta z_r^{(n)}) \leftarrow$ head position change (in $\mathbb{R}^3$)
  $(\Delta\tilde{y}_r^{(n)}, \Delta\tilde{p}_r^{(n)}, \Delta\tilde{r}_r^{(n)}) \leftarrow$ head orientation change (in $[-180, 180)^3$)

  // Set virtual camera state:
  $(x_v^{(n)}, y_v^{(n)}, z_v^{(n)}) \leftarrow (x_v^{(n-1)}, y_v^{(n-1)}, z_v^{(n-1)}) + (\Delta x_r^{(n)}, \Delta y_r^{(n)}, \Delta z_r^{(n)})$ *// position*
  $(\tilde{y}_v^{(n)}, \tilde{p}_v^{(n)}, \tilde{r}_v^{(n)}) \leftarrow (\tilde{y}_v^{(n-1)}, \tilde{p}_v^{(n-1)}, \tilde{r}_v^{(n-1)}) + (\Delta\tilde{y}_r^{(n)}, \Delta\tilde{p}_r^{(n)}, \Delta\tilde{r}_r^{(n)})$ *// orientation*
end for

This approach describes relative transformations from one rendering frame to the next, i.e., it is reasonable to initialize the virtual position and orientation of the user at the beginning of the VR experience with the identical state as in the tracking volume, or to define an initial offset using reference coordinates as described for isometric mappings (see Sect. 10.3.2).

Local Frames of Reference

In contrast to the absolute position and orientation of the user in tracking coordinates, such relative changes are independent of a specific origin defined in the tracking volume. However, the relative changes are not independent of the specific axes defined in the tracking space. In particular, a movement of a user's head is described as a position change for frame $n \in \mathbb{N}$ as $(\Delta x_r^{(n)}, \Delta y_r^{(n)}, \Delta z_r^{(n)}) \in \mathbb{R}^3$ along the x-, y- and z-axes of the tracking volume. Most advanced redirection techniques, however, manipulate position or orientation changes relative to specific coordinates (e.g., determined from the user's head or body state) in the real and virtual world. We can account for such frames of reference by introducing local coordinate transforms, for which the virtual camera transformation at frame $n \in \mathbb{N}$ changes to:

$$\begin{pmatrix} x_v^{(n)} \\ y_v^{(n)} \\ z_v^{(n)} \\ 1 \end{pmatrix} = \begin{pmatrix} 1 & 0 & 0 & x_v^{(n-1)} \\ 0 & 1 & 0 & y_v^{(n-1)} \\ 0 & 0 & 1 & z_v^{(n-1)} \\ 0 & 0 & 0 & 1 \end{pmatrix} \cdot M_v \cdot M_r \cdot \begin{pmatrix} \Delta x_r^{(n)} \\ \Delta y_r^{(n)} \\ \Delta z_r^{(n)} \\ 1 \end{pmatrix},$$

with homogeneous coordinates and 4×4 matrices M_r and M_v defining local coordinate transformations in real and virtual coordinates, respectively. An example of such transformations is discussed in the following section.

10.4.2 Scaling Self-Motions

The most often found redirection techniques are based on nonisometric mappings of user-centric translations or rotations to virtual camera motions. Such mappings can be described by *self-motion gains*, which define ratios between real and virtual self-motions. Two types of self-motion gains are distinguished in immersive virtual environments, i.e., *rotation gains* and *translation gains*.

Rotation Gains

Rotation gains define the ratio between physical head turns and virtual camera rotations [16, 32]. Assuming a relative change in the orientation of the user's head has been determined for frame $n \in \mathbb{N}$ as $(\Delta\tilde{y}_r^{(n)}, \Delta\tilde{p}_r^{(n)}, \Delta\tilde{r}_r^{(n)}) \in \mathbb{R}^3$, rotation gains $g_R = (g_{R[\tilde{y}]}, g_{R[\tilde{p}]}, g_{R[\tilde{r}]}) \in \mathbb{R}^3$ define the resulting virtual camera rotation, which changes to:

$$\begin{pmatrix} \tilde{y}_v^{(n)} \\ \tilde{p}_v^{(n)} \\ \tilde{r}_v^{(n)} \end{pmatrix} = \begin{pmatrix} \tilde{y}_v^{(n-1)} \\ \tilde{p}_v^{(n-1)} \\ \tilde{r}_v^{(n-1)} \end{pmatrix} + \begin{pmatrix} g_{R[\tilde{y}]} & 0 & 0 \\ 0 & g_{R[\tilde{p}]} & 0 \\ 0 & 0 & g_{R[\tilde{r}]} \end{pmatrix} \cdot \begin{pmatrix} \Delta\tilde{y}_r^{(n)} \\ \Delta\tilde{p}_r^{(n)} \\ \Delta\tilde{r}_r^{(n)} \end{pmatrix}$$

Most redirection techniques focus on scaling yaw rotations [13, 22, 23], for which an applied rotation gain $g_{R[\tilde{y}]} \in \mathbb{R}$ causes a tracked real-world head rotation $\Delta\tilde{y}_r^{(n)}$ to cause a virtual camera rotation of $g_{R[\tilde{y}]} \cdot \Delta\tilde{y}_r^{(n)}$, instead of $\Delta\tilde{y}_r^{(n)}$. This means that if $g_{R[\tilde{y}]} = 1$ the virtual scene remains stable considering a user's head orientation change. In case of $g_{R[\tilde{y}]} > 1$ the virtual scene appears to rotate against the direction of the head turn, whereas a gain $g_{R[\tilde{y}]} < 1$ causes the scene to rotate with the direction of the head turn [13]. For instance, if a user rotates the head by a yaw angle of 90°, a gain $g_{R[\tilde{y}]} = 1$ maps this motion one-to-one to a 90° rotation of the virtual camera in the VE. Applying a gain of $g_{R[\tilde{y}]} = 0.5$ results in the user having to rotate the head by 180° physically in order to achieve a 90° virtual rotation. A gain of $g_{R[\tilde{y}]} = 2$ results in the user having to rotate the head by only 45° physically in order to achieve a 90° virtual rotation.

In case such rotation gains cause differences between a user's head orientation in tracking coordinates, and a camera orientation in virtual scene coordinates, this requires us to adapt the direction of subsequent translational movements to the offset between the real and virtual head orientation. We can account for such offsets by introducing user-centric reference coordinates for translational movements in the real and virtual environment (see Sect. 10.4.1). For instance, in the example above, we can account for offsets between real and virtual yaw orientation angles by defining local coordinate transforms for position changes:

$$\begin{pmatrix} x_v^{(n)} \\ y_v^{(n)} \\ z_v^{(n)} \\ 1 \end{pmatrix} = \begin{pmatrix} 1\,0\,0\; x_v^{(n-1)} \\ 0\,1\,0\; y_v^{(n-1)} \\ 0\,0\,1\; z_v^{(n-1)} \\ 0\,0\,0\quad 1 \end{pmatrix} \cdot$$

$$\begin{pmatrix} cos(\tilde{y}_v^{(n-1)}) & 0 & sin(\tilde{y}_v^{(n-1)}) & 0 \\ 0 & 1 & 0 & 0 \\ -sin(\tilde{y}_v^{(n-1)}) & 0 & cos(\tilde{y}_v^{(n-1)}) & 0 \\ 0 & 0 & 0 & 1 \end{pmatrix} \cdot \begin{pmatrix} cos(-\tilde{y}_r^{(n-1)}) & 0 & sin(-\tilde{y}_r^{(n-1)}) & 0 \\ 0 & 1 & 0 & 0 \\ -sin(-\tilde{y}_r^{(n-1)}) & 0 & cos(-\tilde{y}_r^{(n-1)}) & 0 \\ 0 & 0 & 0 & 1 \end{pmatrix} \cdot$$

$$\begin{pmatrix} \Delta x_r^{(n)} \\ \Delta y_r^{(n)} \\ \Delta z_r^{(n)} \\ 1 \end{pmatrix},$$

in which head position changes in tracking coordinates are first transformed into a local coordinate system relative to the yaw orientation angle of the user's head in the previous rendering frame, and then transformed into the local coordinate system relative to the yaw orientation angle of the camera object in virtual coordinates at the previous rendering frame. Using this simple coordinate transformation, we can apply yaw rotation gains without changing the mapping of head translations relative to the user's head orientation. At this point it should be noted that similar transformations can be applied for pitch and roll transformations, e.g., to simulate virtual slopes [17]. However, since pitch and roll angles are usually applied sequentially relative to the virtual camera yaw angle (see Sect. 10.2), in most cases it is not necessary to introduce such coordinate transformations to account for applied pitch or roll gains (cf. [2]).

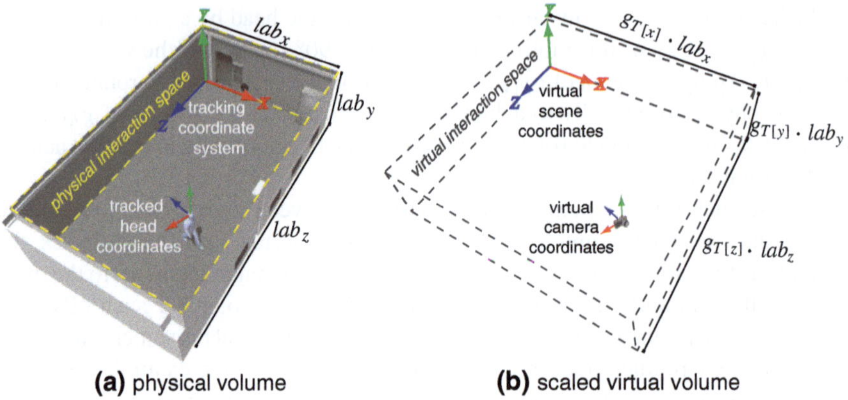

Fig. 10.2 Illustration of translation gains: (**a**) physical interaction volume in the virtual reality laboratory with size $(lab_x, lab_y, lab_z) \in \mathbb{R}^3$, and (**b**) virtual environment interaction volume with size $(g_{T[x]} \cdot lab_x, g_{T[y]} \cdot lab_y, g_{T[z]} \cdot lab_z) \in \mathbb{R}^3$ scaled by translation gains

Translation Gains

Translation gains define the ratio between real and virtual head translations [32]. Similar to rotation gains, scaled translations can be described with translation gains $g_T = (g_{T[x]}, g_{T[y]}, g_{T[z]}) \in \mathbb{R}^3$, which are applied to relative changes in the position of the user's head $(\Delta x_r^{(n)}, \Delta y_r^{(n)}, \Delta z_r^{(n)}) \in \mathbb{R}^3$ for frame $n \in \mathbb{N}$:

$$\begin{pmatrix} x_v^{(n)} \\ y_v^{(n)} \\ z_v^{(n)} \\ 1 \end{pmatrix} = \begin{pmatrix} 1 & 0 & 0 & x_v^{(n-1)} \\ 0 & 1 & 0 & y_v^{(n-1)} \\ 0 & 0 & 1 & z_v^{(n-1)} \\ 0 & 0 & 0 & 1 \end{pmatrix} \cdot \begin{pmatrix} g_{T[x]} & 0 & 0 & 0 \\ 0 & g_{T[y]} & 0 & 0 \\ 0 & 0 & g_{T[z]} & 0 \\ 0 & 0 & 0 & 1 \end{pmatrix} \cdot \begin{pmatrix} \Delta x_r^{(n)} \\ \Delta y_r^{(n)} \\ \Delta z_r^{(n)} \\ 1 \end{pmatrix}$$

For instance, uniform scalings in horizontal walking directions are often applied in immersive virtual environments allowing users to cover a larger distance in the VE when walking in the physical workspace [37], which can be described with translations gains $g_{T[x]} = g_{T[z]} > 1$, and $g_{T[y]} = 1$ (see Fig. 10.2). This causes a position change of the user's head in the real world $(\Delta x_r^{(n)}, \Delta y_r^{(n)}, \Delta z_r^{(n)}) \in \mathbb{R}^3$ to be transferred to the VE as $(g_{T[x]} \cdot \Delta x_r^{(n)}, \Delta y_r^{(n)}, g_{T[x]} \cdot \Delta z_r^{(n)})$, i.e., horizontal movements along the x- and z-axes are scaled uniformly, whereas vertical head bobbing movements along the y-axis are unaffected.

However, this approach still results in the problem that lateral head movements are scaled while a user walks, which can be distracting for the user [11]. Instead of scaling all horizontal motions with a translation gain, Interrante et al. [11] proposed scaling translations only in a user-specified walking direction (i.e., the *seven league boots* metaphor). Using a similar approach, Steinicke et al. [32] proposed using the yaw orientation of the user's head as approximation of walking direction [1] to scale translational movements. The latter approach can be implemented even without additional user instrumentation by changing the mapping to:

$$\begin{pmatrix} x_v^{(n)} \\ y_v^{(n)} \\ z_v^{(n)} \\ 1 \end{pmatrix} = \begin{pmatrix} 1 & 0 & 0 & x_v^{(n-1)} \\ 0 & 1 & 0 & y_v^{(n-1)} \\ 0 & 0 & 1 & z_v^{(n-1)} \\ 0 & 0 & 0 & 1 \end{pmatrix} \cdot \begin{pmatrix} cos(\tilde{y}_v^{(n-1)}) & 0 & sin(\tilde{y}_v^{(n-1)}) & 0 \\ 0 & 1 & 0 & 0 \\ -sin(\tilde{y}_v^{(n-1)}) & 0 & cos(\tilde{y}_v^{(n-1)}) & 0 \\ 0 & 0 & 0 & 1 \end{pmatrix} \cdot$$

$$\begin{pmatrix} g_{T[x]} & 0 & 0 & 0 \\ 0 & g_{T[y]} & 0 & 0 \\ 0 & 0 & g_{T[z]} & 0 \\ 0 & 0 & 0 & 1 \end{pmatrix} \cdot \begin{pmatrix} cos(-\tilde{y}_r^{(n-1)}) & 0 & sin(-\tilde{y}_r^{(n-1)}) & 0 \\ 0 & 1 & 0 & 0 \\ -sin(-\tilde{y}_r^{(n-1)}) & 0 & cos(-\tilde{y}_r^{(n-1)}) & 0 \\ 0 & 0 & 0 & 1 \end{pmatrix} \cdot \begin{pmatrix} \Delta x_r^{(n)} \\ \Delta y_r^{(n)} \\ \Delta z_r^{(n)} \\ 1 \end{pmatrix},$$

which allows to scale head position changes with separate gains relative to the user's locomotion state. In particular, walking distances in the virtual heading direction can be scaled with a gain $g_{T[z]} \in \mathbb{R}$, lateral distances can be scaled with a gain $g_{T[x]} \in \mathbb{R}$, and vertical distances can be scaled with a gain $g_{T[y]} \in \mathbb{R}$.

10.4.3 Redirected Walking

Although scaling self-motions as introduced in Sect. 10.4.2 can be used to redirect a user, e.g., by scaling head rotations to reorient the user away from an obstacle in the physical workspace, the approach has practical limitations. In particular, assuming the user walks straight ahead in the laboratory workspace without performing head rotations, then the virtual travel distance can be scaled relative to the physical walking distance, but at some point the user will eventually reach the end of the physical workspace, and potentially collide with an obstacle. To avoid this problem, researchers proposed various solutions [7, 14, 19, 20, 22, 23, 26, 32, 34, 38, 39], including techniques based on instructing the user to stop walking and start rotating the head, such that rotation gains can be applied to reorient the user away from physical obstacles [22, 38]. However, the most prominent solution for unrestricted walking was presented by Razzaque et al. [23], who proposed to use subtle virtual camera rotations while a user performs translational movements in the physical laboratory workspace. This causes the user to change the heading direction when walking in the real world according to the rotations in the virtual environment. The approach can be implemented with *curvature gains*.

Curvature Gains

Curvature gains define ratios between position changes of the user's head in the real world and virtual camera rotations [32]. For example, when the user walks straight ahead in the physical workspace, a curvature gain that causes reasonably small iterative camera yaw rotations to one side forces the user to walk along a curved path in the opposite direction in the real world in order to stay on a straight path in the virtual world. If the injected manipulations are reasonably small, the user will compensate for the virtual camera rotations without being able to consciously detect the manipulations. Curvature gains $g_C \in \mathbb{R}_0^+$ denote the resulting bending of a user's path in the physical workspace, which is determined as $g_C = 1/r$ for a circular arc with radius $r \in \mathbb{R}^+$. In case no curvature is applied, i.e., $r = \infty$, this corresponds to a curvature gain $g_C = 0$. If an applied curvature gain causes the user to rotate by 90 degrees after $\frac{\Pi}{2}$m walking distance, then the user has covered a quarter circle with radius $r = 1$, which corresponds to a curvature gain $g_C = 1$.

Curvature mappings can be described using the following pseudo code:

Algorithm 4 Curvature mapping from tracked head to camera coordinates

for all rendering frames $n \in \mathbb{N}$ **do**

// Get current head tracking state:

$(x_r^{(n)}, y_r^{(n)}, z_r^{(n)}) \leftarrow$ tracked head position (in $\mathbb{R}^3$)

$(\tilde{y}_r^{(n)}, \tilde{p}_r^{(n)}, \tilde{r}_r^{(n)}) \leftarrow$ tracked head orientation (in $[0, 360)^3$)

// Compute relative changes:

$(\Delta x_r^{(n)}, \Delta y_r^{(n)}, \Delta z_r^{(n)}) \leftarrow$ head position change (in $\mathbb{R}^3$)

$(\Delta \tilde{y}_r^{(n)}, \Delta \tilde{p}_r^{(n)}, \Delta \tilde{r}_r^{(n)}) \leftarrow$ head orientation change (in $[-180, 180)^3$)

// Compute changes relative to curvature:

$(\Delta d_r^{(n)}, \Delta s_r^{(n)}) \leftarrow$ straight and strafe motion (in $\mathbb{R}^2$)

$\Delta \alpha_{\tilde{y}_r}^{(n)} \leftarrow$ arc angle (in $\mathbb{R}$)

// Set virtual camera state:

$$\begin{pmatrix} x_v^{(n)} \\ y_v^{(n)} \\ z_v^{(n)} \end{pmatrix} \leftarrow \begin{pmatrix} x_v^{(n-1)} \\ y_v^{(n-1)} \\ z_v^{(n-1)} \end{pmatrix} + \begin{pmatrix} v_{v_x}^{(n-1)} \cdot \Delta d_r^{(n)} + v_{v_z}^{(n-1)} \cdot \Delta s_r^{(n)} \\ \Delta y_r^{(n)} \\ v_{v_z}^{(n-1)} \cdot \Delta d_r^{(n)} - v_{v_x}^{(n-1)} \cdot \Delta s_r^{(n)} \end{pmatrix} \quad \textit{// position}$$

$$\begin{pmatrix} \tilde{y}_v^{(n)} \\ \tilde{p}_v^{(n)} \\ \tilde{r}_v^{(n)} \end{pmatrix} \leftarrow \begin{pmatrix} \tilde{y}_v^{(n-1)} \\ \tilde{p}_v^{(n-1)} \\ \tilde{r}_v^{(n-1)} \end{pmatrix} + \begin{pmatrix} \Delta \tilde{y}_r^{(n)} - \Delta \alpha_{\tilde{y}_r}^{(n)} \\ \Delta \tilde{p}_r^{(n)} \\ \Delta \tilde{r}_r^{(n)} \end{pmatrix} \quad \textit{// orientation}$$

end for

In the pseudo code, $\Delta d_r^{(n)} \in \mathbb{R}$ denotes the arc length of the traveled circular path along the two-dimensional ground plane in the physical workspace, and $\Delta s_r^{(n)} \in \mathbb{R}$ the strafe distance relative to the center of the circular path, with $\Delta \alpha_{\tilde{y}_r}^{(n)} \in \mathbb{R}$ the corresponding arc angle as shown in Fig. 10.3a. Figure 10.3b illustrates mapping of the arc length to a straight motion in the virtual environment, whereas if the user strays from the circular path in the physical workspace, this is mapped to a strafe motion in the VE. In this example, user movements are mapped relative to the user's real and virtual two-dimensional view direction, denoted as view direction along the xz-plane $(v_{r_x}^{(n)}, v_{r_z}^{(n)}) \in \mathbb{R}^2$ with $\| (v_{r_x}^{(n)}, v_{r_z}^{(n)}) \| = 1$ in the physical workspace shown

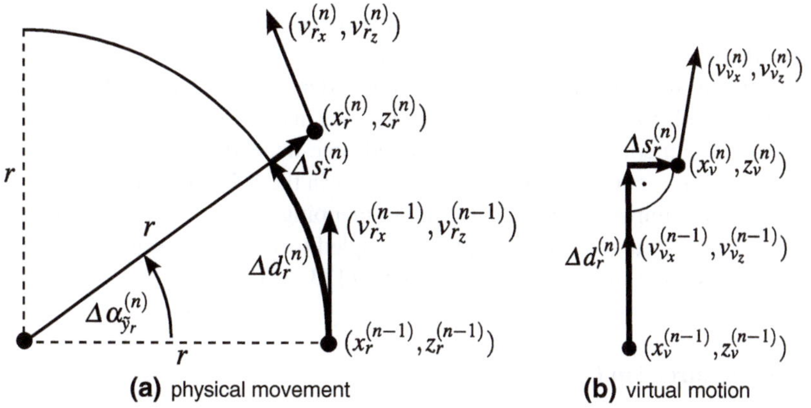

Fig. 10.3 Illustration of two-dimensional mappings in the xz-plane from (**a**) tracking coordinates to (**b**) virtual coordinates for an applied curvature gain $g_C = 1/r$ with radius $r \in \mathbb{R}^+$

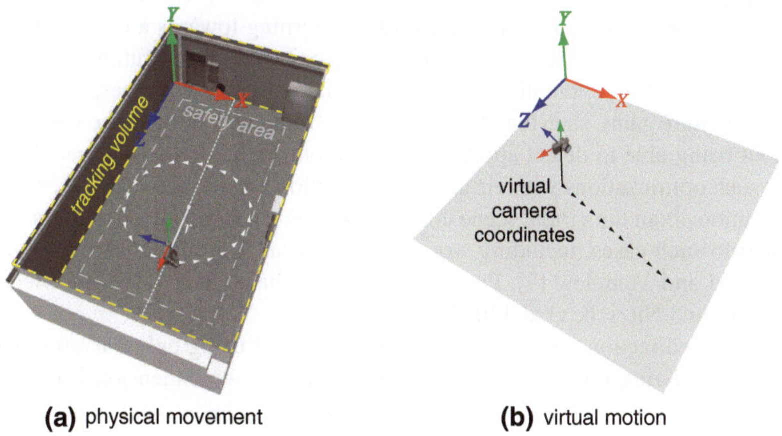

Fig. 10.4 Illustration of (**a**) path redirection with curvature gains in the physical workspace, for (**b**) a predicted virtual straight path

in Fig. 10.3a, as well as $(v_{v_x}^{(n)}, v_{v_z}^{(n)}) \in \mathbb{R}^2$ with $\| (v_{v_x}^{(n)}, v_{v_z}^{(n)}) \| = 1$ in the virtual workspace shown in Fig. 10.3b. Other implementations may use arbitrarily placed curvatures in the physical workspace, e.g., based on real and virtual path planning and transformations [20, 26, 34]. Figure 10.4 shows an example of a predicted straight motion in the VE being mapped to a circular path in the physical workspace.

It is important to note that curvature transformations are based on the assumption that the user will adapt to induced virtual rotations by changing the walking direction in the physical workspace. In particular, if the manipulations are overt, the user has to consciously follow the induced virtual rotations. If the user does not adapt to an induced rotation in the virtual environment, e.g., if the user is walking with eyes closed, or does not have a target in the virtual scene, it is possible that the user may stray off the path planned with curvature gains.

In general, such curvature gains can be applied not only to yaw rotations, but also to pitch and roll rotations, e.g., to simulate slopes in a virtual scene [17]. Moreover, such virtual camera rotations can be applied time-dependently, i.e., not caused by translational or rotational movements of the user in the VR laboratory, which can be described as a simple extension of the above mapping. However, anecdotal evidence suggests that virtual rotations that are not coupled to self-motions are usually easily detectable by users, and potentially distracting [23, 32].

A Basic Redirection Controller

Sophisticated implementations of unrestricted virtual walking with redirection techniques, i.e., *redirection controllers*, are usually based on information about the extents of the physical workspace, the structure of the virtual scene, and assumptions about typical user behavior. For instance, if a user is turning towards a door in a virtual building model, redirection controllers may predict the user's future virtual path to determine how to optimally scale rotations and compress distances, as well as to apply curvature gains, such that the user will be able to walk through the virtual door, without being able to detect applied manipulations [5, 20, 34]. However, in many cases such optimizations with virtual path prediction are not possible, e.g., when no information about the virtual scene is available. Some redirection controllers can be adapted to such cases, including works by the research groups of Razzaque et al. [23], Field and Vamplew [7], Peck et al. [22], Williams et al. [38, 39], Steinicke et al. [34], and Nitzsche et al. [20, 26].

A basic redirection controller can be implemented using only curvature gains. For each rendering frame $n \in \mathbb{N}$ we read the current two-dimensional head position $(x_r^{(n)}, z_r^{(n)}) \in \mathbb{R}^2$, and compute the current two-dimensional view direction $(v_{r_x}^{(n)}, v_{r_z}^{(n)}) \in \mathbb{R}^2$ with $\| (v_{r_x}^{(n)}, v_{r_z}^{(n)}) \| = 1$ in the physical workspace (see Fig. 10.3). Based on the prediction that the user will walk in the virtual view direction [1], we try to map the user's real movements onto a circular path in the physical workspace with largest possible radius, in order to minimize applied curvature manipulations. We accomplish that by computing the strafe view direction $(v_{r_z}^{(n)}, -v_{r_x}^{(n)}) \in \mathbb{R}^2$ in the physical workspace, and solving the optimization problem of finding the point $(x_r^{(n)}, z_r^{(n)}) - r \cdot (v_{r_z}^{(n)}, -v_{r_x}^{(n)})$, with $r \in \mathbb{R}$ that is located within the physical workspace and provides the largest circle through the current user position $(x_r^{(n)}, z_r^{(n)})$, while maintaining at least the same distance to all boundaries of the interaction space, and all obstacles in the laboratory (including a small safety offset, see Fig. 10.4). Mapping user movements onto this computed maximal circle in the physical workspace corresponds to applying a curvature gain of $g_C = 1/r$, using the formulas described above. That means, for each frame the user is redirected onto the optimal circle in the physical workspace, assuming the user will walk straight in the computed view direction. This simple approach allows practitioners to implement a reasonable mapping, which enables users to explore infinite virtual scenes by real walking, and does not require information about the virtual scene. If more information about the user's movements is available, the prediction based on the view direction can be replaced by more sophisticated strategies.

10.5 Conclusion

In this chapter we have described the basic math required to set up real walking user interfaces in immersive VEs. We have shown how isometric and nonisometric transformations can be used to map user movements from a physical workspace to a virtual scene. While isometric transformations provide natural feedback to physical user movements, they limit the virtual space a user can explore by real walking to the size of the tracked physical workspace. We have described how this limitation can be alleviated by combining walking in a limited interaction volume with other traveling techniques (e.g., flying). With nonisometric angular, linear, and curvature transformations we have described how the limitations of interaction space can be broken to support unlimited omnidirectional walking, although this freedom is bought with less natural feedback to physical user movements.

Practitioners interested in implementing real walking user interfaces may follow these rough guidelines:

- If the virtual interaction space is smaller or equal to the tracked physical workspace, isometric transformations should be used, since these will provide optimal self-motion feedback.
- If the virtual places of interest are rather small, but considerably spaced apart in the virtual scene, isometric mappings should be combined with traveling techniques based on additional devices or sensors.
- If the virtual scene consists of one large area of interest that could be explored by walking, then redirected walking with nonisometric mappings is recommended.

As explained above when using nonisometric mappings, the virtual view moves in a different way than the user's head in the tracked physical environment. One interesting question is how much deviation between these motions is tolerated by the user. Recently, several experiments have been reported which have identified detection thresholds for these nonisometric mappings. Interested readers may refer to works by Steinicke et al. [31, 32], Neth et al. [19] and Engel et al. [6].

In summary, movements of a user in immersive VEs have to be transferred to a virtual scene to provide the user with virtual feedback about self-motions, which can be a faithful simulation of real-world movements, or manipulated using different approaches. Since each of the approaches has different advantages and limitations, it depends on the structure of the virtual scene and the application as to which approach is best suited. In the next chapter, these approaches are discussed in more detail.

Acknowledgments The authors of this work are supported by the German Research Foundation (DFG 29160962).

References

1. Berthoz A (2000) The Brain's sense of movement. Harvard University Press, Cambridge
2. Bolte B, Bruder G, Steinicke F, Hinrichs KH, Lappe M (2010) Augmentation techniques for efficient exploration in head-mounted display environments. In: Proceedings of the symposium on virtual reality software and technology (VRST). ACM Press, pp 11–18
3. Bowman D, Koller D, Hodges L (1997) Travel in immersive virtual environments: an evaluation of viewpoint motion control techniques. In: Proceedings of the virtual reality annual international symposium (VRAIS). IEEE Press, pp 45–52
4. Bowman D, Kruijff E, LaViola J, Poupyrev I (2004) 3D user interfaces: theory and practice. Addison-Wesley, Boston
5. Bruder G, Steinicke F, Hinrichs KH (2009) Arch-explore: a natural user interface for immersive architectural walkthroughs. In: Proceedings of the symposium on 3D user interfaces (3DUI). IEEE Press, pp 75–82
6. Engel D, Curio C, Tcheang L, Mohler B, Bülthoff HH (2008) A psychophysically calibrated controller for navigating through large environments in a limited free-walking space. In: Proceedings of the symposium on virtual reality software and technology (VRST). ACM Press, pp 157–164
7. Field T, Vamplew P (2004) Generalised algorithms for redirected walking in virtual environments. In: Proceedings of the international conference on artificial intelligence in science and technology (AISAT), pp 58–63
8. Foley J, Van Dam A (1982) Fundamentals of interactive computer graphics. Addison-Wesley, Reading
9. Hollerbach J (2002) Handbook of virtual environments: design, implementation, and applications, chapter locomotion interfaces. Lawrence Erlbaum Associates, pp 239–254
10. Insko B, Meehan M, Whitton MC, Brooks F (2001) Passive haptics significantly hances virtual environments. In: Proceedings of the annual presence workshop
11. Interrante V, Ries B, Anderson L (2007) Seven league boots: a new metaphor for augmented locomotion through moderately large scale immersive virtual environments. In: Proceedings of the symposium on 3D user interfaces (3DUI). IEEE Press, pp 167–170
12. Interrante V, Ries B, Lindquist J, Anderson L (2007) Elucidating the factors that can facilitate veridical spatial perception in immersive virtual environments. In: Proceedings of virtual reality (VR). IEEE Press, pp 11–18
13. Jerald J, Peck TC, Steinicke F, Whitton MC (2008) Sensitivity to scene motion for phases of head yaws. In: Proceedings of the symposium on applied perception in graphics and visualization (APGV). ACM Press, pp 155–162
14. Kohli L, Burns E, Miller D, Fuchs H (2005) Combining passive haptics with redirected walking. In: Proceedings of the international conference on augmented telexistence. ACM Press, pp 253–254
15. Lappe M, Bremmer F, van den Berg A (1999) Perception of self-motion from visual flow. Trends in Cognitive Sciences 3(9):329–336
16. LaViola J Jr, Katzourin M (2007) An exploration of non-isomorphic 3d rotation in surround screen virtual environments. In: Proceedings of the symposium on 3D user interfaces (3DUI). IEEE Press, pp 49–54
17. Marchal M, Lecuyer A, Cirio G, Bonnet L, Emily M (2010) Walking up and down in immersive virtual worlds: novel interactive techniques based on visual feedback. In: Proceedings of the symposium on 3D user interfaces (3DUI). IEEE Press, pp 19–26

18. Mine M (1995) Virtual environments interaction technqiues. Technical report computer science TR95-018, University of North Carolina at Chapel Hill
19. Neth CT, Souman JL, Engel D, Kloos U, Bülthoff HH, Mohler BJ (2011) Velocity-dependent dynamic curvature gain for redirected walking. In: Proceedings of virtual reality (VR). IEEE Press, pp 1–8
20. Nitzsche N, Hanebeck U, Schmidt G (2002) Extending telepresent walking by motion compression. In: Proceedings of human centered robotic systems (HCRS), pp 83–90
21. Parent R (2007) Computer animation: algorithms and techniques. Morgan Kaufmann, San Francisco
22. Peck TC, Fuchs H, Whitton MC (2010) Improved redirection with distractors: a large-scale-real-walking locomotion interface and its effect on navigation in virtual environments. In: Proceedings of virtual reality (VR). IEEE Press, pp 35–38
23. Razzaque S (2005) Redirected walking. PhD thesis, University of North Carolina at Chapel Hill
24. Robinett W, Holloway R (1992) Implementation of flying, scaling and grabbing in virtual worlds. In: Proceedings of the symposium on interactive 3D graphics (i3D). ACM Press, pp 189–192
25. Robinett W, Holloway R (1994) The visual display transformation for virtual reality. Technical report computer science TR94-031, University of North Carolina at Chapel Hill
26. Rößler P, Beutler F, Hanebeck UD (2005) A framework for telepresent game-play in large virtual environments. In: Proceedings of the international conference on informatics in control, automation and robotics (ICINCO), pp 150–155
27. Ruddle RA, Lessels S (2009) The benefits of using a walking interface to navigate virtual environments. ACM Trans Comput Human Interact (TOCHI) 16:1–18
28. Shreiner D (2009) OpenGL programming guide: the official guide to learning OpenGL, versions 3.0 and 3.1, 7th edn. Addison-Wesley, Reading
29. Slater M (2009) Place illusion and plausibility can lead to realistic behaviour in immersive virtual environments. Philos Trans Royal Soc B Biol Sci 364(1535):3549–3557
30. Speigle JM, Loomis JM (1993) Auditory distance perception by translating observers. In: Proceedings of the symposium on research frontiers in virtual reality, IEEE Press
31. Steinicke F, Bruder G, Jerald J, Frenz H, Lappe M (2008) Analyses of human sensitivity to redirected walking. In: Proceedings of the symposium on virtual reality software and technology (VRST). ACM Press, pp 149–156
32. Steinicke F, Bruder G, Jerald J, Frenz H, Lappe M (2010) Estimation of detection thresholds for redirected walking techniques. IEEE Trans Vis Comput Graph (TVCG) 16(1):17–27
33. Steinicke F, Bruder G, Ries B, Hinrichs KH, Lappe M, Interrante V (2009) Transitional environments enhance distance perception in immersive virtual reality systems. In: Proceedings of the symposium on applied perception in graphics and visualization (APGV). ACM Press, pp 19–26
34. Steinicke F, Weltzel H, Bruder G, Hinrichs KH (2008) A user guidance approach for passive haptic environments. In: Short paper and poster proceedings of the eurographics symposium on virtual environments (EGVE), pp 31–34
35. Taylor II RM, Hudson TC, Seeger A, Weber H, Juliano J, Helser AT (2001) VRPN: a device-independent, network-transparent VR peripheral system. In: Proceedings of the symposium on virtual reality software and technology (VRST). ACM Press, pp 55–61
36. Usoh M, Arthur K, Whitton MC, Bastos R, Steed A, Slater M, Brooks F (1999) Walking > walking-in-place > flying, in virtual environments. In: Proceedings of SIGGRAPH. ACM Press, pp 359–364
37. Williams B, Narasimham G, McNamara TP, Carr TH, Rieser JJ, Bodenheimer B (2006) Updating orientation in large virtual environments using scaled translational gain. In: Proceedings of the symposium on applied perception in graphics and visualization (APGV). ACM Press, pp 21–28
38. Williams B, Narasimham G, Rump B, McNamara TP, Carr TH , Rieser J, Bodenheimer B (2007) Exploring large virtual environments with an HMD when physical space is limited. In:

Proceedings of the symposium on applied perception in graphics and visualization (APGV). ACM Press, pp 41–48

39. Xie X, Lin Q, Wu H, Narasimham G, McNamara TP, Rieser J, Bodenheimer B (2010) A system for exploring large virtual environments that combines scaled translational gain and interventions. In: Proceedings of the symposium on applied perception in graphics and visualization (APGV). ACM Press, pp 65–72
40. Zachmann G (1996) A language for describing behavior of and interaction with virtual worlds. In: Proceedings of the symposium on virtual reality software and technology (VRST). ACM Press, pp 143–150

Chapter 11
Stepping-Driven Locomotion Interfaces

Mary C. Whitton and Tabitha C. Peck

Abstract Walking-in-place and real-walking locomotion interfaces for virtual environment systems are interfaces that are driven by the user's actual stepping motions and do not include treadmills or other mechanical devices. While both walking-in-place and real-walking interfaces compute the user's speed and direction and convert those values into viewpoint movement between frames, they differ in how they enable the user to move to any distant location in very large virtual scenes. Walking-in-place constrains the user's actual movement to a small area and translates stepping-in-place motions into viewpoint movement. Real-walking applies one of several techniques to transform the virtual scene so that the user's physical path stays within the available laboratory space. This chapter discusses implementations of these two types of interfaces with particular regard to how walking-in-place interfaces generate smooth motion and how real-walking interfaces modify the user's view of the scene so deviations from her real motion are less detectable.

11.1 Designing Stepping-Driven Locomotion for Virtual Environment Systems

Arguably, the locomotion interfaces for Immersive Virtual Environment (IVE) systems that are most natural are those that employ a stepping metaphor, i.e., they require that users repeatedly move their feet up and down, just as if walking in the real world.

M. C. Whitton (✉)
Department of Computer Science, The University of North Carolina at Chapel Hill, Chapel Hill, NC, USA
e-mail: whitton@cs.unc.edu

T. C. Peck
Event Lab, Faculty of Psychology, University of Barcelona, Barcelona, Spain
e-mail: tabitha.peck@gmail.com

F. Steinicke et al. (eds.), *Human Walking in Virtual Environments*,
DOI: 10.1007/978-1-4419-8432-6_11, © Springer Science+Business Media New York 2013

Such interfaces give users a locomotion experience that is close to natural walking in the real world. Chapter 9 of this volume, Technologies of Locomotion Interface, describes mechanically-assisted walking interfaces such as treadmills and cycles. This chapter is about stepping-driven interfaces that are not mechanically assisted.

In *walking-in-place* (WIP) *interfaces*, users make stepping motions but do not physically move forward. Sensor data, captured from the user's in-place stepping motions and other sensors, are used to control the movement of the user's viewpoint through the virtual scene. The primary technical challenge in WIP systems is *controlling the user's speed* so that it is both responsive and smooth; direction can be set with any of a number of techniques. Using the taxonomy in Bowman et al. [4], WIP is a hybrid interface: physical because the user makes repeated movements, and virtual because the user does not move through physical space.

In *real-walking interfaces*, a purely physical interface in Bowman et al.'s taxonomy, users really walk to move through the virtual scene and the physical (lab) environment. The easy case is when the virtual scene fits within the lab: There is a one-to-one mapping between the change in the user's tracker-reported pose (position and orientation) and the change in viewpoint for each frame. Speed and direction are controlled by how fast and in what direction the user moves. This is just as in *natural walking*.

The more difficult real-walking case is when the virtual scene is larger than the lab: The mapping between changes in tracker-reported pose and changes in viewpoint can no longer be one-to-one if the user is to travel to areas in the virtual scene that lie outside the confines of the lab. Thus, the primary technical challenge in real-walking interfaces for large scenes is *modifying the transform applied to the viewpoint (or scene)* so that the user changes her real, physical direction in a way that keeps her path through the virtual scene within the physical lab space. Recent locomotion taxonomies have added categories for new real-walking techniques: Arns' taxonomy includes interfaces using *scaled rotation* and/or *scaled translation* [1] and Wendt's taxonomy includes interfaces that *recenter* users via *redirection* techniques [40].

In this chapter we discuss only stepping-driven locomotion interfaces for virtual scenes that are larger than the lab's tracked space. The locomotion interface techniques reported here were developed for IVE systems that use tracked head-mounted display devices (HMDs). With some adaptation, walking-in-place can be used in single- or multi-wall projection display systems. Redirected-walking, one of the techniques for real-walking in large scenes, has also been employed in multi-wall display systems [28]. The interfaces described here do not require stereo-viewing.

Research has shown that locomotion interfaces that require the user to make stepping movements induce a higher sense of presence, are more natural, and enable better user navigation than other interfaces [22, 35]. These benefits make stepping-driven interfaces a worthy subject of study. We conclude this introduction with general goals for locomotion interfaces in IVEs and specific goals for setting locomotion speed and direction. We then discuss walking-in-place and real-walking virtual locomotion interfaces in depth.

General goals for locomotion interfaces. To be widely adopted, a locomotion interface for IVEs has more requirements than simply enabling movement from place to place. Other desirable features of locomotion interfaces include:

- Is easy to learn and easy to use; incurs low cognitive load;
- Leaves the user's hands free so she can use task-related tools;
- Does not increase occurrence or severity of simulator sickness;
- Prevents users from running into real-world obstructions and walls;
- Minimizes encumbrances
  - Is easy to don and doff;
  - Ensures that equipment, including safety equipment, does not interfere with other task-related gear the user may be wearing;
- Minimizes required supporting infrastructure, e.g., tracking systems, for portability and cost.

Goals for setting speed. The notional speed versus time profile (Fig. 11.1a) is a standard against which to compare similar speed/time profiles for our interfaces. Figure 11.1b shows an actual profile generated from (noisy) head-tracker data. The same development, rhythmic, and decay phases are visible in both profiles. We propose four design goals for setting user speed:

- Starting and stopping latency should be minimized. Movement in the virtual scene should begin as soon as the user initiates a step and stop when the user stops stepping. Starting latency is annoying for casual walking and interferes with the timing of quick movements. Stopping latency can result in overshooting the desired stopping location leading to unintended collisions with or interpenetration of objects in the scene.
- Users should be able to adjust their speed continually during a step, as we can with natural walking. If speed is controlled by data measured only once per step, e.g., foot-strike or foot-off speed, continuous control of speed is not possible.

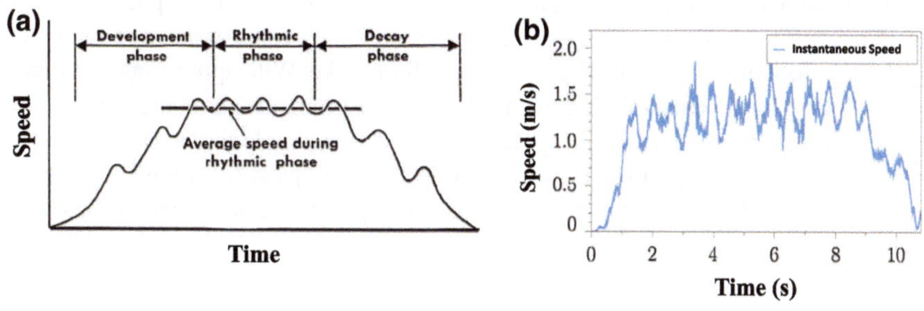

Fig. 11.1 **a** Notional speed versus time plot for one start-to-stop walking event. **b** Speed versus time plot computed from head-tracker data. The higher frequency variations in speed are caused by fore-aft head bob and roughly correspond to steps. Notice that during the rhythmic phase, speed stays well above zero. We try to replicate the general shape of the speed profile with our interfaces. (**a** Adapted from Inman [14]; **b** reproduced from Wendt [40])

- Virtual walking speed should stay relatively constant during the rhythmic phase to avoid detectable variations in optic flow—the change in patterns of light on the retina occurring during movement.
- The system should allow fine positioning or maneuvering steps that do not initiate a full step's movement.

Goals for setting direction. The goals for direction setting are to make it as easy as natural walking and to avoid introducing sensory conflict.

- Users should be able to move in any direction—forward, backward, sideways, or at any angle.
- As in natural walking, the direction of movement should be independent of user's view direction and body orientation. Reinforcing the results reported in Bowman et al. [3], the description of the *Pointman* interface includes a cogent argument for independence of these parameters for tactical movements [37].
- Direction setting should be hands-free, as it is in natural walking, so the hands can be used for application-specific interactions with the environment.

11.2 Walking-in-Place Interfaces

Walking-in-place (WIP) is a locomotion interface technique for Immersive Virtual Environment systems that uses data describing the stepping-in-place gesture to control locomotion speed and uses any one of a number of techniques or input devices to set locomotion direction.

11.2.1 Setting Speed: Interpreting Stepping Gestures

Repeated stepping gestures have several distinct, observable, and measurable phases. Starting from the eight-phase human gait cycle, [41] proposed the six-phase walking-in-place gait cycle shown in Fig. 11.2. There are three events associated with each leg's step: foot off, maximum step height, and foot strike. With appropriate sensors, it is possible to detect each of these events, make measurements about them, and apply time stamps to them. The resulting data are what is available to determine whether the user is moving, and, if she is moving, how fast. The question of whether the user is moving includes both whether the user is starting to move and whether the user is stopping.

11.2.1.1 Detecting Foot-Strike Events

The earliest WIP interfaces computed forward motion based on indirect or direct detection of foot-strikes: each time a foot strike was detected, the user's viewpoint was moved forward by some amount. The faster the foot strikes occurred, the faster the user moved through the virtual scene.

A very early walking-in-place system, called a virtual treadmill, applied a neural network to head tracker data to detect local maxima in stepping-related vertical head-bob [35]. A set amount of forward movement, inserted over several frames, was added between detected steps. The neural network required four positive "step" signals before initiating movement and two "no step" signals before stopping. Starting latency was about two seconds; stopping, about one second.

Other methods of foot-strike detection include pressure sensors in shoes [36], a floor-based array of pressure sensors [2], and head-worn accelerometers [44]. Unlike the first two methods which produce a binary variable when a step is detected, the latter technique generates a stream of accelerometer data in which foot-strikes are detected as local maxima.

Starting latency is a problem for foot-strike techniques: a step is not recognized until the foot has been lifted and returned to the ground. For a casual walking speed of 3 mph and a 24" step length, this latency is around half a second.

Movement can be implemented by choosing a moderate base speed and computing the distance the viewpoint must be moved for each foot strike to achieve that speed through the scene. That incremental distance is added to the viewpoint pose over one or more frames. Stepping faster or slower changes speed, but it is not possible to adjust speed between foot-strikes. Maneuvering is not possible unless the algorithm includes a sensor-signal threshold so that it ignores small foot movements or light floor strikes.

Moving the user forward a set distance for each foot-strike generally does not lead to a relatively constant speed for rhythmic-phase walking even if the total distance to be moved is spread over several frames. In an exaggerated fashion, Fig. 11.3 shows a speed profile for distance (**a**) added uniformly over several frames and (**b**) added in a sawtooth pattern in order to avoid multi-frame pauses in the optic flow occurring when speed goes to zero or near zero between steps. Comparing these profiles to Fig. 11.1 reveals that neither waveform is a good approximation of natural walking. Overcoming the limitations of discrete-step based interfaces—latency, speed variations during rhythmic-phase, inability to maneuver and adjust steed—requires additional data about the user's stepping motion.

11.2.1.2 Continuously Measuring Leg Position

The addition of trackers to the front or back of the user's legs (or knees, shins, ankles, or feet) provides a continuous stream of time-stamped tracker data from which the six events in the walking-in-place cycle can be detected: motion of one leg begins at foot-off, motion reverses direction when the tracker reaches its maximum extent, and motion of that leg stops at foot-strike; then similarly for the other leg. Leg speed can be computed from the tracker data.

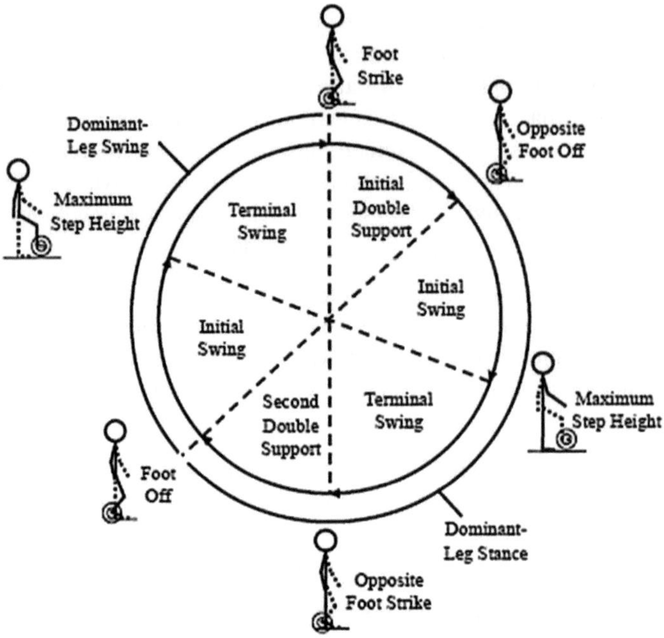

Fig. 11.2 Six-phase walking-in-place gait cycle. Stepping gestures can be quantified by detecting the events, measuring how high the legs are raised, and noting the timing between the events. Note that (1) the phase when both feet are on the floor is called *double support*, and (2) stepping frequency can be calculated from the time stamps of *any* three successive events. (Reproduced from Wendt [40])

Gaiter is a WIP system enabling locomotion in a virtual scene of unlimited size with some limited real-space maneuvering [36]. Knee excursion in the horizontal plane, measured by shin-worn trackers, differentiates virtual and real steps. In a virtual step, i.e., stepping-in-place, the knee moves out (and up) and back again; in a real step the knee moves out and stays out as the user takes the real step. Startup latency is half a step since the system cannot tell if the step is real or virtual until the knee has reached its maximum extent and either stopped or begun to travel back.

Yan et al. designed a system that set locomotion speed based on leg speed during the period of high leg acceleration occurring just after foot-off [45]. Using results from the biomechanics literature and experimentally developed relationships among leg-lift speed, step frequency, and forward velocity for natural walking and for stepping-in-place, the team developed a user-specific linear function relating the stepping-in-place leg-lift speed and forward velocity. Speed was set once per step using this function. Motion did not begin until a leg-lift speed threshold was exceeded, resulting in a starting latency of approximately one-quarter of a step. The threshold prevented false steps and allowed (slow) maneuvering steps. Per-step movement was spread across frames and a Kalman filter was used to smooth forward movement between leg-lifts.

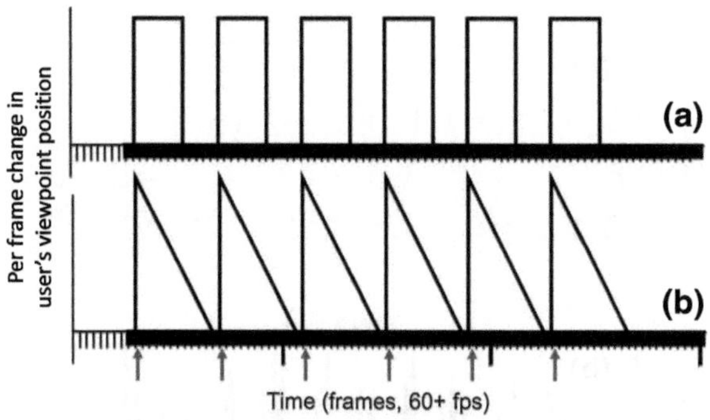

Fig. 11.3 Exaggerated examples of foot-strike driven speed profiles. Time is measured in frames; small arrows below the X-axis indicate foot-strikes every 400 ms (~3.6 mph). **a** User's viewpoint is moved a set distance for each foot-strike and that distance is added uniformly over several frames. **b** To avoid pauses in the optic flow while speed is zero, the set distance is added non-uniformly over several frames. The saw-tooth shape minimizes latency at the start of motion for each step. (After a figure in Feasel et al. [9])

11.2.1.3 Techniques to Smooth Speed Between Foot Strikes

Low-Latency Continuous Motion WIP (LLCM-WIP). LLCM-WIP was developed to reduce starting and stopping latency and to smooth speed during rhythmic-phase walking [9]. LLCM-WIP uses trackers placed just below the user's knees. From the tracker data it finds the location of the user's heel via a rigid body transform and calculates the speed of the user's heel in the vertical axis from that data. LLCM-WIP supports maneuvering by requiring that a heel-speed threshold be exceeded before a full step forward is taken. After some signal processing, vertical heel speeds above the threshold are mapped to locomotion speed. The locomotion speed signal is noisy and dips close to zero during the double support phases of gait. At the cost of approximately 100 ms of latency, filtering smoothes the output speed and reduces, but does not eliminate, those speed dips (Fig. 11.4). Because virtual speed is mapped continuously from heel speed, speed can be changed at any time by speeding or slowing stepping movements.

Gait-Understanding-Driven WIP (GUD-WIP). GUD-WIP addresses the problem of speed variation during rhythmic walking with a technique that updates speed six times during each two-step WIP gait cycle using a quadratic function reported in the biomechanics literature that relates stepping-frequency and speed. Figure 11.5 shows the GUD-WIP system in use.

The timing of events in the WIP gait cycle (Fig. 11.2) is discoverable from time-stamped logs of tracking data from the user's shins. The events occur when tracker position starts changing (foot off), stops changing (foot strike), changes direction (reaching maximum step height). Stepping frequency is computed from the time stamps of the three most recent WIP-cycle events. After startup, step frequency can be (re)computed six times in each two-step cycle. Startup requires three gait events,

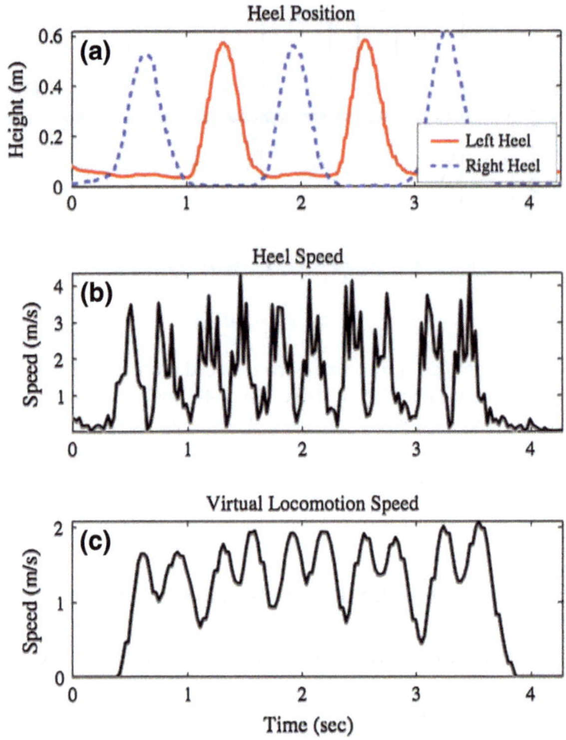

Fig. 11.4 LLCM-WIP system overview. **a** Vertical position of the user's heels; **b** Heel speed obtained through differentiation; **c** Virtual locomotion speed. (Reproduced from Feasel et al. [9])

a latency of one step. The GUD-WIP algorithm consciously traded longer stopping latency (~500 ms) for smoother inter-step motion.

While Yan et al. used a linear relationship between step frequency and speed, the biomechanics literature reports a quadratic relationship between these two values. Wendt used the formula reported by Dean [7] to compute virtual speed six times per 2-step cycle [41]. Figure 11.6 shows Dean's equation, a graph of its curve, and step-frequency to speed data points from other published works. The formula is partially customized with user height, (h).

Figure 11.7 shows LLCM-WIP and GUD-WIP speed profiles computed from the tracker log of the same five-step sequence from the rhythmic phase of a start-to-stop walking event. Note that unlike LLCM-WIP, GUD-WIP speed (and hence optic flow) does not approach zero during double support; however, there are discontinuities when speed is updated (3 times/step). We do not yet know if these discontinuities have perceptual or task-performance consequences.

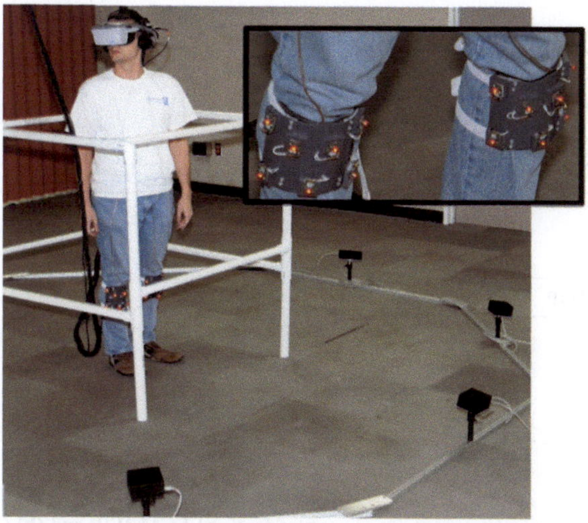

Fig. 11.5 GUD-WIP in use. The user is free to maneuver and rotate within the PVC pipe "cage" that constrains his motion and protects the cameras. The beacons on the user's shins (inset) are sensed by the eight PhaseSpace cameras arrayed around the user. Data from knee tracking is used for both speed and the direction setting. The HMD and head-tracker have wired connections; the PhaseSpace system is wireless. (Reproduced from Wendt [40])

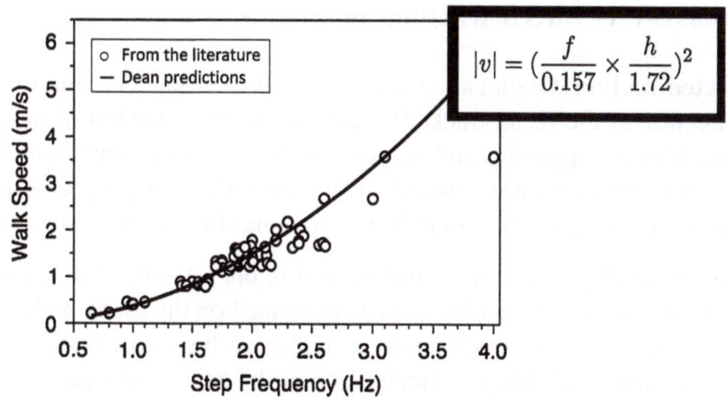

Fig. 11.6 Equation used to compute speed from step frequency and user height. The *solid line* is the curve for $h = 1.67$ m. The *open circles* represent data relating step frequency and speed gathered from the literature and cited in [40]). (Equation from Dean [7]; figure adapted from Wendt [40]

11.2.2 Setting Direction for Walking-in-Place

There is nothing particularly hard about simplistically setting the direction of movement to "forward" in a walking-in-place interface. The difficulty arises when incorporating the goals of allowing the user to move in any direction and keeping the direction of movement independent of view direction and body orientation.

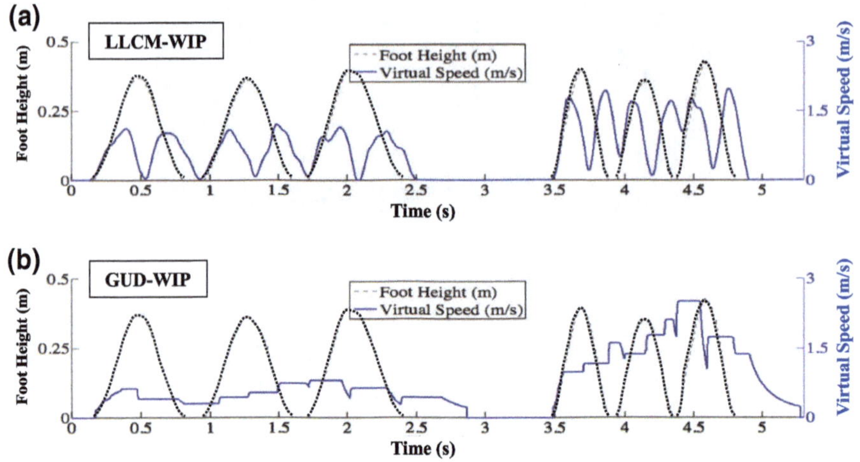

Fig. 11.7 *Solid lines* are the output speed profiles for (**a**) LLCM-WIP and (**b**) GUD-WIP at two walking speeds. *Dotted lines* are the height of *one* ankle. The trace of the other ankle's height would be the same shape, but offset so that its peaks are in the gaps. In **a**, note the dips toward zero in LLCM-WIP output during the double-support event between steps, offset by the ∼100 ms latency. In **b** note the three changes in speed during each step and that the speed does not dip to nearly zero. You can also see the ∼500 ms stopping latency. (Figure adapted from Wendt et al. [41])

11.2.2.1 Hands-Free Direction Setting Techniques

Head-directed motion. Often called *gaze-directed*, head-directed motion uses the forward direction of the (head-tracked) head pose as the direction of motion. This requires no additional apparatus and is easy to implement and learn to use. However, the user cannot move and look around at the same time, as people normally do. Slater's team's neural-network-based WIP system used head-directed motion [35].

Torso-directed motion. Torso-directed motion is one of several direction-setting techniques that depends on data from trackers located on the user's body. A tracker on the user's torso (front or back; chest or hips) can be used to set "forward" to be the direction the user's body is facing. Use of the additional tracker means that torso-directed movement is independent of head orientation, so users can walk and look around at the same time. A limitation of such body-worn tracker techniques is that users cannot move backwards or sideways, as both of those motions require decoupling direction of motion from the direction the body is facing.

Gesture-controlled direction. Gesture-controlled direction setting techniques interpret tracked movements of the user's hands, head, legs, or feet to establish direction of movement. While we would argue that any use of gestures reduces the naturalness of walking-metaphor interfaces, gestures are frequently used. In Gaiter, sideways motion is enabled by swinging the leg to the side from the hip; backward motion is enabled by kicking backward from the knee [36].

11.2.2.2 Hand-Held Direction Setting Devices

The most common hand-held devices for setting direction are tracked wands and joysticks that may or may not be part of a game controller. While the efficacy of these interfaces is well accepted, they come at the cost of limiting how the user can use her hands to interact with the virtual scene in application tasks.

Wands and pointing. Wands typically include a tracker and one or more other input devices such as buttons. Forward direction can be set by a combination of arm gesture and a hand-held three degrees of freedom (3DOF) tracker by using the tracker-measured positions of the user's head and the wand to define the direction vector. If the tracker is 6DOF, direction of movement can be set from the tracker's coordinate system; typically movement is in the direction of the longitudinal axis of the wand. The biomechanics of human shoulders and wrists limit the range of directions that can be set with wands without repositioning the body.

Joysticks/game controllers. Joysticks/game controllers can specify motion in any arbitrary direction, so they are an attractive solution for setting direction. Most often the user wears a 6DOF tracker on her body and the joystick outputs are interpreted in that coordinate system. This means that when the user pushes the joystick perpendicularly away from herself, it causes her move in the direction her body is facing. Note that the tracker data does not restrict the direction of movement; it simply establishes a body-centric coordinate system for the joystick.

Integrated tracker/joystick and task tool. The encumbrance of the hand-held interface devices can be mitigated in part if they are integrated into the task tools used in the IVE system. A well-developed example is the instrumented rifles with integrated thumb-operated joysticks (thumb-sticks) that are used in many military training systems, including the United States Army's relatively new Dismounted Soldier Training System [26]. An evaluation of an earlier system reported both positive and negative aspects of the thumb-sticks [25].

11.2.3 The Future for Walking-in-Place Interfaces

Modeling human walking in ways suitable for use in WIP interfaces is not yet a solved problem. Techniques inspired by biomechanics have addressed setting virtual speed during the rhythmic phase of walking and have tried to minimize starting and stopping latency, but they have not yet addressed the shape of the velocity profile during those two phases of walking, or variations in speed that may result from turning or walking with a heavy load. We do not yet know if the discrete changes in speed that occur in GUD-WIP affect users' perception of the environment or their task performance. We do not know how the mathematical models may change if the user is running.

To be cost effective, walking-in-place techniques have often made do with very little information about the user's actual motion. In some cases, the only data available for use in the locomotion algorithm is from the head tracker. Full body tracking systems provide rich data, but also are costly, encumbering, and inconvenient. Their use has to be carefully balanced against the improvements in naturalness made possible by the richer data.

Consumer products have started to change the landscape. Applications for the Kinect™ range camera can compute and update the 3D pose of a user's skeleton each frame time. The Kinect is inexpensive and does not require the user to wear any additional gear [46]. Small wireless sensors—accelerometers, magnetometers, and gyros—will be an inexpensive and non-encumbering source of data measuring user motion that can be used as inputs to the locomotion algorithm. A proof-of-concept system using such devices is described in Kim et al. [16].

With a richer set of input data, walking-in-place locomotion techniques will be better able to model and simulate the experience of natural walking for users of IVE systems.

11.3 Real-Walking Interfaces

Real-walking interfaces enable HMD-IVE-system users to naturally walk around the virtual scene just as they would in the real world. Because the user must be tracked, restricting the size of the virtual scene to the size of the tracked space is the simplest case for real-walking. If the virtual scene fits in the tracked space, the user can freely walk about in the entire virtual space, the user's real-world speed can be mapped in a one-to-one ratio to her virtual speed, and her direction in the virtual scene can be directly controlled by her direction of motion in the real world.

Complications with real-walking interfaces arise when the virtual scene is larger than the tracked lab area. Mapping the user's actual speed and direction one-to-one with virtual speed and direction no longer enables the user to travel through the entire scene, as to do so would require leaving the tracked area. Numerous techniques, most of which exploit the imprecision of human perception, have been developed to make real-walking a viable locomotion technique for larger-than-tracked-space virtual scenes. Initial implementations focused on transformations of the scene model or the user's motion by manipulating the ratio between the user's real and virtual speeds and directions. A newer technique changes the structure of the scene model [34]. We discuss both approaches.

11.3.1 Manipulating Speed

Manipulating speed in real-walking interfaces can be thought of as altering the ratio between the user's real walking speed and virtual speed so that it is no longer one-to-one.

11.3.1.1 Perceptual Foundation

As people move, their view of their surroundings changes, and information about the layout of the environment and the shape of surfaces, as well as their relative position within the environment, is revealed.

The illusion of self-motion, known as *vection*, can be produced by visual stimulation alone. For example vection can occur when a person is sitting in a stationary car and the adjacent car starts to move forward, causing the person in the stationary car to perceive the sensation of backwards motion.

Movement, essential for accurate perception of the environment, causes *optic flow*, the changing pattern of light on the optic array caused by the relative motion of the observer and environment. Optic flow patterns contain information about self-motion, the motion of objects, and the environment's three-dimensional (3D) structure. If an observer is moving forward, the optic flow will radiate outward from the *center of expansion*—the point toward which the person is moving; if a person is riding in a train and looking out the window, the optic flow will move horizontally across the observer's retina producing *lamellar flow*.

The results of a study by Warren led him to speculate that optical information could be exploited to control locomotion [38]. An experiment by Konczak found that as optic flow slowed, subjects' walking speed slightly increased; however increasing the speed of optic flow appeared to have no effect on participants' real speed [17]. Konczak's results suggest that increasing the ratio between the users' virtual and real walking speeds (i.e., increasing optical flow speed relative to walking speed) could be employed to enable users to travel greater virtual distances in the same number of steps.

11.3.1.2 Interfaces that Manipulate Speed

Real-walking locomotion techniques that alter the ratio between the user's real and virtual speeds, thus altering optic flow, include Seven League Boots [13, 30] and Scaled Translational Gain [42]. Each of these methods maps the user's real translation into increased virtual translation. For example, when the user takes one step in the real world she is translated two or three steps in the virtual world.

Altering the ratio between the user's real and virtual speed enables the size of the virtual scene to be scaled to a multiple of the size of the tracked space, based on the ratio between real and virtual speeds. However problems can occur if the ratio becomes very large. For example, if the user's motion is increased by a factor of 100, then when the user takes one real step she travels 100 steps forward in the virtual scene. This motion, although smooth and in the same direction as the user's motion, may cause disorientation as it places the user far away from their starting location. This rapid change in the user's location is similar to teleportation which is known to disorient the user [3].

An additional problem with speed-scaling methods arises because people move their heads side-to-side as well as forward-to-backward as they walk. When the ratio between real and virtual motion is large, the side-to-side motions are also multiplied and can cause the scene to appear unstable. To eliminate the side-to-side motion, Interrante et al. computed the user's forward direction and scaled user motion only in this predicted direction [13].

Another potential problem with altering user speed is that when the difference between physical and virtual speeds is large, people will be able to notice the discrepancy. A method introduced by Bruder et al. uses change blindness techniques to effectively move the user forward in the VE while the user is unaware of it [5]. Change blindness theory posits that people are unaware of changes made in their view when the changes occur during saccadic eye movements. Change blindness is discussed further in Chap. 14. As is common in change blindness techniques, Bruder et al. display a blank screen that flashes in the HMD for 60–100 ms. While the screen is blanked, the virtual scene is translated in the user's direction, thus altering the ratio between the user's real and virtual speed. Due to change blindness, the user is less aware of the alterations that have occurred.

11.3.2 Manipulating Direction

Manipulating direction for real-walking techniques can be thought of as altering the ratio between real world direction and virtual world directions of movement.

11.3.2.1 Perceptual Foundation

Altering the ratio between real and virtual directions is possible because vision guides *heading direction*, the user's direction of motion. The egocentric direction hypothesis and Gibson's theories about optic flow [10] provide theoretical support for locomotion systems that guide user direction by manipulating the user's view of the virtual scene as generated by the IVE system.

The *egocentric direction hypothesis* states that heading direction is determined by the anterior-posterior axis of the body. This theory was explored by Rushton et al. after observing a subject who suffers from *unilateral visual neglect* (UVN)—damage to one side of the cerebral hemisphere and the inability to respond to stimuli on the side opposite the lesion [31]. UVN is often associated with a misperception of location. Rushton et al. observed the subject walking in curved paths to reach target objects. To simulate the misperception of the target location for individuals without UVN, Rushton et al. had participants wear prisms in front of their eyes and found participants walked a curved path toward the target. The prism translates not only the target object, but also the optic flow produced when the participant walked toward the target (Fig. 11.8).

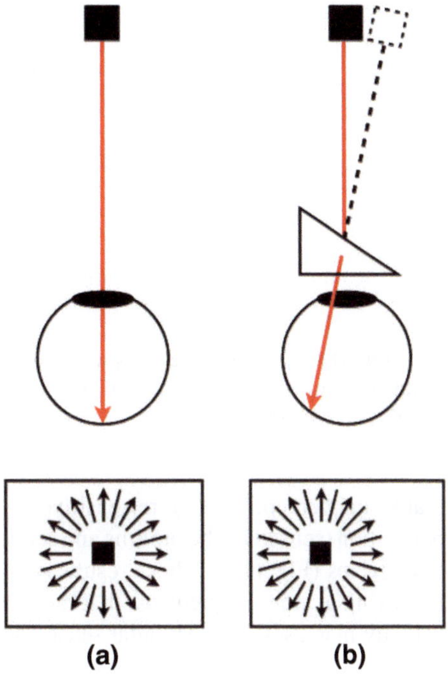

Fig. 11.8 **a** As the person looks directly along the heading vector, optic flow radiates outward from the center of vision. **b** A prism is placed in front of the eye, which shifts the visual location of the goal and the location of the radial optic flow. The optic flow on the retina is the same pattern as on the left but shifted due to the prism. (Adapted from Rushton et al. [31])

Gibson's theories [10] suggest that heading is determined from the center of expansion of optic flow. When people walk toward a target, they adjust their movements to align heading direction with the intended goal. Warren et al. [39] further investigated whether the egocentric direction hypothesis or the optic flow hypothesis dominates. They had people walk through virtual scenes with different textures to create different amounts of optic flow to see if the amount of optic flow affected participants' heading direction as they moved to a target. Their results show that with no optic flow participants followed the egocentric direction hypothesis, however when optic flow was added to the ground plane, participants initially followed the egocentric direction hypothesis, and then after traveling a few meters participants adjusted their heading and used optic flow to aid their guidance.

The results of Warren et al. demonstrated that humans rely on both optic flow and egocentric direction to guide locomotion. These results suggest that manipulations of the visual representation of the scene can guide the user so she walks a straight path in the virtual scene concurrently with walking a curved path in the laboratory.

Slight manipulation of optic flow may go unnoticed by a user; however, extreme changes will be detectable. Studies from aircraft simulation provide further understanding of ways that IVE system and scene designers can manipulate rendered visuals without the user noticing. Research by Hosman and van der Vaart determined the sensitivity of the visual and vestibular senses to different rotation frequencies or speeds, i.e., the frequency response of the two senses [12]. The results suggest

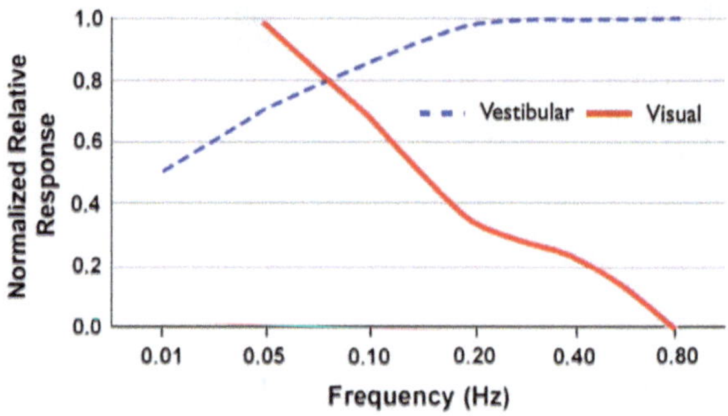

Fig. 11.9 The visual-vestibular crossover. This graph shows, in the frequency domain, the relative contributions of visual and linear vestibular cues to postural stability. (Adapted from Duh et al. [8], reproduced from Razzaque [27])

that visual perception is more sensitive at low frequencies of motion and vestibular perception (sensed by the otoliths and semicircular canals) is more sensitive at higher frequencies (Fig. 11.9). These results suggest that when the head is not moving or is moving at slow frequencies, that the visual system is dominant. As head angular velocity increases, the vestibular sense comes to dominate the visual.

The important outcome of Hosman and van der Vaart's research is the observation that when people turn their heads, the vestibular system dominates and visual manipulation may go unnoticed. Rotation of the virtual scene during head turns is therefore less likely to be detected because when people turn their heads at normal angular velocities the vestibular system dominates the visual system. As a point of reference, an angular rotation of 0.5 Hz corresponds to taking 2 s to rotate your head all the way from one side to the other and back; note that higher angular rotation frequencies (and faster head turns) are further to the right in Fig. 11.9 where vestibular cues almost totally dominate visual.

The egocentric direction hypothesis, Gibson's theories of optic flow, and studies about the visual-vestibular crossover all provide theoretical support for manipulating the views of the virtual scene to cause the user's virtual direction to differ from her real direction. These techniques are employed in the following locomotion interfaces.

11.3.2.2 Interfaces

Motion compression (MC) [19, 33] has a misleading name because it does not in fact compress motion. Instead, MC rotates the virtual scene around the user and remaps areas of the scene that were outside of the tracked-space into the tracked space. The MC algorithm predicts a user's goal location based on points of interest in the scene toward which the user may be walking. The algorithm then maps the straight line of the path from the user to the predicted goal location onto the largest possible arc that will fit into the tracked space. MC continuously updates the goal location and the rotation of the virtual scene relative to the tracked space. It is not a goal of MC to make the rotation undetectable by users.

Redirected walking (RDW) [27–29] is a technique that exploits the imprecision of human perception of self-motion—the motion of humans based on sensory cues other than vision. RDW modifies the direction of the user's gaze by imperceptibly rotating the virtual scene around the user and redirecting the user's (future) path back into the tracked space. Unlike MC, RDW was designed to make rotation undetectable to the user. RDW achieves undetectable rotation by exploiting the visual vestibular crossover described above. The vestibular system is dominant over the visual system at head frequencies greater than 0.07 Hz, approximately one head turn over a 14 s period, causing users to not notice unmatched real and scene rotation while turning their heads at frequencies greater than 0.07 Hz. For this reason, an integral part of the design for RDW was to make users frequently turn their heads.

Razzaque's environments and tasks depended on static waypoints, locations that defined the user's virtual route within the VE, for two reasons. First, a series of waypoints predetermined the user's sequence of goal locations. Knowledge of the future goal locations enables the system to always know what part of the virtual scene should be rotated into the tracked space. Second, waypoints are a mechanism designed to make people look around. That is, users had to turn their heads to find the next waypoint. This enabled the RDW algorithm to rotate the virtual scene (during head turns) and redirect the user's next-path-direction, i.e., the path to the next waypoint, into the tracked space.

Waypoints provided a simple answer for one of the most challenging parts of implementing a redirection system: predicting the user's future direction. Although waypoints enable RDW, they limit applications to those that have predetermined paths and task-related reasons for users to turn their heads.

Newer implementations of redirection have added dynamic controllers: Peck and her colleagues controlled the amount of rotation added to the virtual scene based on the rotation speed of the user's head [21, 22]; Neth et al. controlled the curvature gain based on the user's walking speed [18]; and Hodgson et al. altered the redirection amounts based on both the user's linear and angular velocities [11]. Chapter 10 provides a detailed description of how to modify the view transformation in redirection systems.

Additional studies and techniques have explored determining the appropriate amount of redirection that can be added at any instant [15, 32], how to steer the user within the environment [11, 21, 22, 27], and how to predict the user's future direction [13, 21, 22].

Finally, a method presented by Suma et al. harnesses change blindness techniques by altering part of the scene model when the user is not looking at that part of the scene [34]. For example, the location of a door to a room may change from one wall to another while the user is not looking at it, thus guiding the user to walk in a different direction in the physical space by walking a different direction in the virtual space.

11.3.3 Reorientation Techniques

Many of the locomotion techniques presented in Sects. 11.3.1.1 and 11.3.1.2 use a *reorientation technique* (ROT) to handle the situation when large-area real-walking techniques fail and the user is close to walking out of the tracked space (and possibly into a wall or other obstruction). ROTs discourage the user from leaving the tracked space and rotate the virtual scene around her current virtual location. This moves the user's predicted next-path-direction into the tracked space. The user must also reorient her body by physically turning in the real environment so she can follow her desired path in the newly rotated virtual scene. Some techniques require the user to stop; others do not. As a design goal, ROTs should interfere with the virtual experience as little as possible.

In addition to waypoints, redirected walking [27–29] uses a ROT that employs a loudspeaker in the virtual scene, played through user-worn headphones, that asks the user to stop, turn her head back and forth, and then continue walking in the same direction. During the head turning the virtual world can be undetectably rotated such that the future virtual path lies within the real-world tracked space.

The ROT used in motion compression [19, 33] is built into the motion compression algorithm itself: as the user approaches the edge of the tracked space the arc of minimum curvature grows quite small causing the scene rotation to be large. These large rotations cause the user to feel that the scene is spinning around [19]. This method does not require the user to stop.

In the method presented by Hodgson et al. when the user is about to leave the tracked space the experimenter physically stops the user and physically turns the user back into the tracked area [11]. The HMD visuals are frozen during the turn so that the user can continue walking in the same virtual direction after the turn.

Williams et al. explored three *resetting* methods for manipulating the virtual scene when the user nears the edge of the tracked space [43]. One technique involves turning the HMD off, instructing the user to walk backwards to the middle of the lab, and then turning the HMD back on. The user will then find herself in the same place in the scene but will no longer be near the edge of the laboratory's tracked space. The second technique turns the HMD off, asks the user to turn in place, and then turns the HMD back on. The user will then find herself facing the same direction in the virtual scene, but she is facing a different direction in the tracked space.

Preliminary research suggests that the most promising is a third technique that uses an audio request for the user to stop and turn 360° [43]. The virtual scene rotates at twice the speed of the user and stops rotating after a user turn of 180°. The user is supposed to reorient herself by turning only 180° but should think she has turned 360°. This ROT attempts to trick the user into not noticing the extra rotation; however, results from Peck et al. noticed that few participants were tricked into thinking they turned 360° after only turning 180° [20, 24].

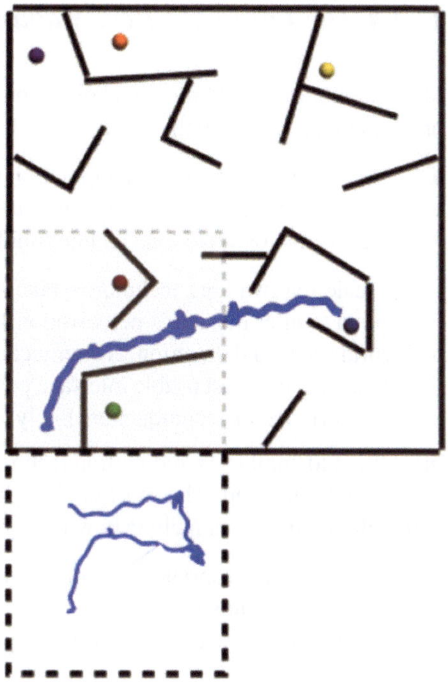

Fig. 11.10 An example of a user's path (*the wiggly blue lines*) through a virtual and real space. The virtual space is represented by the *solid box* and the real space is denoted by the *dashed box*. The *dashed box* shows relative size of real and virtual spaces. (Reproduced from Peck et al. [23])

With reorientation and/or redirection, the paths in the virtual and real world have different shapes and, as is the goal, the real world path covers less area than the virtual. Figure 11.10 shows an example.

Peck et al. introduced *distractors* which are visual objects or sounds in the virtual scene used to stop the user and elicit head rotations. Devoting attention to distractors appears to make people less aware of scene rotation while they are turning their heads [20, 24]. Distractors have been used in conjunction with redirection [21, 22], and users of the combined system scored significantly higher on a variety of navigation metrics than users of walking-in-place and joystick interfaces.

The locomotion interface implemented by Neth et al. used avatars as distractors, and when combined with their implementation of dynamic curvature gain, enabled people to successfully explore a large virtual city [18].

Alternatives to distractors include *deterrents* [22] and *Magic Barrier Tape* [6]. Both techniques display a virtual barrier to mark the real boundaries of the tracked space. The implementation from Cirio et al. uses a joystick method to move the unreachable portions of the virtual scene into the tracked space [6], whereas the implementation by Peck et al. uses distractors and redirection to rotate the unreachable part of the scene back into the tracked space [22].

11.3.4 The Future for Real-Walking Interfaces for IVE Systems

Manipulation of user direction should not be obtrusive to the point that it causes a break in presence. Though not yet studied, it has been proposed that

- For novice users direction manipulation should be undetectable.
- For experienced users direction manipulation should be bounded by the likelihood of increasing cognitive load and/or simulator sickness.

Large-scale real-walking techniques take advantage of the imprecisions of human perception to alter the user's perceived virtual speed and direction compared to the real world speed and direction. Newer techniques are combining multiple manipulations to enable the most usable interface possible. Different combinations of redirection and reorientation techniques are likely to enable different results and experiences.

In addition to combining redirection techniques, the current implementations can be refined and improved. The most challenging and unanswered design decisions for real-walking interfaces include how to:

- Determine an appropriate amount of speed and direction manipulation for both experienced and novice users;
- Determine the most effective way to direct the user away from the edges of the tracked space;
- Predict the user's future virtual direction.

Promising future work would compare different combinations of techniques to guide the VE designer. For training transfer applications where fatigue is important, scaled translational gain methods may not be feasible, however scaled translational gain may be most appropriate for a novice user walking through a virtual city. Possible design goals may include: accurate development of a mental model, usability, user enjoyment, speed of travel, training transfer, and designing for experienced versus novice users.

Acknowledgments Whitton's work on this chapter was supported in part by the NIH National Institute of Biomedical Imaging and Bioengineering and the Renaissance Computing Institute, and Peck's in part by the European grant VERE, an Integrated Project funded under the European Seventh Framework Program. We both thank our colleagues whose work is reported here for the pleasure and stimulation of working with them on this topic.

References

1. Arns L (2002) A new taxonomy for locomotion in virtual environments. Ph.D. Dissertation, Iowa State University, Ames, Iowa, 2002. Accessed 10 Sept 2012 from ProQuest Disseration and Theses
2. Bouguila L, Evequoz F, Courant M, Hirsbrunner B (2004) Walking-pad: a step-in-place locomotion interface for virtual environments. In: Proceedings of the 6th international on multimodal interfaces, ACM, New York, pp 77–81. doi:10.1145/1027933.1027948

3. Bowman DA, Koller D, Hodges, LF (1997) Travel in immersive virtual environments: an evaluation of viewpoint motion control techniques. In: Proceedings of the virtual reality annual international aymposium (VRAIS 1997), IEEE Press, Washington, pp 45–52, 215. doi:10.1109/VRAIS.1997.583043
4. Bowman DA, Kruijff E, LaViola JJ et al (2005) 3D user interface: theory and practice. Addison-Wesley, Boston
5. Bruder G, Steinicke F, Wieland P (2011) Self-motion illusions in immersive virtual reality environments. In: Proceedings of IEEE virtual reality, pp 39–46
6. Cirio G, Marchal M, Regia-Corte T, Lécuyer A (2009) The magic barrier tape: a novel metaphor for infinite navigation in virtual worlds with a restricted walking workspace. In: Proceedings of the ACM symposium on virtual reality software and technology (VRST 2009), pp 155–162
7. Dean GA (1965) An analysis of the energy expenditure in level and grade walking. Ergonomics 8(1):31–47
8. Duh HB, Parker DE, Phillips J, Furness TA (2004) Conflicting" motion cues at the frequency of crossover between the visual and vestibular self-motion systems evoke simulator sickness. Hum Factors 46:142–153
9. Feasel J, Wendt JD, Whitton MC (2008) LLCM-WIP: low-latency, continuous-motion walking-in-place. In: Proceedings of IEEE symposium 3D user interfaces'08, pp 97–104
10. Gibson JJ (1950) The perception of the visual world. Houghton Mifflin, Boston
11. Hodgson E, Bachmann E, Waller D (2011) Redirected walking to explore virtual environments: assessing the potential for spatial Interference. ACM Trans Appl Percept 8(4):1–22 (Article 22)
12. Hosman R, Van der Vaart J (1981) Effects of vestibular and visual motion perception on task performance. Acta Psychol 48(1–3):271–287
13. Interrante V, Ries B, Anderson L (2007) Seven league boots: a new metaphor for augmented locomotion through moderately large scale immersive virtual environments. In: Proceedings of the IEEE symposium on 3D user interfaces, pp 167–170
14. Inman V (1981) Human walking. Williams & Wilkins, Baltimore
15. Jerald J, Peck TC, Steinicke F, Whitton MC (2008) Sensitivity to scene motion for phases of head yaws. In: Proceedings of the ACM symposium on applied perception in graphics and visualization (APGV), pp 155–162
16. Kim J-S, Gracanin D, Quek F (2012) Sensor-fusion walking-in-place interaction technique using mobile devices. In: Proceddings of IEEE virtual reality, pp 39–42
17. Konczak J (1994) Effects of optic flow on the kinematics of human gait—a comparison of young and older adults. J Motor Behav 26:225–236
18. Neth CT, Souman JL, Engle D, Kloos U, Bülthoff HH, Mohler BJ (2011) Velocity-dependent dynamic curvature gain for redirected walking. In: Proceedings of IEEE virtual reality, pp 151–158
19. Nitzsche N, Hanebeck UD, Schmidt G (2004) Motion compression for telepresent walking in large target environments. Presence Teleoper Virtual Environ 13(1):44–60
20. Peck TC, Fuchs H, Whitton MC (2009) Evaluation of reorientation techniques and distractors for walking in large virtual environments. IEEE Trans Vis Comput Graph 15(3):383–394
21. Peck TC, Fuchs H, Whitton MC (2010) Improved redirection with distractors: a large-scale-real-walking locomotion interface and its effect on navigation in virtual environments. In: Proceedings of IEEE virtual reality, pp 35–38
22. Peck TC, Fuchs H, Whitton MC (2011) An evaluation of navigational ability comparing redirected free exploration with distractors to walking-in-place and joystick locomotion interfaces. In: Proceedings of IEEE virtual reality, pp 55–62
23. Peck TC, Fuchs H, Whitton MC (2012) The design and evaluation of a large-scale real-walking locomotion interface. IEEE Trans Vis Comput Graph 18(7):1053–1067
24. Peck T, Whitton M, Fuchs H (2008) Evaluation of reorientation techniques for walking in large virtual environments. In: Proceedings of IEEE virtual reality, pp 121–127
25. Pleban, RJ, Eakin DE, Slater MS et al (2001) Research report 1767: training and assessment of decision-making skills in virtual environments. http://www.dtic.mil/cgi-bin/GetTRDoc?AD=ADA389677. Accessed 19 Aug 2012

26. Quinn K (2011) U.S. Army to get dismounted soldier training system. Def News Train Simul J (Online). http://www.defensenews.com/article/20110825/TSJ01/108250309/U-S-Army-get-dismounted-soldier-training-system. Accessed 19 Aug 2012
27. Razzaque S (2005) Redirected walking. Dissertation and computer science technical report TR05-018, University of North Carolina at Chapel Hill
28. Razzaque S, Kohn Z, Whitton MC (2001) Redirected walking. In: Proceedings of the eurographics workshop on virtual reality, pp 289–294
29. Razzaque S, Swapp D, Slater M, Whitton MC, Steed A (2002) Proceedings of the eighth eurographics workshop on virtual environments, pp 123–130
30. Robinett W, Holloway R (1992) Implementations of flying, scaling and grabbing in virtual worlds. In: ACM Symposium on interactive 3D graphics, pp 189–192
31. Rushton SK, Harris JM, Lloyd MR, Wann JP (1998) Guidance of locomotion on foot uses perceived target location rather than optic flow. Curr Biol 8:1191–1194
32. Steinicke F, Bruder G, Jerald J, Frenz H, Lappe M (2010a) Estimation of detection thresholds for redirected walking techniques. IEEE Trans Vis Comput Graph 16(1):17–27
33. Su J (2007) Motion compression for telepresence locomotion. Presence Teleoper Virtual Environ 16(4):385–398
34. Suma E, Clark S, Krum D, Finkelstein S, Bolas M, Warte Z (2011) Leveraging change blindness for redirection in virtual environments. In: Proceedings of IEEE virtual reality, pp 159–166
35. Slater M, Usoh M, Steed A (1995) Taking steps: the in influence of a walking technique on presence in virtual reality. ACM Trans Comp-Hum Interact (TOCHI) 2(3):201–219
36. Templeman JN, Denbrook PS, Sibert LE (1999) Virtual locomotion: walking in place through virtual environments. Presence Teleoper Virtual Environ 8(6):598–607
37. Templeman JN, Sibert LE et al (2007) Pointman—a new control for simulating tactical infantry movements. Proc IEEE Virtual Real 2007:285–286
38. Warren WH (2004) Optic flow. Chpt 84. MIT Press, Cambridge, pp 1247–1259
39. Warren WHJ, Kay BA, Zosh WD et al (2001) Optic flow is used to control human walking. Nature 4(2):213–216
40. Wendt JD (2010) Real-walking models improve walking-in-place systems. Dissertation and Computer Science Technical Report TR10-009, University of North Carolina at Chapel Hill
41. Wendt JD, Whitton MC, Brooks FP (2010) GUD-WIP: Gait-understanding driven walking-in-place. In: Proceedings of IEEE virtual reality, pp 51–58
42. Williams B, Narasimham G, McNamara TP, Carr TH, Rieser JJ, Bodenheimer B (2006) Updating orientation in large virtual environments using scaled translational gain. In: Proceedings of the 3rd ACM symposium on applied perception in graphics and visualization (APGV 2006), pp 21–28
43. Williams B, Narasimham G, Rump B, McNamara TP, Carr TH, Rieser J, Bodenheimer B (2007) Exploring large virtual environments with an HMD when physical space is limited. In: Proceedings of the 4th ACM symposium on applied perception in graphics and visualization (APGV 2007), pp 41–48
44. Whitton MC, Cohn J, Feasel J et al (2005) Comparing VE locomotion interfaces. In: Proceedings of IEEE virtual reality, pp 123–130
45. Yan L, Allision RS, Rushton SK (2004) New simple virtual walking method-walking on the spot. In: Proceedinds of 9th annual immersive projection technology (IPT) symposium
46. Zheng Y, McCaleb M, Strachan C et al (2012) Exploring a virtual environment by walking in place using the Microsoft Kinect (Poster Abstract). In: Proceedings of the ACM symposium on applied perception, p 131

Chapter 12
Multimodal Rendering of Walking Over Virtual Grounds

Maud Marchal, Gabriel Cirio, Yon Visell, Federico Fontana, Stefania Serafin, Jeremy Cooperstock and Anatole Lécuyer

Abstract The addition of multimodal feedback during navigation in a virtual environment is fundamental when aiming at fully immersive and realistic simulations. Several visual, acoustic, haptic or vibrotactile perceptual cues can be generated when walking over a ground surface. Such sensory feedback can provide crucial and varied information regarding either the ground material itself, the properties of the ground surface such as slope or elasticity, the surrounding environment, the specificities of the foot-floor interaction such as gait phase or forces, or even users' emotions. This chapter addresses the multimodal rendering of walking over virtual ground surfaces, incorporating haptic, acoustic and graphic rendering to enable truly multimodal walking experiences.

M. Marchal (✉) · G. Cirio · A. Lécuyer
INRIA Rennes, France
e-mail: maud.marchal@inria.fr

M. Marchal
INSA Rennes, France

Y. Visell
Drexel University, USA

F. Fontana
University of Udine, Italy

S. Sarafin
Aalborg University, Denmark

J. Cooperstock
Mc Gill University, Canada

F. Steinicke et al. (eds.), *Human Walking in Virtual Environments*,
DOI: 10.1007/978-1-4419-8432-6_12,

12.1 Introduction

Sensations accompanying walking on natural ground surfaces in real world environments are rich, multimodal and highly evocative of the settings in which they occur [110]. For these reasons, foot-based human–computer interaction represents a new means of interacting in Virtual Reality (VR), with potential applications in areas such as architectural visualization, immersive training, rehabilitation or entertainment. However, floor-based multimodal (visual, auditory, tactile) information displays have only recently begun to be investigated [108]. Research work has remained limited as there has been a lack of efficient interfaces and interaction techniques capable of capturing touch via the feet over a distributed display. Related research on virtual and augmented reality environments has mainly focused on the problem of natural navigation in virtual environments [56, 81, 96]. A number of haptic interfaces for enabling omnidirectional in-place locomotion in virtual environments have been developed [46], but known solutions either limit freedom in walking, or are highly complex and costly.

The rendering of multimodal cues combining visual, auditory and haptic feedbacks has rarely been exploited when walking in a virtual environment. Many aspects of touch sensation in the feet have been studied in prior scientific literature, including its roles in the sensorimotor control of balance and locomotion over different terrains. However considerably less is known about how the nature of the ground itself is perceived, and how its different sensory manifestations (touch, sound, visual appearance) and those of the surroundings contribute to the perception of properties of natural ground surfaces, such as their shape, irregularity, or material composition, and our movement upon them. Not surprisingly then, in human–computer interaction and virtual reality communities, little research attention has been devoted to the simulation of multisensory aspects of walking surfaces in ways that could parallel the emerging understanding that has, in recent years, enabled more natural means of human computer interaction with the hands, via direct manipulation, grasping, tool use, and palpation of virtual objects and surfaces.

The present chapter proposes to review the recent interactive techniques that have contributed to develop multimodal rendering of walking in virtual worlds by reproducing virtual experiences of walking on natural ground surfaces. These experiences are enabled primarily through the rendering and presentation of virtual multimodal cues of ground properties, such as texture, inclination, shape, material, or other affordances in the Gibsonian sense [37]. The related work presented in this chapter is organized around the hypothesis that walking, by enabling rich interactions with floor surfaces, consistently conveys enactive information that manifests itself through multimodal cues, and especially via the haptic and auditory channels. In order to better distinguish this investigation from prior work, we adopt a perspective in which vision plays a primarily integrative role linking locomotion to obstacle avoidance, navigation, balance, and the understanding of details occurring at ground level. That is why we will not detail the visual rendering of walking over virtual grounds itself.

This chapter includes the presentation of (1) multimodal rendering techniques for the interactive augmentation of otherwise neutral (i.e., flat, silent, and visually homogeneous) ground surfaces; (2) multisensory effects and cross-modal illusions, involving the senses of touch, kinesthesia, audition, and vision, that were made possible by novel interfaces.

The chapter is organized as follows. Section 12.2 is dedicated to auditory rendering. Section 12.3 begins with the description of haptic rendering approaches, before presenting multimodal systems and cross-modal approaches. Section 12.4 concludes this chapter.

12.2 Auditory Rendering

12.2.1 Introduction

A walking task can be said to be intimately linked to a corresponding auditory task. Not only do walkers constantly hear most of their own footsteps and foot movements, but they are typically also aware of other persons walking in a shared auditory scene. In parallel, the same scene may be populated by passive listeners who, while standing or sitting, and not necessarily visually attending to pedestrians in their surroundings, may nevertheless perceive the footsteps as part of the ambient soundscape.

These simple considerations already say much about the importance of the auditory cues in informing ones perception and action loop during a walking task, and, furthermore, in conveying information that can have social relevance when contributing to form a soundscape that is shared by several listeners.

As with any other type of non visual, ecological feedback, footstep sounds can occupy the periphery of the attention. In other words, we need to make conscious use of this feedback unless it brings to our ears salient cues, either familiar or unexpected. A similar process happens for instance when a car driver's attention may be triggered by an almost imperceptible change in the sound of the engine signaling potential malfunctioning of the car, even after hours on a long trip along a monotonous highway [75]. We do not need much quantitative science to establish these observations in an empirical way: the use of footstep sounds as an auditory warning has long been recognized by movie directors, who used to ask their Foley artist for preparing the right walking sound when a new character entered the movie stage, or for sonifying night-time chase actions that were typical of the "noir" genre.

It should be clear, at this point, that walking sounds are expressive. Through the long familiarity with our own and others' footsteps, we built subjective mental maps linking such sounds to corresponding physical attributes and gestures of the walking person. Some of these links are obvious, and have been exploited for instance in early computer game designs. All vintage electronic game players probably remember the use of iconic footstep sounds to render the number and moving speed of the enemies in Space Invaders™, a popular computer game of the late 1970s: the designers of that game arrived at a successful design by making effective use of extremely simple sound elements, whose staticity expressed well the martial attitude of the adversarial squadron.

12.2.1.1 Psychoacoustic Measurements

The expressivity of footsteps has been analyzed from a scientific perspective as well. On the experimental side, Pastore et al. have adopted an ecological approach to the auditory perception of footsteps. Their experiments investigated the ability of listeners to recognize walkers' gender from walking sounds [61], as well as different kinematics of the gait in people walking with either normal upright or stooped posture [78]. Experiments have been also conducted on the recognition of familiar individuals from their footstep sounds [28]. In all such investigations, an effort has been devoted to identify the acoustic *invariants* that are responsible for the subjective decisions. Arguably such invariants necessarily span a multiplicity of auditory cues. In particular, the demonstrated dependency of these cues on specific spectral features such as spectral slopes, moments, and centroids, can make such perceptual research especially informative for auditory rendering purposes.

A parallel thread in the acoustic analysis of footsteps has concerned their recognition with respect to specific characteristics of the ground. Although starting from an engineering perspective, this thread has introduced even deeper arguments in favor of an ecological approach to these experiments. Cress measured, and hence modeled the acoustic response of outdoor ground sites to individuals who were crawling, walking, and running: not only did he establish the dependence of the response spectra on the ground characteristics of the site; he also showed the relative invariance across frequency of the bands of spectral energy with respect to the walking activities [17]. These conclusions did not contradict earlier assessments made by Watters, who had found dependence on the floor type of impact force values measured from a single hard-heeled female footstep on various floors [113]. Stimulated by these experiences, Ekimov and Sabatier searched broad-band components of footstep sound *signatures* for different floor materials and walking styles: although the high-frequency band of these signatures contains most of the information about the frictional (i.e. tangential force) components giving rise to the footstep sounds, the same band has been shown to be relatively invariant with respect to changes in both floor covering and walking styles [25]. Irrespectively of their conclusions, overall these studies have called for introducing the floor dimension in the psychophysics of footstep recognition.

Research in this area has, consequently, begun to reveal the mechanisms underlying the active recognition of footsteps over different grounds. In such cases, subjects are engaged in a perception and action (walking) task, i.e., they are not just passive listeners, and thus the recognition process involves also use of the tactile sensory channel. In another investigation, by masking the tactile channel using active shoes capable of generating vibrational noise at sole level, Giordano et al. were able to study walkers' abilities to identify different ground surfaces comprising both solid materials (e.g., marble, wood) and granular media (e.g., gravel, sand) when alternately auditory, haptic, or audio-haptic information was available. The authors found that walkers could perform this perceptual task through a variety of different sensory modalities [39].

12.2.1.2 Premises for an Auditory Rendering of Grounds

The latest experiment is even more interesting, since the walkers' experience was *augmented* with elements of synthetic feedback, specifically to mask tactile cues of real ground materials. This design strategy opens new scenarios, in which the non visual "ground display" (as it is perceived by walkers) is contaminated with synthetic cues that mix with the rest of the floor feedback. Although the previous experiment is clear in posing limits to the salience of the auditory feedback when it does not match with the simultaneous (in that case noise-masked) tactile cues, yet it leaves room to sound as a mean for enriching the information brought by these cues. Specifically, one may think to mould an otherwise neutral tactile feedback, such as that experienced while walking on a silent, homogeneous flat and solid floor, using auditory cues reporting about a different type of ground; likewise, one may try to bias a multimodal stream of ground cues by altering some of their auditory parameters through the use of virtual sounds, without breaking the coherence of the feedback overall. In both such cases, however, an artificial perturbation of the auditory feedback has a chance to shape the recognition of a floor without disrupting the perceived realism of the multimodal percept, only if this perturbation elicits some form of cross-modal (specifically, audio-tactile) illusion.

Several cross-modal tactile effects induced by auditory cues have been discovered [8, 47]. In the following, we will report on recent studies that have investigated partial or total sensory substitutions of ground attributes in walkers, who were presented virtual auditory cues of the ground using different techniques, reproduction methods, as well as experimental setups, methodologies and tasks. Preliminary to these studies, a state of the art of the models for the rendering of footstep sounds is surveyed starting from the early experiences, until current developments. The section concludes by providing guidelines to the sound designer, who is interested in the realization of interactive floors including the auditory modality as part of their multimodal feedback.

12.2.2 Footstep Sound Synthesis

For what we have previously seen, the acoustic reproduction of walking requires one to render at least at two levels the auditory information: a low level, accounting for the sonic signatures of a single footstep, and a high level, that conversely reports about the frequency of the walking cycle and its fluctuations across time. Further cues would be needed to render the spatial movement across a walking area: although necessary to define a realistic soundscape, such cues are closely related to the *spatialization* features of sound reproduction, an issue that raises questions of 3D audio, a research and application field whose specific links to the rendering of walking sounds will be treated later in Sect. 12.2.3.

High-level cues are intuitively not too difficult to be rendered, as soon as a sufficiently large collection of data is put available for inferring a convenient statistical model for the walking cycle of a homogeneous population. More interesting are the constraints among instances of such cycles taking place in collective contexts, giving rise to *entrainment* effects [104]: for these effects the exact role of sound is currently unknown, in spite of a conspicuous number of works dealing with the relationships existing between gait cycle and rhythmic (especially musical/dance) sonic patterns [93].

Low-level cues represent an even more challenging design issue. By bringing information on the interactions taking place during the contact between the foot and the ground, they mainly report about the materials the floor and the shoes are made of. For this reason, the accuracy of their reproduction depends on the ability to embed this information within a sound synthesis model. Normally, these models must keep parametric control of the temporal as well as spectral features of the synthesis: as we will see in Sect. 12.2.2.2, the former are especially important for determining the correct *particle density* during the reproduction of aggregate grounds such as those made of crumples, ice, snow, creaking wood; conversely, the latter provide a unique *color* to the contact events, hence becoming crucial in interactions with solid floors, where the entire footstep sound is represented by one or very few contact events.

Further information, concerning several characteristics of a walker (weight, height, age, sex) results from the interplay of low- and high-level cues, and the information they provide about foot gesture, postural habits and locomotion style of the walking person: a credible rendering of footstep sounds must account also for this interplay, for which a comprehensive collection of kinematic and biomechanical data is not available yet [22]. This and other knowledge gaps currently make the design of interactive walking sound synthesizers a difficult task.

12.2.2.1 Early Models

The first systematic attempt to synthesize walking sounds was proposed by Cook in 2002 [14]. In this pioneering system, engineered on an STK-based sound engine known as Bill's Gait, the author introduced research elements that are still stimulating nowadays. In particular, Bill's Gait successfully implemented a number of solutions that are still largely state-of-the art in the realm of real-time sound processing: he detailed an analysis procedure which included Linear Predictive Coding for the extraction of footstep color, a Wavelet analysis for estimating the particle density, and an envelope following of the gait sequence for informing the higher-level statistics on amplitude and frequency of the walking cycle. His model could store data on footstep signatures from sound signals, which were recorded during foot interactions with diverse floors. The same signatures could be reproduced online essentially by reversing this procedure, i.e., by mapping the predictor onto the coefficients of a parametric re-synthesis filter, and by feeding this filter with signals having the temporal density and envelopes calculated during the analysis.

Especially innovative and rewarding, in this modeling approach, was its tight interactivity with non musical sound events. Not only did this system allow straightforward connection of floor interfaces like sensing mats; it also put a palette of controls available to the users, who could manipulate the synthesis parameters for trimming the results of the analysis, and furthermore introduce their own taste to the footstep sounds. A similar interaction design approach was followed by Fontana and Bresin one year later in form of C external code for the Puredata realtime environment, limitedly to the interactive simulation of aggregate grounds [30]: as opposed to Cook, their model was completely independent of pre-recorded material, instead relying on a physics-based impact model simulating a point-wise mass colliding against a resonant object through a nonlinear spring. This model was employed to generate bursts of micro impacts in real time, whose individual amplitude and temporal density followed stochastic processes taken from respective physical descriptions of *crumpling* events. Such descriptions expose macro parameters (respectively of amplitude and temporal density) that, for the purpose of this model, could be used for user control. Finally, an amount of potential energy could be set which was progressively consumed by the micro impacts during every footstep: this feature made it possible to trigger a footstep on a specific floor directly, i.e. with no further information needed, and allowed the authors to reproduce slow-downs taking place at the end of a run, based on assumptions on human movement having links to musical performance.

Both such models have imposed the closed-loop interaction paradigm to the specific area of interactive walking simulation. This paradigm is even more constraining in the case of acoustic rendering, as only few milliseconds are allowed to the system for displaying the response signal in front of an action of the foot in contact with a sensing floor, or wearing an instrumented shoe. From there, further experiences have aimed at refining the maps linking foot actions to the synthesized sound. In particular, an attempt to integrate some biomechanical parameters of locomotion, particularly the ground reaction force, in a real time footstep sound synthesizer was made by Farnell in 2007 [27]. The result was a patch for Puredata that was furthermore intended to provide an audio engine for computer games, in which walking is interactively sonified.

12.2.2.2 Current Approaches to Walking Sound Synthesis

The synthesis of walking sounds has been recently centering around multimodal, interactive contexts where users are engaged in a perception and action task. In fact, for the mentioned lack of robust maps linking biomechanical data of human walking to dynamic contact laws between the foot and grounds having different properties, if the listener is not physically walking then the synthesis model can be conveniently resolved by a good dataset of footstep sounds recorded over a multiplicity of grounds, that is managed by an intelligent agent capable of understanding the

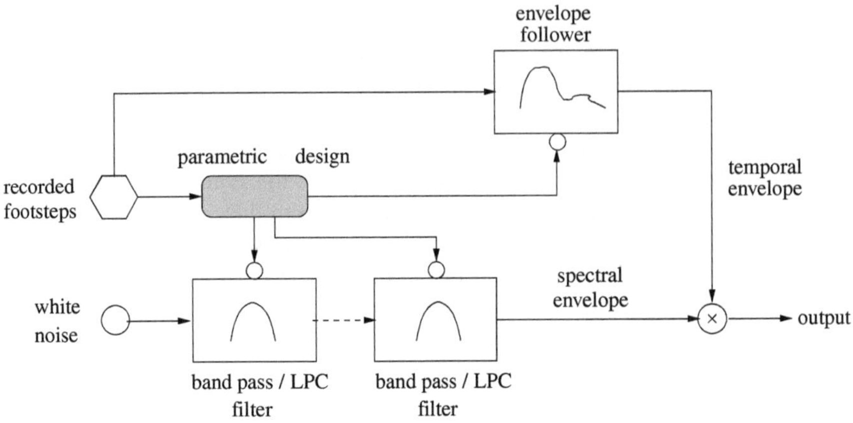

Fig. 12.1 Hybrid synthesis of footstep sounds [31]

context: this happens e.g. in recent videogames, where the scenarios and situations in which the game characters are engaged provide the ground parameters and the kinematic data enabling the selection of appropriate elements of a knowledge base. Extremely accurate collections of walking sounds exist, especially in commercial repositories like http://sounddogs.com, for creating such a dataset.

Somehow closer to an interactive synthesis paradigm, a hybrid model has been recently proposed based on a simplified version of Cook's method, relying on the temporal envelope control of filtered noise [31]. As Fig. 12.1 shows, every footstep results by weighing the output of a series of linear filters through a temporal envelope function. Both such filters coefficients and this function report of a characteristic locomotion style on a specific ground material, whose sonic signature is extracted from a set of recorded samples: the former obtained by Linear Predictive Coding of these samples, the latter created by defining a force-dependent stochastic process on top of the same recorded information.

The approach based on datasets or hybrid generation has fewer points when the auditory feedback must tightly follow the locomotion and foot gestures of the walkers. As we previously said, in this situation users are not passive listeners, conversely they are engaged in a perception and action task. However in this case, also on the light of the psychoacoustic experiments previously described, the applicability of interactive sound rendering is necessarily limited since real walking cannot be substituted with a virtual experience, nor can the auditory cues contradict the tactile perception through the feet. For this reason, the synthesis models which are currently receiving most attention are those capable of rendering aggregate grounds. The related cues, in fact, can conveniently "overwrite" the feedback provided by flat and homogeneous, sufficiently silent floors such as those covering normal buildings and other urban spaces. For these floors, interesting augmentations can be realized especially if companion vibrotactile cues of aggregate ground material are provided underfoot, simultaneously with the corresponding auditory feedback.

The Natural Interactive Walking EU project, active until fall 2011, has put major emphasis on the audio-tactile augmentation of otherwise neutral floors through the use of active tiles as well as instrumented shoes. Both such interfaces, detailed in Sect. 12.3.3, were designed based on the fundamental hypothesis that a credible, however informative augmentation of a flat, solid floor could be realized via the superposition of virtual audio-tactile cues. As noted in Sect. 12.2.3, in practice these cues had to guarantee an especially "strong" characterization to walkers having normal sensory abilities, mainly to counterbalance the unavoidable bias caused by the visual appearance of a ground surface: silent floors, then, were augmented so to sound either as aggregate grounds, or strongly coloring (such as wooden) surfaces.

Effective audio-tactile simulations of aggregate and resonant ground categories have been obtained through physically-based sound synthesizers, whose low-level core made use of the same dynamic impact model as that used by Fontana and Bresin [30]. In phenomenological sense, physics-based models have the fundamental advantage to provide a coherent multimodal feedback: since they reproduce force and velocity signals, then their response can be inherently used to mechanically excite the resonant body, in our case a floor; once this excitation is known, along with the resonance properties of the same floor, then it is not difficult to get sounds as well as vibrations from it. Specifically, a footstep sound can be considered to be the result of multiple microimpacts between a shoe and a floor. Either they converge to form a unique percept consisting of a single impact, in the case of solid materials, or conversely they result in a more or less dispersed, however coherent burst of impulsive sounds in the case of aggregate materials.

An impact involves the interaction between two bodies: an active exciter, i.e., the impactor, and a passive resonator. Sonic impacts between solid surfaces have been extensively investigated, and results are available which describe relationships between physical and perceptual parameters of the objects in contact [52, 103]. The most simple approach to the synthesis of such sounds is based on a lumped source-filter model, in which a signal $s(t)$ modeling the excitation is passed through a linear filter with impulse response $h(t)$ modeling the resonator, and resulting in an output expressed by the linear convolution of these two signals: $y(t) = s(t) * h(t)$. A more accurate reproduction of the contact between two bodies can be obtained by simulating the nonlinear dynamics of this contact: a widely adopted description considers the force f between them to be a function of the compression x of the exciter and velocity of impact $\dot{x}$, depending on the parameters of elasticity of the materials, masses, and local geometry around the contact surface [3]:

$$f(x, \dot{x}) = \begin{cases} -kx^{\alpha} - \lambda x^{\alpha}\dot{x}\,, & x > 0 \\ 0\,, & x \leq 0 \end{cases} \tag{12.1}$$

where k accounts for the material stiffness, λ represents the force dissipation due to internal friction during the impact, α depends on the local geometry around the contact surface. When $x \leq 0$ the two bodies are not in contact.

Friction is another crucial category at the base of footstep sound generation [36]. This phenomenon has been synthesized as well, by means of a dynamic model in which the relationship between relative velocity v of the bodies in contact and friction force f is represented as a differential problem [4]. Assuming that friction results from a large number of microscopic elastic bonds, also called bristles, the velocity-to-force $f(\ldots, v, \ldots)$ relationship is expressed as:

$$f(z, \dot{z}, v, w) = \sigma_0 z + \sigma_1 \dot{z} + \sigma_2 v + \sigma_3 w \tag{12.2}$$

where z is the average bristle deflection, the coefficient σ_0 is the bristle stiffness, σ_1 the bristle damping, and the term $\sigma_2 v$ accounts for linear viscous friction. The fourth component $\sigma_3 w$ relates to surface roughness, and is simulated as fractal noise.

12.2.3 Walking Sounds and Soundscape Reproduction

The algorithms described in the previous section provide faithful simulations of walking sound on different surfaces. In order to achieve realistic simulations of virtual environments, it is important to provide a context to such sounds, i.e., to be able to render them as delivered in specific locations.

"Spaces speak, are you listening?" asks the title of a book by Blesser and Salter, which explores the topic of aural architecture from an interdisciplinary perspective considering audio engineering, anthropology, human perception and cognitive psychology [7]. Indeed listening to a soundscape can provide useful information regarding the size of the space, the location, the events happening. The sounds associated to a place can also evoke emotions and memories. Moreover, when exploring a place by walking, at least two categories of sounds can be identified: the person's own footsteps and the surrounding soundscape. Studies on soundscape originated with the work of Murray Schafer [87]. Among other ideas, Schafer proposed soundwalks as empirical methods for identifying a soundscape for a specific location. During a soundwalk it is important to pay attention to the surrounding environment from an auditory perspective, while physically blocking the input from strong sensorial modality like vision, by walking blindfolded. Schafer claimed that each place has a soundmark, i.e., sounds which one identifies a place with.

When reproducing real soundscapes in laboratory settings several challenges are present, both from the designer's point of view and from the technologist's point of view. From the designer's point of view, the main challenge is how to select the different sonic events that combined together produce a specific soundscape. From this perspective the scientific literature is rather scarce. The approach usually adopted is merely based on the artistic skills and intuitions of the sound designer. However, an exception is the work of Chueng [11], who suggested to design soundscapes based on users' expectations. Her methodology consists of asking people which sounds they associate to specific places, and then use their answers as a starting point to create soundscapes. Chueng also proposes discrimination as an important parameters in soundscape design. Discrimination is defined as the ability of a soundscape to present few easily identifiable soundmarks. In her approach, this is also called minimal ecological sound design.

Studies have shown the importance of auditory cues in virtual reality simulation, and how they can lead to measurable enhancement in what is called the feeling of presence. In [86] it is reported how sound contributes to user's sense of presence, as evidenced by electrodermal activity and temperature measurements, as well as questionnaire's scores. Moreover, significant differences were noticed when measuring delivered sound through headphones or surround sound (5.1) using loudspeakers. Other studies show how ratings of presence are enhanced by either the addition of bass or the increase of volume. On the other hand, an increase on number of channels does not increase ratings of presence [32]. The role of self-produced sounds to enhance sense of presence in virtual environments has also been investigated. By combining different kinds of auditory feedback consisting of interactive footstep sounds created by ego-motion using the techniques described in the previous section with static soundscapes, it was shown how a person's motion in a virtual reality environment is significantly enhanced when moving sound sources and ego-motion are rendered [74].

Concerning delivery of footstep sounds, they can be conveyed to the walker by means of different hardware devices, such as headphones, loudspeakers or through bone conduction. The choice of delivery methods depends on several factors, for example if the soundscape has to be part of a mobile or augmented reality installation, or if it is part of a virtual reality laboratory setting. An ecologically valid solution consists of placing loudspeakers at the shoes' level, since this faithfully reproduces the equivalent situation in real life, where footstep sounds come at the level of the interaction between a shoe and a floor. As an alternative, sounds can be conveyed by means of a system of multichannel loudspeakers. In this case a problem arises regarding how footstep sounds can be rendered in a 3D space, and how many loudspeakers should be used and where they should be placed.

Sound rendering for virtual environments has reached a level of sophistication that it is possible to render in realtime most of the phenomena which appear in the real world [34]. 3D spatialized audio in immersive virtual environments remains however still challenging. In delivering through multichannel speakers, the choice of rendering algorithms is fundamental. As a matter of fact, various typologies of soundscapes can be classified: static soundscapes, dynamic soundscapes and interactive soundscapes. Static soundscapes are those composed without rendering the appropriate spatial position of the sound sources. In static soundscapes the same content is delivered to every channel of the surround sound system. The main advantage of this approach is the fact that the user exploring the virtual environment does not need to be tracked, since the same content is displayed to every speaker no matter where the user is placed. The main disadvantage is the fact that the simulation does not represent a real life scenario, where different sonic cues are received depending on where a person is placed. Dynamic soundscapes are those where the spatial position of each sound source is taken into account, as well as their eventual movements along three-dimensional trajectories. Finally, interactive soundscapes are based on the dynamic ones where in addition the user can interact with the simulated environment generating an auditory feedback as result of actions. This last situation ideally represents the scenario with augmented footstep sounds, where each step of the user must be tracked and rendered while the user is walking in the virtual environment, without any perceivable latency, in order to recreate for example the illusion of walking on a surface different from the one the user is actually stepping upon, or to allow the user to interact with objects of the virtual environment.

Sound delivery using headphones can also be performed using two general approaches: the simple mono or stereo delivery and a solution based on binaural synthesis. One of the main issue in combining footstep sounds and soundscape design is to find the right amplitude balance between the two. One approach can be empirical, by asking subjects to walk freely while interactively producing the simulated footstep sounds and hearing the reproduced soundscape through multichannel speakers. Subjects are then able to adjust the volume of footstep sounds until they find a level which they considered satisfactory. After describing the possibilities offered by hardware technologies, the next section describes available software packages for footstep sound design.

12.2.4 Footstep Sound Design Toolkits

A specific treatment on the use of the above models for foot-floor interaction purposes has been presented by Serafin et al. [88], along with pointers to sources of software, sound, and other documentation material. Implementing such models is not straightforward, but real-time software modules realizing impact and friction interactions are available, that are open and flexible enough for inclusion in more general architectures for the synthesis of footstep sounds. In particular, the Sound Design Toolkit[1] (SDT) [21] contains a set of physically-consistent tools for designing, synthesizing and manipulating ecological sounds [36] in real time. SDT consists of a collection of visual programs (or *patches*) and dynamic libraries (or *externals*) for the software Puredata, which is publicly available, and Max/MSP, which is easier to work with although commercial. SDT provides also examples, allowing users to launch these patches and see them at work in both such visual environments.

Public software is also available, which implements footstep sound synthesis models that are ready for use. Farnell accompanied his work with a patch and an external for Puredata, both referenced in the related paper [27]. Fontana's crumpling model for Puredata has been integrated in SDT: examples of this model at work can be found, among others, in the Natural Interactive Walking project website.[2] The same website collects sound examples resulting from alternative instantiations of the physically-based approach, based on a sound synthesis engine that has not been put available in the public domain [102]. Furthermore, it contains footstep sounds that have been generated using the aforementioned hybrid model descending from Cook's synthesis technique.

[1] http://www.soundobject.org/SDT/

[2] http://niw.soundobject.org

12.3 From Haptic to Multimodal Rendering

12.3.1 Introduction

12.3.1.1 Walking and Haptic Feedback in Virtual Environments

Virtual reality applications aim at simulating digital environments with which users can interact and, as a result, perceive through different modalities the effects of their actions in real time. Current VR applications draw primarily on vision and hearing. Haptic feedback—which aims to reproduce forces, movements and other cutaneous sensations felt via the sense of touch—is rarely incorporated, especially in those VR applications where users are enabled to walk.

A heightened sense of presence can be achieved in a VR simulation via the addition of even low-fidelity tactile feedback to an existing visual and auditory environment, and the potential gains can, in some cases, be larger than those obtained by improving feedback received from a single existing modality, such as the visual display [91].

High-frequency information in mechanical signals often closely links the haptic and auditory modalities, since both types of stimuli have their origin in the same physical contact interactions. Thus, during walking, individuals can be said to be performing simultaneous auditory and haptic probing of the ground surface and environment. As demonstrated in recent literature, walkers are capable of perceptually distinguishing ground surfaces using either discriminative touch via the feet or audition [39]. Thus, approaches to haptic and auditory rendering like those reviewed in this chapter share common features, while the two types of display can be said to be partially interchangeable.

An important component of haptic sensation is movement. Walking is arguably the most intuitive means of self-motion within a real or virtual environment. In most research on virtual environments, users are constrained to remain seated or to stand in place, which can have a negative impact on the sense of immersion [90]. Consequently, there has been much recent interest in enabling users of such environments to navigate by walking. One feasible, but potentially cumbersome and costly, solution to this problem is to develop motorized interfaces that allow the use of normal walking movements to change position within a virtual world. Motorized treadmills have been extensively used to enable movement in one-dimension, and this paradigm has been extended to allow for omnidirectional locomotion through an array of treadmills revolving around a larger one [49]. Another configuration consists of a pair of robotic platforms beneath the feet that are controlled so as to provide support during virtual foot-ground contact, while keeping the user in place. Another configuration consists of a spherical cage that rotates as a user walks inside of it [46]. The reader could refer to the chapter by Iwata in this volume for further discussion of these scenarios. The range of motion, forces, and speeds that are required to simulate omnidirectional motion make these devices intrinsically large, challenging to engineer, and costly to produce. In addition, while they are able to simulate the support and traction supplied by the ground, they cannot reproduce the feeling of walking on different materials.

Lower-cost methods for walking in virtual environments have been widely pursued in the VR research community. Passive sensing interfaces have been used to allow for the control of position via locomotion-like movements without force feedback [94]. Walking in place is another simple technique, in which movements of the body are sensed, and used to infer an intended movement trajectory [96]. For virtual environments that are experienced via an audiovisual head mounted display, a user's locomotion can be directly mapped to movements in a virtual environment. The real walkable workspace is typically much smaller than the virtual environment, and this has led to the development of techniques, such as redirected walking [81], that can engender the perceptual illusion that one is walking in a large virtual space.

The auditory and tactile experience of walking on virtual materials can be simulated by augmenting foot-ground interactions with appropriate sounds or vibrations. Although vibrotactile interfaces are simpler and lower in cost to implement than haptic force feedback devices [62], they have only recently been used in relation to walking in virtual environments. Auditory displays have been more widely investigated, and walking sounds are commonly used to accompany first-person movements in immersive games, although they are rarely accompanied by real foot movements. Cook developed a floor interface (the Pholiemat), for controlling synthesized walking sounds via the feet, inspired by foley practice in film [14, 15], and other researchers have experimented with acoustically augmented shoes [77]. Research on the use of vibrotactile displays for simulating virtual walking experiences via instrumented shoes [89] or floor surfaces [107] is still in its infancy.

Although tactile displays have, to date, been integrated in very few foot-based interfaces for human–computer interaction, several researchers have investigated the use of simple forms of tactile feedback for passive information conveyance to the feet. Actuated shoe soles used to provide tactile indicators related to meaningful computing events [85, 105], and rhythmic cues supplied to the feet via a stair climber have been found to be effective at maintaining a user's activity level when exercising. In automotive settings, tactile warning cues delivered via the accelerator pedal have been studied for many years [67], and eventually appeared in production vehicles. Tactile stimulation to the feet has also been explored as an additional feedback modality in computer music performance [84].

12.3.1.2 Haptic and Acoustic Signals Generated by Walking Interactions

During walking interactions, several touch interactions are involved with the virtual ground. Stepping onto a natural or man-made surface produces rich multimodal information, including mechanical vibrations that are indicative of the actions and types of materials involved. Stepping on solid floors in hard-soled shoes is typified by transient signals associated with the strike of the heel or toe against the floor, while sliding can produce signals such as high-pitched squeaking (when surfaces are clean) or textured noise. In indoor environments, the operation of common foot operated switches, used for lamps, dental equipment, or other machines, is often accompanied by transient clicks accompanying the engagement solid mechanical elements. The discrete quality of these mechanical signals contrasts with the more continuous nature of those generated by a step onto natural ground coverings, such as gravel, dry sand, or branches. Here, discrete impacts may not be as apparent, and can be accompanied by both viscoelastic deformation and complex transient, oscillatory, or noise-like vibrations generated through the inelastic displacement of heterogeneous materials [45]. A few of the processes that can be involved include brittle fracture and the production of in-solid acoustic bursts during rapid micro-fracture growth [1, 2, 45], stress fluctuations during shear sliding on granular media [5, 19, 72, 73], and the collapse of air pockets in soil or sand.

A series of mechanical events can be said to accompany the contact of a shoed foot with the ground. There may be an impact, or merely a soft landing, according to the type of shoe, the type of ground, and the stride of the walker. Once the initial transitory effects have vanished and until the foot lifts off the ground, there may be crushing, fracturing, or little movement at all if the ground is stiff. There may also be slipping if the ground is solid, or soil displacement if the ground is granular. There may be other mechanical effects, such as the compacting of a compressible ground material (e.g., soil, sod, snow). The question of what form of haptic signal to reproduce in virtual reality applications is therefore not so simple to answer. The sense of touch is nearly as refined in the foot as it is in the hand. It has, in fact, great discriminative acumen, even through a shoe sole [39]. However, like vision or audition, in accordance to the perceptual task, it may be satisfied by relatively little input. In the case of foot, our habit to wear shoes plays in our favor since shoes filter out most of the distributed aspects of the haptic interaction with the ground, save perhaps for a distinction between the front and back of the foot at the moment of the impact. In that sense, wearing a shoe is a bit like interacting with an object through a hand-tool. The later case, as is well known, is immeasurably easier to simulate in virtual reality than direct interaction with the hand. When it comes to stimulating the foot, the options are intrinsically limited by the environmental circumstances. While it is tempting to think of simulating the foot by the same methods as those used to stimulate the hand [43, 44], this option must be discarded in favor of approaches that are specific to the foot. In particular, options involving treadmills, robot arms and other heavy equipment will remain confined to applications where the motor aspects dominate over the perceptual aspects of interacting with a ground surface [10, 20, 48, 79].

Broadly speaking, then, foot-ground interactions can be said to be commonly accompanied by mechanical vibrations with energy distributed over a broad range of frequencies (see Fig. 12.2). High-frequency vibrations can originate with a few different categories of physical interaction, including impacts, fracture, and sliding friction. The physics involved is relatively easy to characterize in restricted settings, such as those involving homogeneous solids, but becomes more complex to describe when disordered, heterogeneous materials are involved.

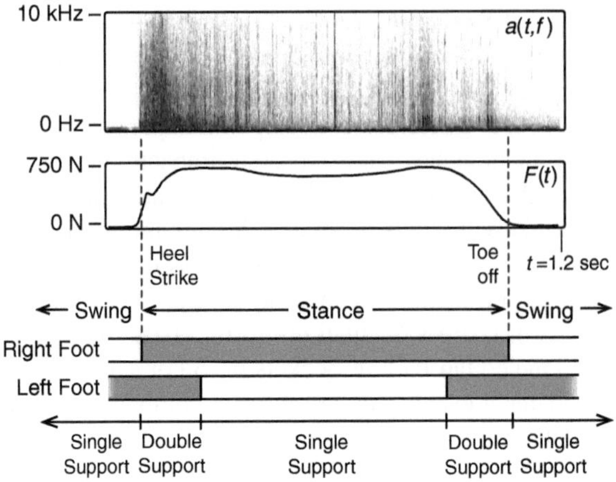

Fig. 12.2 Walking in real environments produces rich, step-dependent vibromechanical information. Shown: vibration spectrogram $a(t, f)$ and low-frequency normal foot-ground force $F(t)$ measured at the hard sole of a men's shoe during one footstep of a walker onto rock gravel, together with the corresponding foot contact states within the gait cycle (author's measurements). The *dark vertical stripes* in the spectrogram correspond to discrete impact or dislocation events that are characteristic of dynamic loading of a complex, granular medium

12.3.2 Touch Sensation in the Feet

The sense of touch in the human foot is highly evolved, and is physiologically highly similar to that in the hand, with the same types of tactile receptor populations as are present in the former, including the fast-adapting (FA) type I and II and slow-adapting (SA) type I and II cutaneous mechanoreceptors [50, 100], in addition to proprioceptive receptors including Golgi organs, muscle spindles, and joint capsule receptors in the muscles, tendons, and joints. The sole is sensitive to vibrotactile stimuli over a broad range of frequencies, up to nearly 1000 Hz [109], with FA receptors comprising about 70 % of the cutaneous population. Several differences between tactile sensation in the foot and hand have been found, including an enlargement and more even distribution of receptive fields in the foot, and higher physiological and psychophysical thresholds for vibrotactile stimuli [50, 114], possibly related to biomechanical differences between the skin of the hands and feet [115]. Further comparisons of the vibrotactile sensitivity of the hand and foot were performed by Morioka et al. [68].

Self motion is the key function of walking, and most of the scientific research in this area is related to the biomechanics of human locomotion, and to the systems and processes underlying motor behavior on foot, including the integration of multisensory information. During locomotion, sensory input and muscular responses are coordinated by reflexes in the lower appendages [83, 95, 116], and prior literature has characterized the dependence of muscular responses on both stimulus properties and gait phase. The vibrotactile sense in the foot has been less studied in this regard, presumably because it is not a primary channel for directly acquiring information about forces and displacements that are required for the control of locomotion and balance.

Perceptual abilities of the foot are essential to the sensorimotor loop involved in the control of locomotion, but have been less studied than those of the hand. Prior literature has emphasized perceptual-motor abilities related to the regulation of locomotion and balance on slippery, compliant, or slanted surfaces [23, 29, 41, 51, 63, 66, 69, 70]. The stepping foot is able to discriminate materials distinguished by elasticity [53, 82] or by raised tactile patterns [16, 54], as demonstrated in research aimed at evaluating the utility of these features for aiding visually impaired people in walking or navigating safely and effectively.

Although walking on natural ground surfaces generates rich haptic information [25, 35, 110], little research exists on the perception of such materials during locomotion. Giordano et al. investigated a setting in which walkers were tasked with identifying man-made and natural walking surface materials in different non-visual sensory conditions, while wearing shoes [38]. Better than chance performance was observed in all conditions in which tactile information was unmodified. Performance was worse when tactile information was degraded by a vibrotactile masking signal supplied to the foot sole. Although the latter could have affected haptic information in multiple ways (by perturbing high- and low-frequency cutaneous tactile information and/or information from deeper joint and muscle proprioceptors) subsequent analyses indicated that this information was highly relevant for discriminating walking grounds. Furthermore, the results suggested that similar high frequency information was communicated through both auditory and tactile channels.

12.3.2.1 Vibrotactile Rendering of Footsteps

Due to the highly interactive nature of the generation of haptic stimuli in response to foot-applied pressure, the display of haptic textures, in the form of high frequency vibrations simulating the feel of stepping onto heterogeneous solid ground materials [107], is a significant challenge to be overcome in the multimodal rendering of walking on virtual ground surfaces. During a step onto quasi-brittle porous natural materials (e.g., sand or gravel), one evokes physical interaction forces that include viscoelastic components, describing the recoverable deformation of the volume of the ground surrounding the contact interface; transient shocks from the impact of foot against the ground; and plastic components from the collapse of brittle structures or granular force chains, resulting in unrecoverable deformations [24, 92]. Combinations of such effects give rise to the high frequency, texture-like vibrations characteristic of the feel of walking on different surfaces [26]. Figure 12.3 presents an example of force and vibration data acquired by the authors from one footstep on a gravel surface.

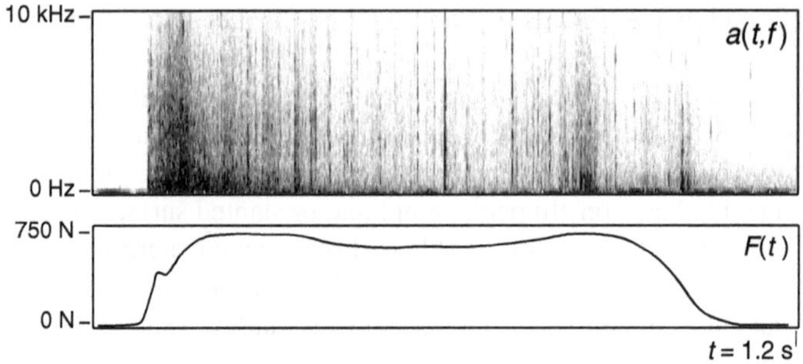

Fig. 12.3 Vibration spectrogram $a(t, f)$ and normal force $F(t)$ measured from one footstep onto rock gravel (Authors' recording). Note the discrete (impulsive) broadband impact events evidenced by *vertical lines* in the spectrogram

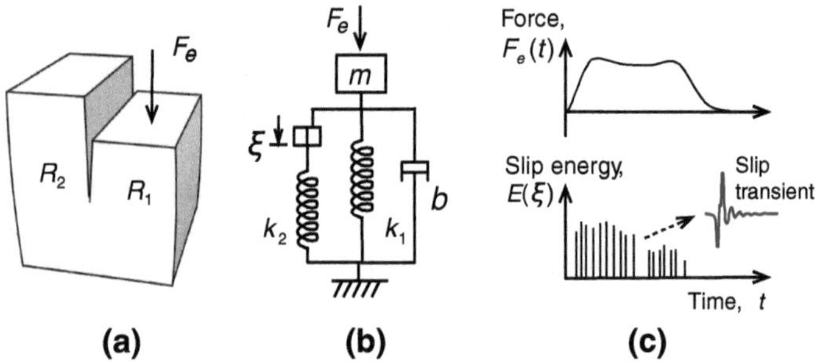

Fig. 12.4 Normal force texture synthesis. **a** A fracture mechanics approach is adopted. A visco-elasto-plastic body undergoes shear sliding fracture due to applied force F_e. **b** A simple mechanical analog for the generation of slip events $\xi(t)$ in response to F_e. **c** For vibrotactile display, each slip event is rendered as an impulsive transient using an event-based approach [108]

12.3.2.2 Stepping on Disordered Heterogeneous Materials

Due to the continuous coupling of acoustic and vibromechanical signals with force input in examples such as that described above, there is no straightforward way to convincingly use recorded footsteps for acoustic or vibrotactile rendering, although more flexible granular sound-synthesis methods could be used [6, 18]. For the modeling of simpler interactions, involving impulsive contact with solid materials, recorded transient playback techniques could be used [55].

A simple yet physically-motivated approach that can be used in the haptic synthesis of interaction with complex, compressible surfaces is based on a minimal fracture mechanics model [108]. Similar approaches have proved useful for modeling other types of haptic interaction involving damage [42, 64]. Figure 12.4 illustrates the continuum model and a simple mechanical analog used for synthesis.

In the stuck state, the surface has stiffness $K = k_1 + k_2$, effective mass m and damping constant b. It undergoes a displacement x in response to a force F, as

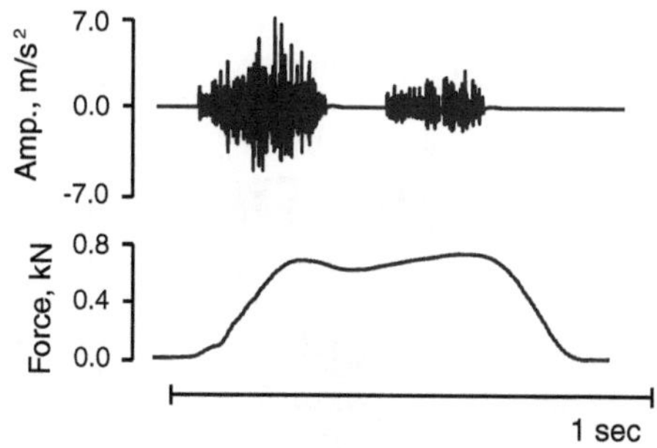

Fig. 12.5 Example footstep normal force and synthesized waveform using the simple normal force texture algorithm described in the text. The respective signals were captured through force and acceleration sensors integrated in the vibrotactile display device [108]

governed by:

$$F(t) = m\ddot{x} + b\dot{x} + K(x - x_0), \quad x_0 = k_2\xi(t)/K \tag{12.3}$$

In the stuck state, virtual surface admittance $Y(s) = \dot{x}(s)/F(s)$ is given, in the Laplace-transformed (s) domain, by:

$$Y(s) = s(ms^2 + bs + K)^{-1}, \quad K = k_1k_2\xi/(k_1 + k_2) \tag{12.4}$$

where $\xi(t)$ represents the net plastic displacement up to time t. A Mohr-Coulomb yield criterion is applied to determine slip onset: when the force on the plastic unit exceeds a threshold value (which may be constant or noise-dependent), a slip event generates an incremental displacement $\Delta\xi(t)$, along with an energy loss of ΔW representing the inelastic work of fracture growth.

Slip displacements are rendered as discrete transient signals, using an event-based approach [55]. High frequency components of such transient mechanical events are known to depend on the materials and forces of interaction, and we model some of these dependencies when synthesizing the transients [110]. An example of normal force texture resulting from a footstep load during walking is shown in Fig. 12.5.

12.3.3 Multimodal Displays

Several issues arise in the rendering of multimodal walking interactions, including combinations of visual, auditory and haptic rendering to enable truly multimodal experiences. A model of the global rendering loop for interactive multimodal experiences is summarized in Fig. 12.6. Walking over virtual grounds requires the use of specific hardware devices that can coherently present visual, vibrotactile and acoustic signals. Several devices dedicated to multimodal rendering of walking over virtual grounds are described below, in Sect. 12.3.4, while examples of multimodal scenarios are discussed in Sect. 12.3.5. Multimodal rendering is often complicated due to hardware and software constraints. In some cases, crossmodal perceptual effects

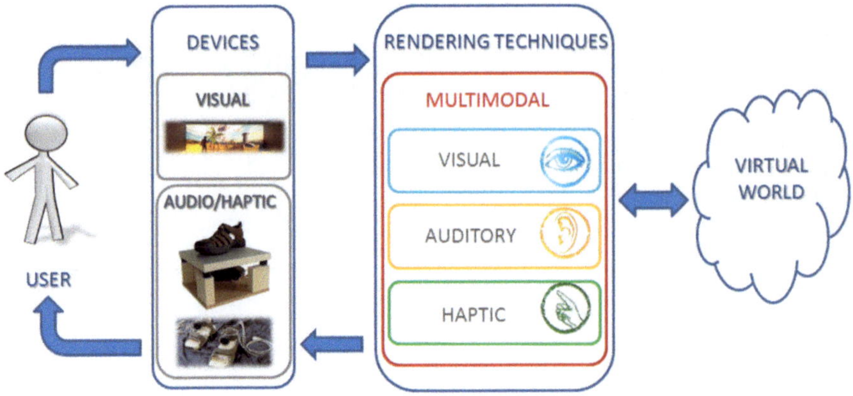

Fig. 12.6 Global loop for multimodal rendering approaches of walking over virtual grounds. The user (*left*) is interacting with the virtual world (*right*) through specific hardware devices. Rendering techniques allows multimodal feedback by taking into account the user's input and the virtual environment. The feedback are provided through visual, acoustic and haptic interfaces

can be exploited to allow one modality (for example, vision) to render sensations that would normally be presented via another modality that may not be feasible to reproduce (e.g., via haptic force feedback). Some of these approaches are described at the end of the Sect. 12.3.5.

12.3.4 Display Configurations

In this paragraph, we discuss two types of devices capable of the generation of multimodal cues for the interaction with virtual grounds, and corresponding to two different approaches: actuated floors, an array of sensors an actuators laid out on a given space transmitting the different cues to the user stepping on them; and actuated shoes, mobile devices worn by the user with sensors and actuators embedded in the shoes.

The two approaches stimulate the foot with the simulated high-frequency mechanical feedback, viz. 30–800 Hz, from foot-ground interactions. As it turns out, a wide variety of sensations can be produced this way, including those that would normally be ascribed to kinesthesia [109]. Auditory feedback is also generated by the resulting prototypes, as a by-product of the vibrotactile actuators aboard them, or via associated loudspeaker arrays, and visual feedback may be supplied via top-down video projection systems. One approach is to tile a floor and actuate each tile according to the movement and interaction of the walker or the user. Another approach is to provide the walker with shoes augmented with appropriate transducers. In addition to the devices described in other sections of this chapter, the vibrotactile augmentation of touch surfaces has been widely investigated for HCI applications

[33, 71, 80], although design issues affecting their perceptual transparency have often been neglected. As case studies, two approaches are described below, starting with floor-based stimulator and continuing with a shoe-based stimulator.

12.3.4.1 Actuated Floors

Floor-based systems for providing multimodal feedback to the foot offer the advantage of easy accessibility, since users are not required to wear any special footwear or equipment in order to use them. Furthermore, they can be readily designed with an extensible architecture, which allows them to be networked and powered easily, as they can be integrated within existing room infrastructures. However, on the negative side, such systems can be said to be somewhat invasive, since they require modifications to the existing floor infrastructure of a building, thus requiring a comparatively permanent installation space. The workspace available to users—that is, the amount of real space within which they can interact—depends on the size of the actuated floor, with a larger workspace inevitably entailing higher costs and complexity.

The vibrotactile floor tile interface developed by Visell et al. [106, 107, 110, 111] represents the first systematically designed device of its type for haptic human–computer interaction. Passive floor-based vibrotactile actuation has been used to present low frequency information in audiovisual display applications, for special effects (e.g., vehicle rumble), in immersive cinema or VR settings [99]. The fidelity requirements that must be met by an *interactive* haptic display are, however, higher, since users are able to actively sample its response to actions of the feet. The device of Visell et al. is based on a high fidelity vibrotactile interface integrated in a rigid surface, with visual feedback from top-down video projection and a spatialized, eight-louspeaker auditory display. The main application for which it was envisioned is the vibrotactile display of virtual ground surface material properties for immersive environments. The device consists of an actuated composite plate mounted on an elastic suspension, with integrated force sensors. The structural dynamics of the device was designed to enable it to accurately reproduce vibrations felt during stepping on virtual ground materials over a wide range of frequencies. Measurements demonstrated that it is capable of reproducing forces of more than 40 N across a usable frequency band from 50 to 750 Hz. In a broader sense, potential applications of such a device include the simulation of ground textures for virtual and augmented reality simulation [112] or telepresence (e.g., for remote planetary simulation), the rendering of abstract effects or other ecological cues for rehabilitation, the presentation of tactile feedback to accompany the operation of virtual foot controls, control surfaces, or other interfaces [111], and to study human perception. In light of the latter, an effort was undertaken to ensure a high fidelity response that would avoid the reproduction of vibromechanical stimuli.

The interface of the device (Fig. 12.7) consists of a rigid plate that supplies vibrations in response to forces exerted by a user's foot, via the shoe. The total normal force applied to the plate by a user is measured. It can be assumed to consist of two components: isolated transients with high frequency content, generated by foot

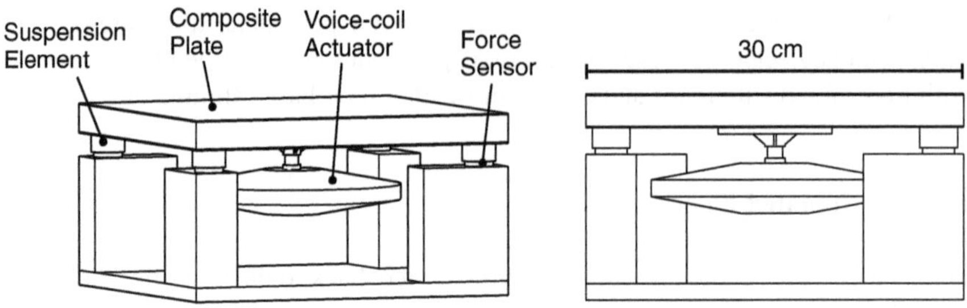

Fig. 12.7 Vibrotactile floor interface hardware for a single tile unit. *Left* (View) showing main components. *Right* Side view with top dimension

impacts with the plate, and low-frequency forces generated by active human motions, limited in bandwidth to no more than 10 Hz [9, 110]. A haptic simulation provides feedback approximating the vibration response felt during interaction with a virtual object. The rendering algorithms are of admittance type, computing displacements (or their time derivatives) in response to forces applied to the virtual object. Force sensing is performed via four load cell force transducers (Measurement Systems model FX19) located below the vibration mount located under each corner of the plate. Although the cost for outfitting a single-plate device with these sensors is not prohibitive, potential applications of this device to interaction across distributed floor surface areas may involve two dimensional $m \times n$ arrays of tiles, requiring a number $N = 4mn$ of sensors. As a result, in a second configuration, four low-cost resistive force sensors are used in place of load cells. After conditioning, the response of these sensors to an applied force is nonlinear, and varies up to 25 % from part to part (according to manufacturer ratings). A linearization and calibration of force sensing is performed [112] ensuring a response accurate to within a few percent. Analog data from the force sensors is conditioned, amplified, and digitized, and used as an input to drive a physically-based simulation of a ground surface such as sand, snow, or ice. Vibromechanical feedback is provided by a single Lorentz force type inertial motor (Clark Synthesis model TST429) with a usable bandwidth of about 25 Hz to 20 kHz, which is driven using standard digital and analog audio hardware. The Fig. 12.8 provides an overview of the system.

12.3.4.2 Actuated Shoes

Actuated shoes provide a mobile solution to foot-floor interaction setups, not requiring the use of large floors laid out in specific spaces. However, the realization of a mobile device delivering the same cues as a static actuated floor poses serious technical questions. While the size of the device needs to remain small enough not to impair the natural walking gait of the user, the intensity of the signals it delivers must allow the rendering of perceivable interaction cues. Power supply as well as

Fig. 12.8 Photo of an actuated tile with large mens' shoe, showing representative size. The model shown is based on the low-cost force sensing resistor option. The cable in the foreground interfaces the sensors with the data acquisition unit

computation units might be too large and cumbersome to be located on the wearable device itself, requiring their offload to other parts of the body.

Papetti et al. [77] addressed the design of multimodal actuated shoes through a first prototype delivering vibrotactile and acoustic signals. This device is illustrated in Fig. 12.9. Force data acquisition is made through two force sensing resistors (Interlink FSR model 402) located under the insole, one at the toe and one at the heel position. Vibrotactile feedback is produced by two vibrotactile transducers embedded in the front and the rear of the shoe sole respectively [16] (Haptuator, Tactile Labs Inc., Deux-Montagnes, Qc, Canada). Two cavities were made in the soles to accommodate these broadband vibrotactile actuators. These electromagnetic recoil-type actuators have an operational, linear bandwidth of 50–500 Hz and can provide up to 3G of acceleration when connected to light loads. They were bonded in place to ensure good transmission of the vibrations inside the soles. When activated, vibrations propagated

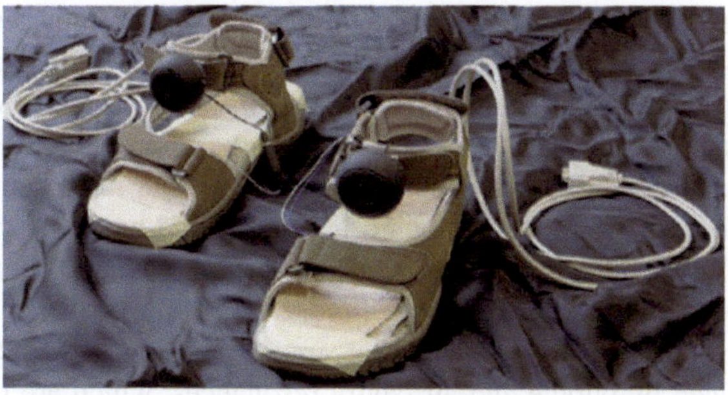

Fig. 12.9 Photo of the actuated shoes [77] with loudspeakers mounted on the *top*. The cable in the foreground interfaces the sensors in the shoes with the data acquisition unit

well in the light, stiff foam. In addition to vibrations, each shoe emits sounds from one Goobay Soundball Mobile battery loudspeaker mounted on the top buckle. These devices are provided with on-board micro-amplifiers, hence they can be connected directly to the audio card. As any small, low-power loudspeaker device, they exhibit unavoidable performance limits both in the emitted sound pressure level (2.4 W RMS) and low frequency cutoff (about 200 Hz).

An evolution of such shoes concept has made use of vibrotactile exciters, such as those capable of making an entire desk sound and vibrate like a musical soundboard once they are firmly attached to it. In the case of the actuated shoes, two Dayton Audio DAEX32 exciters were secured inside the sole of each sandal, respectively under the toes and the heel: together, they provided a more coherent audio-tactile feedback beneath the respective areas of the feet, furthermore eliciting some low resonance energy from the floor that was otherwise impossible to obtain using small speakers such as those mentioned previously. Moreover, by employing lightweight power amplification (in this case a pair of Class T battery-powered digital stereo amplifiers) and a low latency connection to and from the host, respectively to transmit force data and to receive the audiotactile signals, a good compromise between realism of the feedback and wearability of the prototype could be achieved at least for some materials such as frozen ponds, muddy soil, aggregate grounds and, if supported by headphones providing the necessary auditory spaciousness to a walking listener, also metal grates [76].

12.3.5 Interactive Scenarios

12.3.5.1 Description

We will now briefly present examples of multimodal rendering of ground materials. The examples correspond to two categories of ground materials that exhibit strong high-frequency components: granular materials and fluids. Footsteps onto granular (aggregate) ground materials, such as sand, snow, or ice fragments belie a common temporal process originating with the transition toward a minimum-energy configuration of an ensemble of microscopic systems, via a sequence of transient events. The latter are characterized by energies and transition times that depend on the characteristics of the system and the amount of power it absorbs while changing configuration. They dynamically capture macroscopic information about the resulting composite system through time. On the other hand, liquid-covered ground surfaces, such as water puddles and shallow pools, have an important kinesthetic component due to pressure and viscosity forces within the fluid, and may, at first, seem to lack high frequency mechanical responses. However, important high frequency components exist, as generated by bubble and air cavity resonances, which are responsible for the characteristic sound of moving fluids.

The two examples presented in this section utilize the fact that vibrotactile and acoustic phenomena share a common physical source by designing the vibrotactile models based on existing knowledge of fluid sound rendering. Both types of ground materials exhibit very interesting high frequency features adequate for their restitution through an actuated vibrotactile floor: as opposed to rigid surfaces, the overall signal is not reduced to transients at the moment of impact, but can produce a signal during the entire foot-floor contact duration. Although mainly focused on the vibrotactile modality, the approaches described here are multimodal. The synthesis models are also capable of generating acoustic feedback, due to common generation mechanisms and physical sources. The visual modality is an absolute requirement on its own, since interacting with virtual environments without visual feedback is of little interest, except in very specific cases.

12.3.5.2 Frozen Pond and Snow Field

In a multimodal scenario, Law et al. [57] designed a virtual frozen pond demonstration in which users may walk on the frozen surface, producing patterns of surface cracks that are rendered and displayed via audio, visual and vibrotactile channels. Audio and vibrotactile feedback accompany the fracture of the virtual ice sheet underfoot. The two are derived from a simplified mechanical model analogous to that used for rendering basic footstep sensations (see Sect. 12.3.2.2).

Based on the floor tile interface described in Sect. 12.3.4.1, the authors designed a virtual frozen pond demonstration that users may walk on, producing patterns of surface cracks that are rendered and displayed via audio, visual, and vibrotactile channels. The advantage of this scenario is that plausibly realistic visual feedback could be rendered without detailed knowledge of foot-floor contact conditions, which would require a more complex sensing configuration.

Vibrotactile and acoustic feedback are generated through the simplified fracture model described in Sect. 12.3. The visual rendering of crack surfaces on the ice is generated with sequences of line primitives on the ice texture. Cracks originate at seed locations determined by foot-floor contact, as illustrated in Fig. 12.10. In another application [57], using the same interface, the authors simulated a snow field, as also shown in Fig. 12.10. Users were enabled to leave footsteps onto virtual snow, with acoustic and vibrotactile similar to the feeling of stepping onto real snow.

12.3.5.3 Walking on Fluids

Cirio et al. [13] proposed a physically-based vibrotactile fluid rendering model for solid-fluid interaction, allowing "splashing on the beach" scenarios. Since fluid sound is generated mainly through bubble and air cavity resonance, they developed a physically-based simulator generating real-time bubble creation and solid-fluid interaction and synthesizing vibrotactile feedback from interaction and simulation events. The vibrotactile model proposed by Cirio et al. is divided in three components, following the physical processes that generate sound during solid-fluid interaction [12]: (1) an initial high frequency impact, (2) small bubble harmonics and (2) a main cavity

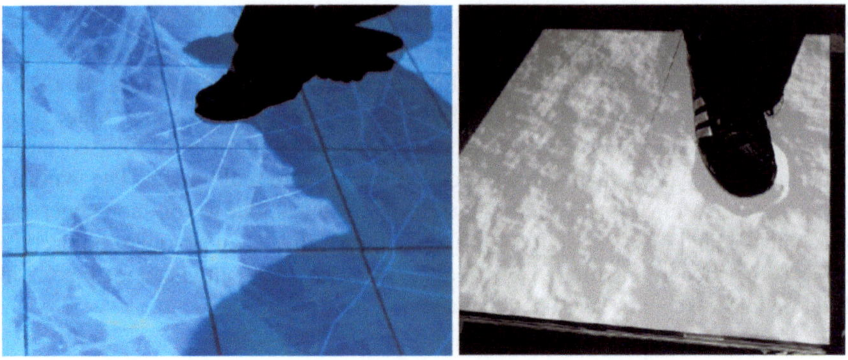

Fig. 12.10 Example of a multimodal foot-floor interaction. (*Left*) The frozen pond scenario generates vibrotactile, acoustic and visual feedback. (*Right*) The snow field is modified according to the user steps, providing multimodal feedback [57]

oscillation. A real-time fluid simulator based on Smoothed-Particle Hydrodynamics and enhanced with bubble synthesis gathers the physical simulation process.

Based on the fluid vibrotactile model [12] and using the floor tile array described in Sect. 12.3.4, Cirio et al. [13] designed two multimodal scenarios generating haptic, acoustic and visual feedback. An active shallow pool scenario allowed the user to walk about a virtual pool with 20 cm of water, splashing water as they stepped on the pool. A passive beach front scenario allowed users to stand still and feel waves washing up at their feet on a sandy beach. The floor rendered the vibrotactile feedback to the user's feet through the appropriate vibrotactile transducers. Acoustic feedback was also be provided through speakers or headphones. The user's feet were modeled as parallelepiped rigid bodies and tracked through the floor pressure sensors. Visual feedback was generated by a GPU meshless screen-based technique optimized for high frequency rendering [12] appropriate to the underlying particle based simulation. The Fig. 12.11 shows the two scenarios.

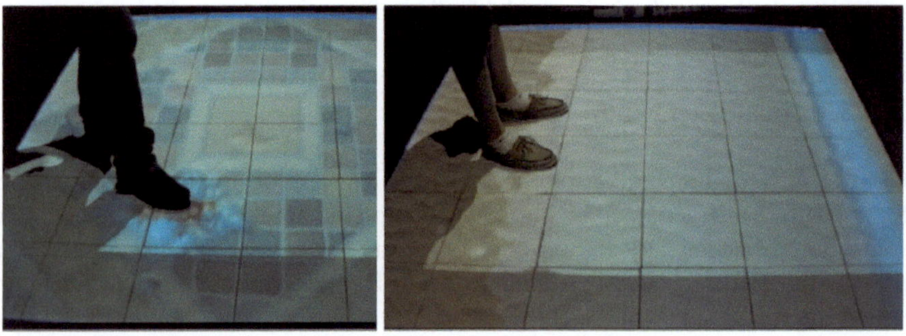

Fig. 12.11 Interacting with water with multimodal feedback. (*Left*) a shallow water pool. (*Right*) a wave washing up on a beach [13]

12.3.5.4 Augmenting Footsteps with Simulated Multimodal Feedback

The enhancement of walking sensations over virtual grounds is not necessarily limited to immersive virtual reality setups. Some applications should be able to run in desktop mode, i.e. when the user is seated and is using a basic computer. This includes training applications that need to be massively deployed, or video games. To give the sensation of walking, video games use auditory feedback intensively and footstep sounds to simulate steps. Visual information can also be used to enhance the sensation of walking.

In this desktop VR context, Terziman et al. [98] introduced a set of cues to augment virtual footsteps with artificial (exaggerated) multimodal feedback, called "King-Kong Effects". These sensory cues are inspired by special effects in movies in which the incoming of a gigantic creature is suggested by adding visual vibrations/pulses to the camera at each of its steps. Visual, tactile and acoustic signals artificially enhance each footstep detected (or simulated) during the virtual walk of the user sitting in front of the computer. The system leverages the tiles presented in Sect. 12.3.4.2 located under the user's feet, for vibrotactile rendering of foot-floor impact, in addition to the visual camera vibrations and acoustic rendering of footsteps. The authors studied the use of different kinds of feedback cues based on vertical or lateral oscillations, physical or metaphorical patterns, and one or two peaks for heal-toe contacts simulation. They showed that for a seated user, the sensation of walking is increased when the different modalities are taken together, and strongly recommend the use of such a multimodal simulation for an improved user immersion.

12.3.5.5 Pseudo-Haptic Rendering of Virtual Grounds

Pseudo-haptic feedback leverages the crossmodal integration of visual and kinesthetic cues giving rise to an illusion of force feedback [58]. Pseudo-haptic feedback was initially obtained by combining the use of a passive input device with visual feedback, simulating haptic properties such as stiffness or friction [59]. For example, to simulate the friction occurring when inserting an object inside a narrow passage, researchers proposed to artificially reduce the speed of the manipulated object during the insertion. Assuming that the object is manipulated with an isometric input device, the user will have to increase his pressure on the device to make the object advance inside the passage. The coupling between the slowing down of the object on the screen and the increasing reaction force coming from the device gives the user the illusion of a force feedback as if a friction force was generated.

Marchal et al. [65] brought the concept of pseudo-hatic feedback to walking interaction in immersive VR, inspired by the use of virtual camera motions [60, 97] to improve the sensation of walking in a virtual environment. The modification of the subjective visual feedback of the user, combined to the real kinesthetic cues of the user walking on the real ground surface, gives rise to the illusion of walking on uneven terrain. The authors base their study on the modification of the user viewpoint by changing height, speed and orientation of the virtual subjective camera according to the slope of the virtual ground. While the user walks on the flat real ground, these camera effects are injected in the virtual environment and rendered through a head-mounted display. Experimental results showed that these visual effects are very efficient for the simulation of two canonical shapes: bumps and holes located on the ground. Interestingly, a strong "orientation-height illusion" is found, as changes in pitch viewing orientation produce perception of height changes (although camera's height remains strictly the same in this case).

Other pseudo-haptic effects could be envisioned to improve the sensation of walking over virtual grounds. One promising direction would be the simulation of pseudo-haptic materials with the King-Kong effects: the current simple visual vibration patterns could give way to physically based patterns representing the impact on different materials (wood, rubber, metal), as demonstrated in previous work in a hand-based interaction context [40]. Extension of the pseudo-haptic walking has also been performed for auditory rendering by Turchet et al. [101] for simulating bumps and holes on different ground surfaces.

12.4 Conclusion

The present chapter proposed to review interactive techniques related to multimodal rendering of walking over virtual ground surfaces. We successively detailed existing auditory, vibrotactile and then multimodal rendering approaches. As for today, high-end VR setups and devices dedicated to multi-sensory walking in virtual environments could succeed in providing realistic acoustic and haptic feedback corresponding to complex scenarios. It becomes indeed possible to walk over snow, beaches, or dead leaves, and hear and feel the corresponding walking sensations using sonic shoes or haptic floors. Besides, some cross-modal effects enable to fool the senses and perceive changing ground properties.

Through the description of different rendering approaches, the chapter provided some concrete examples of how sensations accompanying walking on natural ground surfaces could be rich, multimodal and highly evocative of the settings in which they occur. We believe including multimodal cues when exploring virtual environments could bring major benefits in various applications, such as for medical rehabilitation for gait and postural exercises, training simulations, and entertainment, for an improved immersion within rich virtual environments and compelling interaction with realistic virtual grounds.

References

1. Adjémian F, Evesque P (2002) Different regimes of stick-slip in granular matter: from quasi-periodicity to randomness. In: EGS XXVII general assembly
2. Alava MJ, Nukala PKVV, Zapperi S (2006) Statistical models of fracture. Adv Phys 55(3):349–476

3. Avanzini F, Rocchesso D (2001) Modeling collision sounds: non-linear contact force. In: Proceedings of COST-G6 conference on digital audio effects, pp 61–66
4. Avanzini F, Serafin S, Rocchesso D (2005) Interactive simulation of rigid body interaction with friction-induced sound generation. IEEE Trans Speech Audio Process 13(5):1073–1081
5. Baldassarri A, Dalton F, Petri A, Zapperi S, Pontuale G, Pietronero L (2006) Brownian forces in sheared granular matter. Phys Rev Lett 96(11):118002
6. Barrass S, Adcock M (2002) Interactive granular synthesis of haptic contact sounds. In: Proceedings of AES. 22nd international conference on virtual, synthetic and entertainment audio
7. Blesser B (2001) An interdisciplinary synthesis of reverberation viewpoints. J Audio Eng Soc 49(10):867–903
8. Bresciani J-P, Ernst MO, Drewing K, Bouyer G, Maury V, Kheddar A (2005) Feeling what you hear: auditory signals can modulate tactile tap perception. Exp Brain Res 162:172–180
9. Burdea GC, Akay M (1996) Force and touch feedback for virtual reality. Wiley, New York
10. Christensen RR, Hollerbach JM, Xu Y, Meek SG (2000) Inertial-force feedback for the treadport locomotion interface. Presence: Teleoper Virtual Environ 9(1):1–14
11. Chueng P (2002) Designing sound canvas: the role of expectation and discrimination. In: Proceedings of ACM CHI extended abstracts on human factors in computing systems. pp 848–849
12. Cirio G, Marchal M, Hillaire S, Lécuyer A (2011) Six degrees-of-freedom haptic interaction with fluids. IEEE Trans Visual Comput Graph 17(11):1714–1727
13. Cirio G, Marchal M, Lécuyer A, Cooperstock J (2012) Vibrotactile rendering of splashing fluids. IEEE Trans Haptics 99:1
14. Cook PR (2002) Modeling Bill's gait: analysis and parametric synthesis of walking sounds. In: Proceedings of audio engineering society 22nd conference on virtual, synthetic and entertainment audio, Espoo, Finland
15. Cook PR (2002) Real sound synthesis for interactive applications. AK Peters, Natick
16. Courtney A, Chow HM (2000) A study of tile design for tactile guide pathways. Int J Ind Ergon 25(6):693–698
17. Cress DH (1978) Terrain considerations and data base development for the design and testing of devices to detect intruder-induced ground motion. Technical report M-78-1, U.S. Army Engineer Waterways Experiment Station, Vicksburg, Mississippi
18. Crossan A, Williamson J, Murray-Smith R (2004) Haptic granular synthesis: targeting, visualisation and texturing. In: Proceedings of international symposium on non-visual & multimodal visualization, pp 527–532
19. Dalton F, Farrelly F, Petri A, Pietronero L, Pitolli L, Pontuale G (2005) Shear stress fluctuations in the granular liquid and solid phases. Phys Rev Lett 95(13):138001
20. Darken RP, Cockayne WR, Carmein D (1997) The omni-directional treadmill: a locomotion device for virtual worlds. In: Proceedings of the 10th annual ACM symposium on user interface software and technology. pp 213–221
21. Delle Monache S, Polotti P, Rocchesso D (2010) A toolkit for explorations in sonic interaction design. In: Proceedings of the ACM 5th audio mostly conference. pp 1–7
22. Dickstein R, Plax M (2012) Metronome rate and walking foot contact time in young adults. Percept Mot Skills 114(1):21–28
23. Dixon SJ, Collop AC, Batt ME (2000) Surface effects on ground reaction forces and lower extremity kinematics in running. Med Sci Sports Exerc 32(11):1919–1926
24. Dixon SJ, Cooke A (2004) Shoe-surface interaction in tennis. In: Hung GK, Pallis JM (eds) Biomedical engineering principles in sports. Kluwer Academic/Plenum Publishers, New York, p 125
25. Ekimov A, Sabatier J (2006) Vibration and sound signatures of human footsteps in buildings. J Acoust Soc Am 120:762
26. Ekimov A, Sabatier J (2008) A review of human signatures in urban environments using acoustic and seismic methods. In: Proceedings of IEEE technologies for homeland security

27. Farnell AJ (2007) Marching onwards—procedural synthetic footsteps for video games and animation. In: Pd convention
28. Farrington SM (1998) Invariant acoustic cues for determining source characteristics of footstep sounds. Master's thesis, State University of New York
29. Ferris DP, Louie M, Farley CT (1998) Running in the real world: adjusting leg stiffness for different surfaces. Proc R Soc B: Biol Sci 265(1400):989
30. Fontana F, Bresin R (2003) Physics-based sound synthesis and control: crushing, walking and running by crumpling sounds. In: Proceedings of colloquium on musical informatics, Florence, Italy, pp 109–114
31. Fontana F, Morreale F, Regia-Corte T, Lécuyer A, Marchal M (2011) Auditory recognition of floor surfaces by temporal and spectral cues of walking. In: Proceedings of international conference on auditory display
32. Freeman J, Lessiter J (2001) Hear there & everywhere: the effects of multi-channel audio on presence. In: Proceedings of ICAD 2001, pp 231–234
33. Fukumoto M, Sugimura T (2001) Active click: tactile feedback for touch panels. In: Proceedings of ACM CHI extended abstracts. pp 121–122
34. Funkhouser T, Tsingos N, Jot JM (2003) Survey of methods for modeling sound propagation in interactive virtual environment systems. Presence
35. Galbraith F, Barton M (1970) Ground loading from footsteps. J Acoust Soc Am 48:1288–1292
36. Gaver WW (1993) What in the world do we hear? An ecological approach to auditory event perception. Ecol Psychol 5(1):1–29
37. Gibson JJ (1977) The theory of affordances. In: Erlbaum L (eds) Perceiving, Acting, and Knowing. Erlbaum, Hillsdale
38. Giordano BL, McAdams S, Rocchesso D (2010) Integration of acoustical information in the perception of impacted sound sources: the role of information accuracy and exploitability. J Exp Psychol: Hum Percept Perform 36(2):462–476
39. Giordano BL, McAdams S, Visell Y, Cooperstock J, Yao H-Y, Hayward V (2008) Non-visual identification of walking grounds. J Acous Soc Am 123(5):3412
40. Hachisu T, Cirio G, Marchal M, Lécuyer A, Kajimoto H (2011) Pseudo-haptic feedback augmented with visual and tactile vibrations. In: Proceedings of international symposium on VR innovations, pp 331–332
41. Hardin EC, van den Bogert AJ, Hamill J (2004) Kinematic adaptations during running: effects of footwear, surface, and duration. Med Sci Sports Exerc 36(5):838
42. Hayward V (2007) Physically-based haptic synthesis. In: Lin M, Otaduy M (eds) Haptic rendering: foundations, algorithms and applications. AK Peters, Natick
43. Hayward V, Astley OR, Cruz-Hernandez M, Grant D, Robles-De-La-Torre G (2004) Haptic interfaces and devices. Sens Rev 24(1):16–29
44. Hayward V, Maclean K (2007) Do it yourself haptics, part I. IEEE Robot Autom 14:88–104
45. Herrmann HJ, Roux S (1990) Statistical models for the fracture of disordered media. North Holland, Amsterdam
46. Hollerbach J (2008) Locomotion interfaces and rendering. In: Lin M, Otaduy M (eds) Haptic rendering: foundations, algorithms and applications. AK Peters, Natick
47. Hötting K, Röder B (2004) Hearing cheats touch, but less in congenitally blind than in sighted individuals. Psychol Sci 15:60–64
48. Iwata H (1999) Walking about virtual environments on an infinite floor. In: Proceedings of IEEE international conference on virtual reality, pp 286–293
49. Iwata H (2008) Haptic interface. In: Sears A, Jacko JA (eds) The human-computer interaction handbook, 2nd edn. Lawrence Erlbaum Assoc, New York
50. Kennedy PM, Inglis JT (2002) Distribution and behaviour of glabrous cutaneous receptors in the human foot sole. J Physiol 583(3):731–742
51. Kinsella-Shaw JM, Shaw B, Turvey MT (1992) Perceiving "walk-on-able" slopes. Ecol Psychol 4(4):223–239
52. Klatzky RL, Pai DK, Krotkov EP (2000) Perception of material from contact sounds. Presence: Teleoper Virtual Environ 9(4):399–410

53. Kobayashi Y, Osaka R, Hara T, Fujimoto H (2008) How accurately people can discriminate the differences of floor materials with various elasticities. IEEE Trans Neural Syst Rehabil Eng 16(1):99–105
54. Kobayashi Y, Takashima T, Hayashi M, Fujimoto H (2005) Gait analysis of people walking on tactile ground surface indicators. IEEE Trans Neural Syst Rehabil Eng 13(1):53–59
55. Kuchenbecker KJ, Fiene J, Niemeyer G (2006) Improving contact realism through event-based haptic feedback. IEEE Trans Visual Comput Graph 12(2):219–230
56. LaViola JJ, Feliz DA, Keefe DF, Zeleznik RC (2001) Hands-free multi-scale navigation in virtual environments. In: Proceedings of the ACM symposium on interactive 3D graphics. pp 9–15
57. Law AW, Ip J, Peck B, Visell Y, Kry PG, Cooperstock JR (2009) Multimodal floor for immersive environments. In: ACM SIGGRAPH emerging technologies
58. Lécuyer A (2009) Simulating haptic feedback using vision: a survey of research and applications of pseudo-haptic feedback. Presence: Teleoper Virtual Environ 18(1):39–53
59. Lécuyer A, Burkhardt J-M, Etienne L (2004) Feeling bumps and holes without a haptic interface: the perception of pseudo-haptic textures. In: Proceedings of SIGCHI conference on human factors in computing systems, pp 239–246
60. Lécuyer A, Burkhardt J-M, Henaff J-M, Donikian S (2006) Camera motions improve sensation of walking in virtual environments. In: Proceedings of IEEE international conference on virtual reality, pp 11–18
61. Li X, Logan RJ, Pastore RE (1991) Perception of acoustic source characteristics: walking sounds. J Acoust Soc Am 90(6):3036–3049
62. MacLean KE (2008) Foundations of transparency in tactile information design. IEEE Trans Haptics 1(2):84–95
63. MacLellan MJ, Patla AE (2006) Adaptations of walking pattern on a compliant surface to regulate dynamic stability. Exp Brain Res 173(3):521–530
64. Mahvash M, Hayward V (2001) Haptic rendering of cutting: a fracture mechanics approach. Haptics-e 2(3):1–12
65. Marchal M, Lécuyer A, Cirio G, Bonnet L, Emily M (2010) Walking up and down in immersive virtual worlds: novel interactive techniques based on visual feedback. In: Proceedings of IEEE symposium on 3D user interface, pp 19–26
66. Marigold DS, Patla AE (2005) Adapting locomotion to different surface compliances: neuromuscular responses and changes in movement dynamics. J Neurophysiol 94(3):1733
67. Michon JA (ed) (1993) Generic intelligent driver support. CRC Press, Boca Raton
68. Morioka M, Whitehouse DJ, Griffin MJ (2008) Vibrotactile thresholds at the fingertip, volar forearm, large toe, and heel. Somatosens Mot Res 25(2):101–112
69. Moritz CT, Farley CT (2004) Passive dynamics change leg mechanics for an unexpected surface during human hopping. J Appl Physiol 97(4):1313
70. Moritz CT, Greene SM, Farley CT (2004) Neuromuscular changes for hopping on a range of damped surfaces. J Appl Physiol 96(5):1996
71. Nashel A, Razzaque S (2003) Tactile virtual buttons for mobile devices. In: Proceedings of ACM CHI. pp 854–855
72. Nasuno S, Kudrolli A, Bak A, Gollub JP (1998) Time-resolved studies of stick-slip friction in sheared granular layers. Phys Rev E 58(2):2161–2171
73. Nasuno S, Kudrolli A, Gollub JP (1997) Friction in granular layers: hysteresis and precursors. Phys Rev Lett 79(5):949–952
74. Nordahl R (2006) Increasing the motion of users in photorealistic virtual environments by utilizing auditory rendering of the environment and ego-motion. In: Proceedings of presence, pp 57–62
75. Norman D (2007) The design of future things. Basic Books, New York
76. Papetti S, Civolani M, Fontana F (2011) Rhythmnshoes: a wearable foot tapping interface with audio-tactile feedback. In: Proceedings of international conference of new interfaces for musical expression

77. Papetti S, Fontana F, Civolani M, Berrezag A, Hayward V (2010) Audio-tactile display of ground properties using interactive shoes. In: Haptic and audio interaction design, pp 117–128
78. Pastore RE, Flint JD, Gaston JR, Solomon MJ (2008) Auditory event perception: the source-perception loop for posture in human gait. Percept Psychophys 70(1):13–29
79. Perreault S, Gosselin CM (2008) Cable-driven parallel mechanisms: application to a locomotion interface. J Mech Des 130:102301
80. Poupyrev I, Maruyama S, Rekimoto J (2002) Ambient touch: designing tactile interfaces for handheld devices. In: Proceedings of ACM UIST. pp 51–60
81. Razzaque S, Kohn Z, Whitton MC (2001) Redirected walking. In: Proceedings of Eurographics, pp 289–294
82. Rosburg T (2008) Tactile ground surface indicators in public places. In: Grunwald M (ed) Human haptic perception: basics and applications. Birkhäuser, Basel, pp 491–499
83. Rossignol S, Dubuc R, Gossard JP (2006) Dynamic sensorimotor interactions in locomotion. Physiol Rev 86(1):89
84. Rovan J, Hayward V (2000) Typology of tactile sounds and their synthesis in gesture-driven computer music performance. In: Wanderley M, Battier M (eds) Trends in gestural control of music. Editions IRCAM, Paris
85. Rovers AF, van Essen H (2006) Guidelines for haptic interpersonal communication applications. Virtual Real 9:177–191
86. Sanders RD, Scorgie MA (2002) The effect of sound delivery methods on a user's sense of presence in a virtual environment
87. Schafer RM (1977) The tuning of the world. Random House, New York
88. Serafin S, Fontana F, Turchet L, Papetti S (2012) Auditory rendering and display of interactive floor cues. In: Fontana F, Visell Y (eds) Walking with the senses, Chap 7. Logos Verlag, Berlin, pp 123–152
89. Serafin S, Turchet L, Nordahl R, Dimitrov S, Berrezag A, Hayward V (2010) Identification of virtual grounds using virtual reality haptic shoes and sound synthesis. In: Proceedings of Eurohaptics, p 61
90. Slater M, Usoh M, Steed A (1995) Taking steps: the influence of a walking technique on presence in virtual reality. ACM Trans Comput-Hum Inter 2(3):201–219
91. Srinivasan MA, Basdogan C (1997) Haptics in virtual environments: taxonomy, research status, and challenges. Comput Graph 21(4):393–404
92. Stiles VH, James IT, Dixon SJ, Guisasola IN (2009) Natural turf surfaces: the case for continued research. Sports Med 39(1):65
93. Styns F, van Noorden L, Moelants D, Leman M (2007) Walking on music. Hum Movem Sci 26(5):769–785
94. Swapp D, Williams J, Steed A (2010) The implementation of a novel walking interface within an immersive display. In: Proceedings of IEEE Symposium on 3D user interfaces, pp 71–74
95. Tax A, van Wezel B, Dietz V (1995) Bipedal reflex coordination to tactile stimulation of the sural nerve during human running. J Neurophysiol 73(5):1947–1964
96. Templeman JN, Denbrook PS, Sibert LE (1999) Virtual locomotion: walking in place through virtual environments. Presence: Teleoper Virtual Environ 8(6):598–617
97. Terziman L, Lécuyer A, Hillaire S, Wiener JM (2009) Can camera motions improve the perception of traveled distance in virtual environments? In: Proceedings of IEEE international conference on virtual reality, pp 131–134
98. Terziman L, Marchal M, Multon F, Arnaldi B, Lécuyer A (2012) The king-kong effects: improving sensation of walking in VR with visual and tactile vibrations at each step. In: Proceedings of IEEE symposium on 3D user interfaces, pp 19–26
99. Tramberend H, Hasenbrink F, Eckel G, Lechner U, Goebel M (1997) CyberStage—an advanced virtual environment. ERCIM News 31:22–23
100. Trulsson M (2001) Mechanoreceptive afferents in the human sural nerve. Exp Brain Res 137(1):111–116
101. Turchet L, Marchal M, Lécuyer A, Serafin S, Nordahl R (2010) Influence of visual feedback for perceiving walking over bumps and holes in desktop VR. In: Proceedings of 17th ACM symposium on virtual reality software and technology, pp 139–142

102. Turchet L, Serafin S, Nordahl R (2010) Physically based sound synthesis and control of footsteps sounds. In: Proceedings of conference on digital audio effects
103. van den Doel K, Kry PG, Pai DK (2001) FoleyAutomatic: physically-based sound effects for interactive simulation and animation. In: Proceedings of the 28th annual conference on computer graphics and interactive techniques pp 537–544
104. van Ulzen NR, Lamoth CJC, Daffertshofer A, Semin GR, Beek PJ (2008) Characteristics of instructed and uninstructed interpersonal coordination while walking side-by-side. Neurosci Lett 432(2):88–93
105. Velazquez R (2008) On-shoe tactile display. In: Proceedings of the IEEE international workshop on haptic audio visual environment and their applications
106. Visell Y, Cooperstock J (2010) Design of a vibrotactile display via a rigid surface. In: Proceedings of IEEE haptics symposium, pp 133–140
107. Visell Y, Cooperstock J, Giordano BL, Franinovic K, Law A, McAdams S, Jathal K, Fontana F (2008) A vibrotactile device for display of virtual ground materials in walking. In: Proceedings of Eurohaptics 2008, pp 420–426
108. Visell Y, Fontana F, Giordano BL, Nordahl R, Serafin S, Bresin R (2009) Sound design and perception in walking interactions. Int J Hum-Comput Studies 67:947–959
109. Visell Y, Giordano BL, Millet G, Cooperstock JR (2011) Vibration influences haptic perception of surface compliance during walking. PLoS one 6(3):e17697
110. Visell Y, Law A, Cooperstock JR (2009) Touch is everywhere: floor surfaces as ambient haptic interfaces. IEEE Trans Haptics 2:148–159
111. Visell Y, Law A, Smith S, Rajalingham R, Cooperstock JR (2010) Contact sensing and interaction techniques for a distributed multimodal floor display. In: Proceedings of IEEE Symposium on 3D user interfaces. pp 75–78
112. Visell Y, Law A, Ip J, Smith S, Cooperstock JR (2010) Interaction capture in immersive virtual environments via an intelligent floor surface. In: Proceedings of IEEE international conference on virtual reality, pp 313–314
113. Watters BG (1965) Impact noise characteristics of female hard-heeled foot traffic. J Acoust Soc Am 37:619–630
114. Weinstein S (1968) Intensive and extensive aspects of tactile sensitivity as a function of body part, sex, and laterality. In: Kenshalo DR (ed) The skin senses. Thomas, Springfield
115. Wells C, Ward LM, Chua R, Inglis JT (2003) Regional variation and changes with ageing in vibrotactile sensitivity in the human footsole. J Gerontol A Biol Sci Med Sci 58(8):B680–B686
116. Zehr EP, Stein RB, Komiyama T (1998) Function of sural nerve reflexes during human walking. J Physiol 507:305–314

Part III
Applications and Interactive Techniques

Chapter 13
Displacements in Virtual Reality for Sports Performance Analysis

Richard Kulpa, Benoit Bideau and Sébastien Brault

Abstract In real situations, analyzing the contribution of different parameters on sports performance is a difficult task. In a duel for example, an athlete needs to anticipate his opponent's actions to win. To evaluate the relationship between perception and action in such a duel, the parameters used to anticipate the opponent's action must then be determined. Only a fully standardized and controllable environment such as virtual reality can allow this analysis. Nevertheless, movement is inherent in sports and only a system providing a complete freedom of movements of the immersed subject (including displacements) would allow the study of the link between visual information uptake and action, that is related to performance. Two case studies are described to illustrate such use of virtual reality to better understand sports performance. Finally, we discuss how the introduction of new displacement devices can extend the range of applications in sports.

13.1 Introduction

Sport is gaining an increasing place in nowadays societies, not only as an entertainment but also as a socio-economical matter. There is now overwhelming evidence that regular physical activity has important and wide-ranging benefits on health [14]. Moreover, sport offers great pedagogical values for young people. It conveys moral values and requires appropriate behaviors in groups. As a consequence, studying sports becomes more and more important, not only to better understand performance

R. Kulpa (✉) · B. Bideau · S. Brault
M2S Lab, University of Rennes 2, Avenue Robert Schuman, 35170Bruz, France
e-mail: richard.kulpa@univ-rennes2.fr

B. Bideau
e-mail: benoit.bideau@univ-rennes2.fr

S. Brault
e-mail: sebastien.brault@univ-rennes2.fr

F. Steinicke et al. (eds.), *Human Walking in Virtual Environments*,
DOI: 10.1007/978-1-4419-8432-6_13, © Springer Science+Business Media New York 2013

and improve it for high-level athletes but also to better train people and encourage them to practice more.

Sport is characterized by complex displacements and movements. These movements are dependent on visual information that the athlete gathers from his environment, including the opponent's actions. Perception is thus fundamental to the performance. Indeed, a sportive action, unique, complex and often limited in time, requires a selective gathering of information. While everyone is able to perform simple tasks, everyone cannot be top athlete and return service tennis reaching nearly 200 km/h, stop handball throw or prevent a rugby attacker to pass with a deceptive movement. For many sports, performance is dependent on the ability to correctly react under time pressure; everything being often played in a few milliseconds, it is important to be at the right place at the right time.

If we all access to the same information during the action, why do not we all react in the same way? The reason is probably that time constraint requires a selection of perceptual information, that is to say eliciting some information to get right to the point. This perception, often seen as a prerogative for action, takes the role of a passive collector of information. However, the perception-action relationship should not be considered uniquely but rather as a coupling: we perceive to act but we must act to perceive [27]. There would thus be laws of coupling between the informational variables available in the environment and the motor responses of a subject.

In sport, this framework has already inspired Farrow and Abernethy who preserved the perception-action coupling as close to the real situation as possible during a video-based experiment [24]. In this work, the authors tried to reproduce a realistic viewpoint by capturing video sequence from within the field of basketball. The authors tested two conditions: a coupled (perception and action) and uncoupled one. The results attest that participants were better in prediction accuracy when perception and action were coupled and when the ball flight was available. In other words, it is necessary that top athletes can act to better perceive the opportunities of action from the environment. Whichever school of thought considered, Virtual Reality offers new, more pertinent and more accurate perspectives to address these concepts.

13.1.1 Why Virtual Reality for Sports?

In sport, top athletes develop important skills and perceptual motor coordination adapted to the situation. As a consequence, many studies have focused on the analysis of perceptual information of athletes [61]. For example, in sports duel, it was shown that experts have better skills than novices in using visual information to guide their early response [6, 58]. However the analysis of duels is relatively complex. Indeed, the player's movement is modulated by his opponent's one. Understanding how and on which criteria players adjust their movements requires the development of specific methodologies. To better understand why Virtual Reality can be used for such studies, it is important to describe the limitations of previous approaches used to investigate sports performance, and to understand why it can overcome them.

One of the first methods used to analyze the visual information gathered by an athlete consists in interviewing him [21]. The use of questionnaires allows athletes to formalize and transcribe their impressions about the actions they performed and the perceptual information necessary for decision. Nevertheless, this methodology is based on the feelings of the subject after the experiment, without any time pressure. The relative importance of kinematic variables in the decision-making is then very difficult to quantify. In general, this technique, based on subjective measures of perception, does not accurately characterize the visual information gathered by the subject.

Methodologies based on video uses a video clip in which an opponent is currently conducting an action. In this setup, a camcorder is placed at subject's eye position during a real game situation. For example, to study the visual information gathered by a tennis player when his opponent serves, the sequence is performed by placing the camera on the bottom line of the short side of the court. If one is interested now in analyzing visual information of a football goalkeeper during a penalty kick, the video camera is positioned at the center of the goal line. Although it is not always respected during the experiments, the perspective of the athlete is fundamental for a realistic movie. Once the video is acquired, it is projected onto a screen and the subject has then to predict the final outcome of the action displayed. Two techniques can then be used to evaluate the visual information uptake: (i) the reaction time (ii) the temporal and spatial masking. The technique of the reaction time is based on the time taken by an observer to respond correctly to a task. In other words, when using this technique, the subject has as much time as he wants to perform his task. The duration of the response is then correlated with the accuracy of the response [1]. Studies using this method have demonstrated the superiority of experts over novices in the brevity and precision of the response [5, 59, 63]. But this technique does not know the precise time and location of the visual information gathered. The second technique based on occultation is more frequent. In this methodology, two procedures are used. The first one consists in using temporal occlusion and implies to cut the video at different times to get several key sequences, each with different visual information (motion shooting, ball flight, etc.) [2, 33, 34, 45]. The second procedure is a spatial occlusion where certain visual cues are masked in the video clip in order to see if they are used or not by the participant for performing the action. In both conditions (temporal or spatial occlusion) the instructions given to subjects viewing these sequences (repeated several times in random order) are then to predict the final outcome of the observed action [40, 47]. The way the subject responds may vary from one study to another: verbal responses [7, 45], written directions [17, 38, 39, 58] or button pressed [25, 26, 54]. Means of response do not modify the response once it has been given. More elaborated means were employed such as using a joystick to save the changes made by the subject during the experiment [46, 47]. The advantage of temporal occultation is to identify a relation between the perceptual processes involved and some key instants or visual cues of the video clip. It is thus possible to determine precisely at which key moment the subject effectively predicts the outcome of the action or which visual cues are used during the decision-making process. The visual information present at this key moment or in this visual area is then considered as

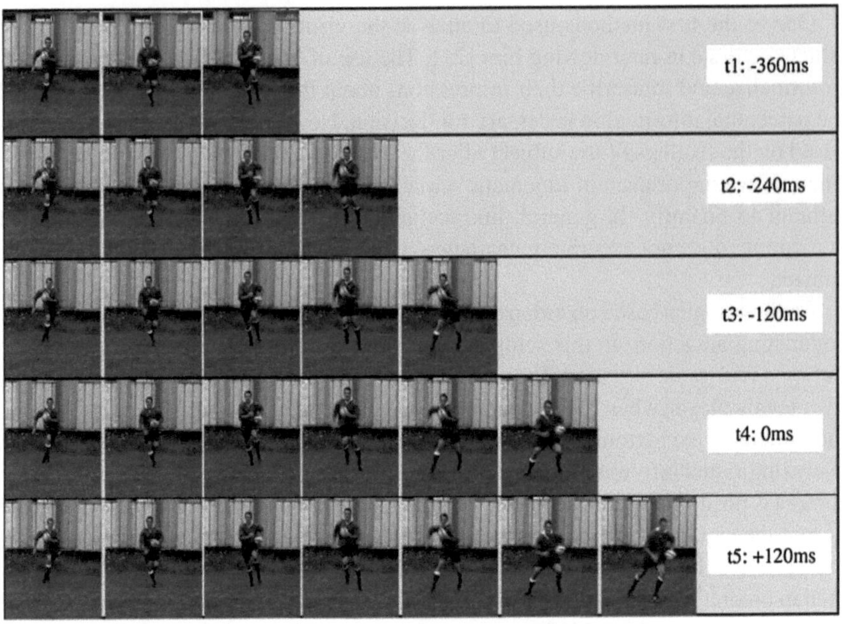

Fig. 13.1 Occlusion times used in the video-based work of Jackson et al. [33]

relevant for the subject. The occlusion-based methodology relies on the link between perceptual capabilities and performance of athletes [60] (Fig. 13.1).

However, some criticisms and drawbacks inherent to the use of video can be formulated:

- The video is a two-dimensional display. Therefore, it does not provide access to all the visual information of a real game situation [4, 61]. But the stereoscopic information can be critical to performance since it provides depth information [31, 37]. The three-dimensional video nevertheless allows to overcome this limitation.
- Another disadvantage of the video is that the viewpoint is fixed and the subject cannot change the way he sees the environment. Therefore, it is always the same information that is displayed. When an individual moves in reality, it can observe new visual information according to its viewpoint. By changing his angle of view, a soccer goalkeeper for example can obtain information about the orientation of the foot of the kicker or about the displacement of other player behind this kicker.
- The making of video clips can also be a problem. The presence of the camera indeed can force the players to adapt their behavior to avoid it. For example, as part of the study of deceptive movements in rugby, researchers filmed a player simulating this type of movement [33]. But it raises the question of the relevance of such a movement: would this deceptive movement have been effective in real one-against-one duel? Does his deceptive movement be the same if it has been done against a real opponent?

- The size of the screens can also be a problem. Sometimes, simple television screens were indeed used to view the video clips. The field of view is then limited and it reduces the access to depth information [21]. For example, in tennis, Fery and Crognier [25] have shown that the presentation of video clips on a TV screen lead to overestimation of the length of short ball trajectories and underestimation of long ball trajectories. These errors were not made in real situations.
- Video-based methodology mainly provides the instant when visual information are important but not the location of these information. This lack can be overcome by combining the technique of time occlusion with an occlusion of spatial visual information presented to subjects [1, 6]. However, few studies have used this coupling between time and space occlusions because the modification of the video is not an easy task. Other methods can be used to determine the position of pertinent visual information. The most used tool is the eye-tracker: this technique consists in analyzing the subject's point of gaze during an action. In duels for example, the subject must focus his attention on the most relevant sources of visual information on his opponent's gesture to react accordingly. It is therefore important to know where an athlete looks at [61]. It was indeed demonstrated that experts used specific visual search strategies during different actions [8]. This spatial dimension of the visual information gathered by athletes in front of a video clip has been analyzed with an eye-tracker [50]. The focus of the subject is then deduced from the position, the duration and the frequency of the fixations [47]. Many studies using this coupling between eye-tracking and video clips techniques have shown differences between expert and novice players, in soccer [29, 46, 47, 59, 63], tennis [28, 50], baseball [49], badminton [6] and combat sport [44, 62]. However, conflicting conclusions emerged about the visual strategies used by the same group level [61]. There are indeed several limitations to the use of eye-trackers. The main problem is that observing a visual fixation during an action does not necessarily mean that it is linked to pertinent information used to react [61]. Thus, the location of the gaze is not necessarily related to the visual information uptake [3, 22, 58].

If these two methodologies can be used to define the relationship between the decision of the player and his level of expertise, or to define the moment when the player makes his decision, it seems very difficult to differentiate the influence of one kinematic parameter compared to another. In the case of sports without human interactions, it is easier to standardize situations and to control their modifications, even in the field by the means of throwing machines for example. But in the case of human interactive sports, it becomes very difficult to standardize situations since the athlete cannot perform twice the same movement and further less modify only one part of his motion without modifying the other parameters. For example in the case of duel between a goalkeeper and a thrower in handball, it is very difficult to demonstrate the influence of the wrist of the thrower on the reaction of the goalkeeper. Indeed, the thrower cannot change the position of his wrist at ball release without modifying the rest of his movement. The experimental conditions are thus not controlled. To understand and determine the weight of each kinematical parameter of the movement of an opponent in the decision of a player, it is necessary to control all these parameters. Virtual reality meets the requirements of standardization of the situation through the control of synthetic humanoids. All the motion editing techniques can indeed be applied to modify only one part of the motion up to completely control the virtual character to react differently, such as by modifying the direction of his run, change his stances, etc. [36, 41]. Moreover, it overcomes the limitations of previous methodologies. It indeed allows three-dimensional display, viewpoint adaptations, and modifications of the visual information provided to the immersed subject.

13.1.2 Requirements for Using Virtual Reality for Sports

13.1.2.1 A Sufficient Level of Presence

Using virtual reality for sports analysis must nevertheless be done with caution. Some requirements must indeed be observed. The most important is to provide an environment that generates a high level of presence, the subject then has the feeling of being in the virtual environment [10]. The sense of presence is related to various factors. Slater and Usoh [53] distinguish two types of factors: external factors related to technology and internal factors related to psychological aspects. Internal factors are the way to internalize the experience of an individual. External factors are the types of technologies and materials used to display and interact with the virtual environment. Evaluation of presence is thus fundamental to consider that a virtual environment can be used to study physical activity and sports performance.

Measuring presence is very complex since it results from a set of parameters difficult to control. Hendrix divides this measurement into objective and subjective evaluations [30]. Objective evaluation depends on several categories of indicators [9]. These indicators are physiological, they are function of muscle tension, eye and cardiovascular responses to virtual events. They are also linked to the achievement of one or more tasks in the synthetic world, the precision of movement and speed of response [52]. Subjective measures correspond to a psychological evaluation, usually conducted using questionnaires [51, 55, 64]. Given the intrinsic nature and complexity of the presence, validate subjective measures of presence is not obvious. As Hendrix highlighted [30], "Evaluation of the presence requires both the use of subjective and objective measures. This is the most appropriate measure.". In the case of physical activities, it is necessary that the subject can reproduce gestures as close as possible to reality. This requirement is in addition to other constraints that determine the presence. To assess the degree of presence of an athlete, it seems necessary to use an objective method in connection with the completion of the task in a virtual environment. This achievement must be compared to that encountered in the real world. Comparison of the kinematics of athletes between a real and a virtual situation is then an additional quantitative assessment of presence.

This kind of objective kinematical validation of the presence has been done by Bideau et al. [11]. They focused on the duel between a handball goalkeeper and a thrower. To this end, they defined an experiment divided into two steps: a motion capture of duels in real situation and an experiment of the same duels in virtual environment between a real goalkeeper and a virtual thrower. At last, they compared the goalkeeper's gestures in the real and in the virtual experiments to determine if virtual reality engendered the same movements in reaction to the same throws. Their results showed that these movements were similar between the real and virtual environments.

13.1.2.2 Displacements and Freedom of Movements

In the case of this last work, the high level of presence is also due to the freedom of movements of the immersed subjects. The freedom of movement is indeed an important requirement for the use of virtual reality for sport applications since it is inherent in the physical activities. However it is not the only requirement. The displacement of the immersed player is also important. In the example of duel between attacker and defender in rugby (first case study), the displacement of the immersed player is limited comparatively to other situations. The second case study shows an example of such a situation: the soccer goalkeeper's action in front of a free kick.

In real sport situation the displacement of a player influences the displacement of his opponent. In virtual reality, the coupling of the virtual and real environments should be done to handle the interaction between the immersed subject and his virtual opponent. A real time motion capture system should then be used. Moreover, it can be used to concurrently handle the viewpoint of the subject. As mentioned before, having an egocentric viewpoint is important to gather all the visual information. Moreover, the motion capture of the immersed subject allows the biomechanical analysis necessary to evaluate his reactions.

13.1.3 Some Applications of Virtual Reality for Sports

Virtual reality is more and more used for sports applications. In 1997, Noser et al. already proposed an interactive situation in tennis [42]. If this study highlighted the technological tool, the situation was not standardized: two real players confronted each other through two virtual reality systems. Each real player then played against the avatar of the other player. Since the situation was not standardized, no link can be done between the decision made by one player and the actions of the other. Many other simulators in virtual environments have been developed such as for rowing [66], bobsleigh [35], etc.

Other studies have used virtual reality to study decision making of sports athletes. For example, Craig et al. [19, 23] used virtual reality to study the perception of soccer goalkeepers depending on the effects of ball. Their study demonstrated that the Magnus effect was hard to perceive by goalkeepers, even experts. Others have studied baseball [43], rugby [57] or handball [12, 56] to analyze the decision-making of sports players.

Two case studies are presented below to illustrate this kind of studies but implying moreover large displacements of the immersed subject. The first example shows a duel between a real rugby defender and a virtual attacker who makes deceptive movements (case study 1). The defender must thus perform medio-lateral displacements to intercept the attacker. This study is based on an HMD technology to allow the defender to move. The second case study examines the performance of a soccer goalkeeper against a free kick depending on the configuration of the wall (case study 2). The goalkeeper has to dive as in real situation to intercept the ball.

13.2 Case Study 1: Deceptive Movements in Rugby

In rugby, the aim is to progress with a ball, toward the opposite team for scoring a try after the goal-line. In order to win, each team must thus develop individual and collective displacements to avoid being intercepted by opponents. The main difficulty for the defender is to intercept a human and not an object (such as a ball) that could follow a predefined trajectory. Indeed, a rugby attacker has the opportunity to suddenly change his running direction at every moment. This is precisely what happens during a deceptive movement of an attacker. If such motor strategies are very used during sports interaction, few studies investigated them [18, 33, 48].

The wealth of the rugby duel is this opportunity for an attacker to play with deceptive and non-deceptive movements to take the advantage over a defender. The goal of this case study is to explore this complex interaction and more precisely how a defender expert, compared to a novice one, reacts to a deceptive movement of an attacker? For analyzing such strategies, a controllable and repeatable stimuli (attacker movement) as well as a system allowing free displacements of the real defender is necessary. Virtual reality is the solution.

13.2.1 Setup

In this experiment, deceptive (DM) and non-deceptive (NDM) movements were presented to novice and expert defenders. The goal of the immersed defender is to stop the attacker. The virtual reality system described in this section allows the displacement and freedom of movement of the defender to achieve his task (Fig. 13.2).

Real 1 attacker versus 1 defender actions were recorded using the optoelectronic motion capture Vicon MX system. Eight rugby players (age 21.38 ± 1.18 years) who played in the French national league took part in the motion capture session. The aim for the attacking player was the same as in the game of rugby: the attacker had to try to beat the defender whilst the defender had to try to stop the attacker. Both deceptive (DM—changing running direction to beat the defender) and non-deceptive (NDM—not changing running direction to beat the defender) were recorded [15]. These captured movements were then used to animate a virtual rugby player by using

Fig. 13.2 Setup of the experiment

the animation engine MKM (Manageable Kinematic Motion) [36, 41]. MKM has already been used and validated in other experiments involving different sporting duels [11, 13]. The 3D development software Virtools manages the final virtual reality solution and integrates all the developed components (3D rugby pitch, humanoid animation, management of the interface).

13.2.2 Method

12 expert rugby players (age 23.9 $\pm$ 2.9 years) from the professional club of Ulster rugby (Belfast, UK) and 12 non-rugby players (age 22.6 $\pm$ 2.6 years) took part in the study. 12 attacking runs were selected from the real 1 vs 1 motion capture session: four in which the attacker passed to the left of the defender by performing an effective DM to the right and four in which he passed to the right of the defender by performing an effective DM to the left. The other four attacking runs did not involve any DMs and were made up of two simple changes in running direction (NDM): two to the left and two to the right of the defender.

The virtual attacking movements (8 DM, 4 NDM) were pseudo-randomly presented five times (total of 60 trials). Participants were asked to act as in a real match situation by stopping the virtual attacker. Defenders' movements were recorded with a Qualisys ProReflex motion capture cameras. Thirty-eight reflective markers were placed on key anatomical landmarks on the defenders' body. The external markers attached to the body of the participants were used to compute the 3D positions of the joint centers and then to obtain the 3D position of the global center of mass [65]. This latter was used to determine if the defender initiated his action by an early movement bias in the wrong direction. The initiation of the displacement was taken as effective when thc COM mediolateral linear velocity get over a 0.5 m.s-1 threshold. Four parameters were then calculated to compare novices and experts: (i) movement initiation time (ms) of the defending action, (ii) percentage of early movement bias, (iii) displacement amplitude (cm) of the early bias, and (iv) the minimal distance (cm) observed between the defender and the attacker during the duel.

The real defender is also equipped with a head mounted display (Cybermind Visette pro, 45deg field of view, resolution 1280 * 1024) to have stereoscopic vision. His viewpoint is changed in real time (120 Hz) thanks to an Intersense head tracker sensor mounted on the front of the headset.

13.2.3 Results

The different results illustrate the defending strategies differences between experts and novices. Firstly, results show that experts wait significantly longer before initiating movement (Experts 267.74 $\pm$ 36.18 ms vs. Novices 192.71 $\pm$ 63.82 ms; $t(22) = 3.54$, $p = .002$) (Fig. 13.3). Secondly, regarding the percentage of early bias, authors note that novices initiates more often in the wrong direction (Novices 41.9 $\pm$ 20.5% vs. Experts 14.62 $\pm$ 9.8%; $t(12) = -4.219$, $p = .01$). These two results can be logically linked. The third parameter, namely the amplitude of the early bias, has significantly greater values for novices (14.99 $\pm$ 2.68 cm) compared to experts (11.74 $\pm$ 3.81 cm) ($t(22) = 2.41$; $p = .025$, $d = .98$) (Fig. 13.3). Finally, the minimal distance between defender and attacker in front of DM, which highlights the real performance of the defender, appears to be logically smaller for experts compared to novices (Novices 70.8 $\pm$ 7.6 cm; Experts 49.2 $\pm$ 11.4 cm, $p < .001$).

13.2.4 Discussion

The aim of this study was to explore the differences of displacement strategies between novices and experts defender in front of DM. The results highlight the fact that experts, compared to novices, wait much longer before initiating a displacement to intercept the virtual attacker. Consequently, experts are able to make a significantly lower number of early bias movement in the wrong direction as well as a significantly lower amplitude in the wrong direction during an early bias. This type of results has already been observed in the literature [23]. Dessing and Craig

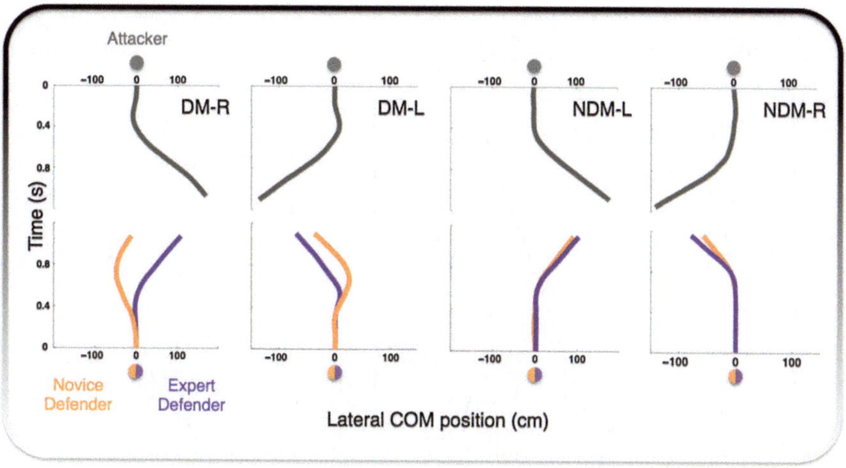

Fig. 13.3 Effects of expertise on movement initiation and displacement. Four examples of the virtual attacker's movements: *dark grey*—DM-R (deceptive movement *right*), DM-L (deceptive movement left), NDM-L (non-deceptive movement *left*) and NDM-R (non-deceptive movement *right*) and the corresponding influence on movements of an expert (*purple*) and a novice (*orange*) defender. Displacements represent the lateral movement (cm) of the COM (center of mass) over time (s) [15]

have indeed immersed soccer goalkeepers (10 novices and 2 experts) in a virtual environment and observed their motor strategies in front of different conditions of free kicks. Their results have demonstrated that the most experienced goalkeepers wait longer before initiating movement. This allows him to observe the ball trajectory during a longer period and thus allows him to better interpret the curvature of the trajectory before initiating movement. In our case, waiting longer would also allow the defender to obtain more reliable information about the true direction of the attacker and finally be closer for intercepting him.

Such an exploration is very interesting for understanding novices vs. experts differences in terms of displacement strategies. For researchers, a permanent difficulty is to be able to explore performance during a situation that is as close as possible to the real and complex situation and that keeps a high level of perception-action coupling. In this case study, virtual reality presents the opportunity to recreate a situation in which the defender can move freely as in a real game situation. It allows researchers and coaches to access to quality and pertinent knowledges about expert strategies. Indeed, the results previously presented can be used on the field for training attacker's ability to perform deceptive movements and attacker's ability to intercept them.

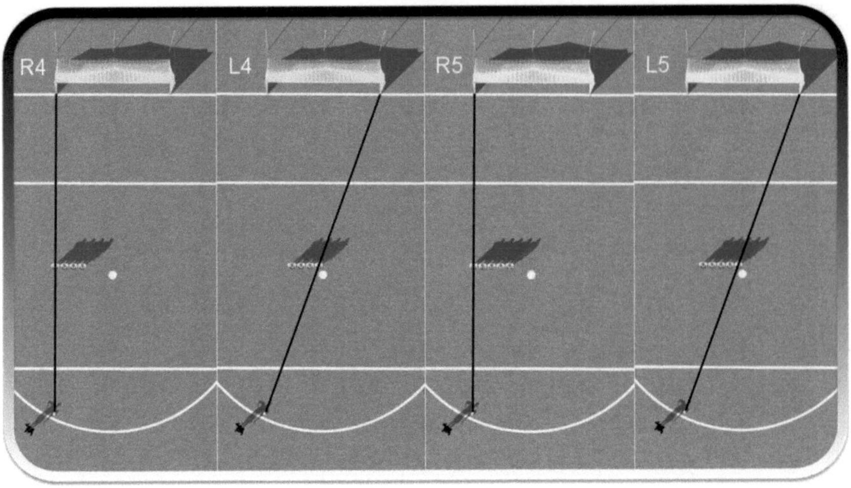

Fig. 13.4 Wall configurations: 4 or 5 players, aligned on right or left

13.3 Case Study 2: Wall Configuration for Soccer Free Kicks

During a soccer match, when a foul is committed near the penalty area and when the attacking team has a free kick, the defending team has the opportunity to set up a defensive wall between the kicker and the goalkeeper. The way this wall is configured is basically based on two parameters: its position and the numbers of players composing it. Concerning the position of the wall, one of the two external players is placed between the ball and the goalpost (Fig. 13.4). This ensures that at least one side of the goal is protected by the wall. Commonly, if the kick comes from the right and if the kicker is right-hander, the external player of the right side of the wall will be aligned with the ball and the right goalpost. As pointed out by Hugues [32], the aim for the goalkeeper is to cover two-thirds of the goal with the wall. To help covering the goal, the goalkeeper can also change the numbers of players composing the wall, usually 4 or 5. If the wall is composed of 5 players then the goal is well covered by the wall and the free zone (not covered by the wall) is then reduced. Nevertheless, the goalkeeper that is covering the free zone of the goal is far from the opposite goalpost. If the ball is kicked over the wall (it is not unusual to observe such a kick and a goal scored on the side normally covered by the wall), the goalkeeper has only a few chance to intercept the ball. On the opposite, 4 players in the wall means a larger free zone. It means it is easier to score a goal. But it also allows a better placement of the goalkeeper. Indeed, he generally places himself in the alignment of the external player of the wall and the ball. He is then closer to the middle of the goal and can thus be more efficient if the ball is kicked over the wall. The configuration of the wall is then a dilemma for the goalkeepers. The aim of this study is precisely to analyze the influence of different wall configurations on the goalkeeper performance.

13.3.1 Setup

As described in the introduction of this chapter, virtual reality systems are the only way to control all the parameters of the displayed situation. Indeed, the animation of the kicker as well as the trajectory of the ball must be completely controlled to change only one parameter at a time. Nevertheless, in that study, the displacements of the goalkeeper must be the core of the Virtual Reality system designed for this experiment. Indeed, the goalkeeper must be able to see the situation from different viewpoints going from the two goalposts to any position in the goal. Moreover, when the ball is kicked, the system must allow the goalkeeper to dive to intercept the ball and to evaluate his performance.

13.3.1.1 Hardware Setup

To allow a complete freedom of movement of the goalkeeper, the stereoscopic vision was ensured by the Nvidia 3D vision system at 120 Hz. This system is composed of wireless glasses and USB infrared emitter. These glasses are similar to normal ones and are maintained with a headband attached to the arms of the glasses. Thus, participants are fully free of their movements. In addition, contrary to HMD devices for example, the goalkeepers can also see their body segments such as their arms that are really important for such intercepting tasks. These glasses are synchronized with an Acer high frequency H5360 video projector. Rear projection was done to avoid shadows since the goalkeeper was placed near the screen to have a very large field of view. In front of this screen, a real goal was placed on synthetic grass to allow the immersed goalkeeper to feel the pitch and to dive as realistically as possible (Fig. 13.5).

The viewpoint used in the simulation was modified according to the head movement. Five markers were indeed placed on a hair band to record the head's position via the Vicon MX motion capture system. This latter was coupled with the Autodesk Motionbuilder software in order to update in real time the player's viewpoint as well as handling the stereoscopic vision.

13.3.1.2 Virtual Kicker and Ball Animations

Kicker animation was based on captured motions (see [16] for detailed information on the motion capture session). Real free kick situations were captured using the optoelectronic motion capture Vicon MX camera system at 300 Hz. The participants were mid-level football players. They had to kick the ball in a zone of 1.7 m * 1 m in the top corner of the net avoiding a plastic wall of five players. The ball was positioned at 20 meters from the goal and aligned with the left goal post (Fig. 13.4). Successful motions were kept for animation.

Concurrently, the initial linear and angular velocities of the ball at kick were captured by using reflective tapes on the ball. In addition to the final position of

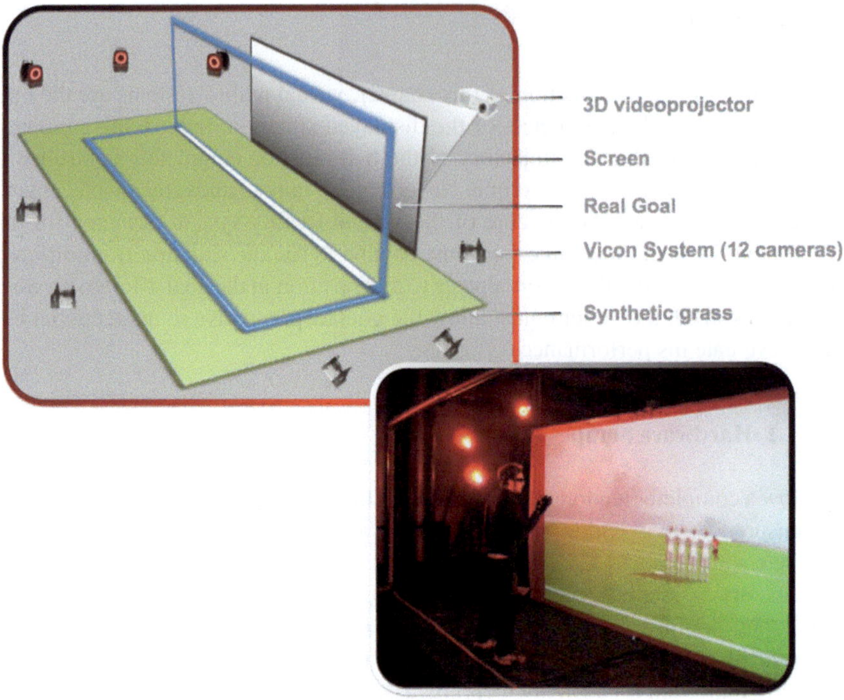

Fig. 13.5 Setup of the experiment

the ball in the goal, these velocities allowed the definition of the full ball trajectory thanks to a dynamic ball model (Fig. 13.6 and [16] for details).

All the developed components of the final virtual reality solution were integrated in the 3D development software Motionbuilder (3D soccer pitch, humanoid animation, ball animation, management of the interface).

13.3.2 Methods

11 mid-level goalkeepers who play at a departmental or regional level (mean age 26.1 $\pm$ 4.1 years) volunteered to take part in the experiment. The mean playing experience for these experts was 16.4 $\pm$ 5.4 years.Participants had to intercept different free kick conditions as in a real situation. These following conditions were mixed: 2 free kicks (in the top left and right corners), 2 wall positions (aligned with the right or left goalpost) and 2 numbers of players in the wall (4 or 5). Moreover, four other free kicks were added in order to keep a sufficient level of uncertainty. Note that all the left and right sides refer to the kicker's viewpoint. The 48 experimental trials were presented in four randomized blocks of 12 trials (2 free kicks * 2 wall positions * 2 numbers of players + 4 random free kicks). The participants then saw 8 different free

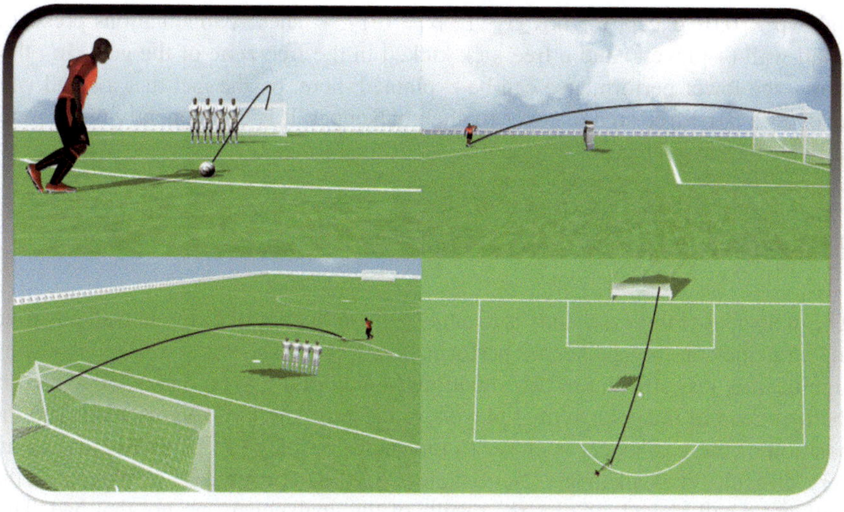

Fig. 13.6 Virtual environment

kick conditions (noted i.e. fkL_4_wL with fkL: freekick to the Left, 4: 4 players in the wall, and wL: wall aligned with the Left goalpost). A training period was carried out before the experiment to familiarize subjects with the virtual environment and the task. Five situations were randomly selected and presented to participants that had to react as in the real experiment.

13.3.3 Results

The distance between the center of the ball and the center of the closest goalkeeper's hand was calculated in real time and its minimal value is kept for each trial. A three-way analysis of variance (2 free kicks (right, left) × 2 walls (right, left) × 2 numbers of players (4, 5)) allows us to highlight significant differences. One can indeed observe a significant main effect of the number of players in the wall ($F(1,352) = 11.05$, $p < .001$, $\eta^2 = 1069.49$) with a significant better performance for situations with 5 players in the wall (minimal distance ball/hand for 5 players $= 14.39 \pm 11.14$; for 4 players $= 17.88 \pm 10.01$). Another main effect concerns the wall position condition ($F(1,352) = 11.05$, $p < .001$, $\eta^2 = 1069.49$) with a significant better performance for situations with a wall aligned with the right post (minimal distance ball/hand for right wall $= 14.9 \pm 10.21$; for left wall $= 17.35 \pm 11.09$). No difference appeared between the free kick kicked to the right or the left of the goal.

About the pairwise comparison post-hoc test (Tuckey HSD test), results highlight two conditions with a significantly lower performance, namely, "fkR_4_wL": free kick to the right, 4 players in the wall, wall on the left ($p < 0.001$ except for fkR_5_wL, $p < 0.01$) and "fkL_4_wR": free kick to the left, 4 players in the wall, wall on the right ($p < 0.01$ except for fkR_5_wL, $p = 0.175$). These two situations present a small wall (4 players) and a free kick kicked in the free zone of the goal. In other terms, the lower performance can be explained, here, by the fact that a small wall involves a large free area to cover by the goalkeeper. It is consequently much more difficult for the goalkeeper to stop it.

13.3.4 Discussion

The main goal of this case study is to analyze the influence of different wall positioning strategies on the goalkeeper performance. Results show that goalkeepers were significantly less efficient in two situations: a wall of 4 players and a ball kicked in the free zone of the goal. With 4 players in the wall, the free zone is wider and the position of the goalkeeper is closer to the center of the goal. This situation is thus very difficult for goalkeepers. Nevertheless, in the opposite case (5 players in the wall and a ball kicked over it), the results of the goalkeepers are not that bad. This result is startling since the free zone is smaller and the goalkeeper is farthest from the center of the goal and even more from the opposite goalpost. Finally, looking at all the situations, it thus seems that the goalkeeper should choose to place 5 players in the wall. Notice that the side where the wall is aligned does not influence the ability of the goalkeeper to intercept the ball.

Such an analysis of the goalkeeper's performance in front of different wall configurations would not be possible without a Virtual Reality system that ensures a complete freedom of movement of the goalkeeper. If a lot of studies have already investigated this very popular sport, the large space of the pitch and the complexity of the interaction make the experimentations difficult. If numerous studies have explored the goalkeeper's performance during penalties [46, 47], very few have analyzed the goalkeeper performance during free kick. Some interesting studies such as those of Craig et al. [19, 20] were focused on free kick in order to explore the perceptual skills of goalkeeper and to identify the optical variables that underlie judgments. In these studies, the aim was to evaluate the influence of ball's effect on perception. In the case study presented here, the goal is to evaluate the goalkeeper's performance on the perception-action coupling. His movements are thus fundamental, to intercept the ball of course but also to place the wall. The goalkeeper indeed must change his viewpoint from one goalpost to the other to be able to align the wall depending on the number of players in the wall and the position of the kicker. This kind of studies can then not be done if no displacement in the virtual reality system is provided.

13.4 Conclusion

The two case studies presented in this chapter show how virtual reality can be used to analyze the visual information uptake in sports duel. The first example deals with the interception of a virtual attacker doing deceptive movements, by a real defender in rugby. The second one concerns the influence of the wall configuration during free kick in soccer on the goalkeeper's performance. Both examples emphasize the importance of freedom of movements and displacements for the analysis of the performance of the immersed athletes. These features are indeed fundamental when studying sports. It allows the immersed subject to react realistically without constraints, to act as in real situation.

Thus, as we have illustrated, virtual reality is a fundamental tool for sports applications. It indeed offers standardized and fully controlled situations. The influence of one parameter on the performance of an athlete can then be analyzed by only modifying it and by observing the reaction of the immersed subject. Virtual reality thus offers a new way to understand performance and to increase the knowledge on sports. Another important application of the use of virtual reality for sports concerns the training. These simulators can indeed also be used to train the athletes. It offers an environment that controls exactly the information provided and allows the training of specific situations. Moreover, information can be added to the virtual environment to focus the attention of the athlete on important features of the opponents' movement for example. Indeed it is possible to use the knowledge obtained in virtual reality experiments on sports to create a new generation of training systems.

Nevertheless, in this kind of studies in virtual reality, the situations are always chosen with limited displacements. Having displacement devices would largely extend the range of studies that can be done. In the duel of the first case study for example, the defender is real and the attacker virtual. In the opposite duel between a virtual defender and a real attacker, the displacements are mainly done toward the screen. The use of a locomotion device is then necessary to handle these displacements. A specific device should obviously be developed nevertheless because the displacements are fast and jerky. In the same way, in soccer, some studies [19, 20] worked on the perception of aerial balls with effects (such as the Magnus one). To go further and study the interception of these aerial balls (for example lobs in soccer, fly balls in baseball), it is then necessary to have a displacement device that allows to move at least forward and backward. Indeed, having a large CAVE for example is not a solution since the immersed subject must be near the screen to see the balls that can be very high. When the subject is too far from the screen, the vertical field of view is then too limited and the subject looses the ball. A displacement device is then necessary to walk around the virtual field while staying near the screen of the virtual reality system.

References

1. Abernethy B (1985) Cue usage in open motor skills: a review of available procedures. Motor memory and control. Human Performance Associates, Otago
2. Abernethy B (1987) Anticipation in sport: a review. Phys educ rev 10(1):5–16
3. Abernethy B (1988) Dual-task methodology and motor skills research: Some applications and methodological constraints. J Hum Mov Study 14:101–132

4. Abernethy B, Gill D, Parks S, Packer S (2001) Expertise and the perception of kinematic and situational probability information. Perception 30:233–252
5. Abernethy B, Russell D (1984) Advance cue utilisation by skilled cricket batsmen. Aust J Sci Med Sport 16(2):2–10
6. Abernethy B, Russell D (1987) The relationship between expertise and visual search strategy in a racquet sport. Hum mov sci 3:283–319
7. Allard F, Starkes J (1980) Perception in sport: volleyball. J Sport Psychol 2:22–33
8. Bard C, Fleury M (1981) Considering eye movement as a predictor of attainment. Vision sport pp 28–41
9. Barfield W, Weghorst S (1993) The sense of presence within virtual environments: A conceptual framework. Proc Fifth Int Conf on Hum-Comput Inter 2:699–704
10. Barfield W, Zeltzer D, Sheridan T, Slater M (1995) Presence and performance within virtual environments. Oxford University Press, Inc, New York, NY, USA, pp 473–513
11. Bideau B, Kulpa R, Menardais S, Fradet L, Multon F, Delamarche P, Arnaldi B (2003) Presence: Teleoperators & Virtual Environments. Real Handball Goalkeeper vs. Virtual Handball Thrower 12(4):411–421
12. Bideau B, Kulpa R, Vignais N, Brault S, Multon F, Craig C (2010) Using virtual reality to analyze sports performance. IEEE Comput Graphics Appl 30(2):14–21
13. Bideau B, Multon F, Kulpa R, Fradet L, Arnaldi B, Delamarche P (2004) Using virtual reality to analyze links between handball thrower kinematics and goalkeeper's reactions. Neurosci Lett 372(1–2):119–122
14. Blair SN (2009) Physical inactivity: the biggest public health problem of the 21st century. Br J Sports Med 43(1):1–2
15. Brault S, Bideau B, Kulpa R, Craig C (2012) Detecting deception in movement: The case of the side-step in rugby. PlosOne 7(6):e37494
16. Brault S, Bideau N, Bideau B, Duliscouët L, Marin A, Multon F, Kulpa R Virtual kicker vs. real goalkeeper in soccer: the influence of the wall positioning strategy on the goalkeeper's performance. J Sports Sci (submitted)
17. Buckolz E, Prapavesis H, Fairs J (1988) Advance cues and their use in predicting tennis passing shots. Canadian journal of sport sciences=. Journal canadien des sciences du sport 13(1):20
18. Cañal-Bruland R, Schmidt M (2009) Response bias in judging deceptive movements. Acta psychol 130(3):235–240
19. Craig C, Berton E, Rao G, Fernandez L, Bootsma R (2006) Judging where a ball will go: the case of curved free kicks in football. Naturwissenschaften 93(2):97–101
20. Craig C, Goulon C, Berton E, Rao G, Fernandez L, Bootsma R (2009) Optic variables used to judge future ball arrival position in expert and novice soccer players. Attention, Perception, & Psychophysics 71(3):515–522
21. Crognier L, Fery Y (2007) 40 years of research on anticipation in tennis: A critical review. Science et Motricit 62:9
22. Davids K (1984) The role of peripheral vision in ball games: Some theoretical and practical notions. Phys Edu Rev (PER) 7(1):26–40
23. Dessing JC, Craig CM (2010) Bending it like beckham: how to visually fool the goalkeeper. PLoS ONE 5(10), e13,161
24. Farrow D, Abernethy B (2003) Do expertise and the degree of perception: action coupling affect natural anticipatory performance? Perception 32(9):1127–1139
25. Fery Y, Crognier L (2001) On the tactical significance of game situations in anticipating ball trajectories in tennis. Res Q Exerc sport 72(2):143
26. Franks I, Harvey I (1997) Cues for goalkeepers-high-tech methods used to measure penalty shot response. Soccer Journal-Binghatmon-National Soccer Coaches Association of. America 42:30–33
27. Gibson J (1979) The Ecological Approach to Perception. Haughton Mifflin, Boston
28. Goulet C, Bard C, Fleury M (1989) Expertise differences in preparing to return a tennis serve: A visual information processing approach. J Sport Exerc Psychol 11(4):382–398

29. Helsen W, Pauwels J (1993) The relationship between expertise and visual information processing in sport. Adv psychol 102:109–134
30. Hendrix C (1994) Exploratory studies on the sense of presence in virtual environments as a function of visual and auditory display parameters. Ph.D. thesis, University of Washington
31. Hendrix C, Barfield W (1996) Presence within virtual environments as a function of visual display parameters. Presence: Teleoperators and Virtual Environments 5(3), 274–289
32. Hugues C (1999) The F.A. coaching book of soccer tactics and skills. Harpenden: Queen Anne Press
33. Jackson R, Warren S, Abernethy B (2006) Anticipation skill and susceptibility to deceptive movement. Acta Psychol 123(3):355–371
34. Jones C, Miles T (1978) Use of advance cues in predicting the flight of a lawn tennis ball. J hum mov stud 4:231–235
35. Kelly A, Hubbard M (2000) Design and construction of a bobsled driver training simulator. Sports Eng 3(1):13–24
36. Kulpa R, Multon F, Arnaldi B (2005) Morphology-independent representation of motions for interactive human-like, animation. pp. 343–352 .
37. Mazyn L, Lenoir M, Montagne G, Savelsbergh G (2004) The contribution of stereo vision to one-handed catching. Exp Brain Res 157(3):383–390
38. McMorris T, Colenso S (1996) Anticipation of professional soccer goalkeepers when facing right-and left-footed penalty kicks. Percept mot skills 82(3):931–934
39. McMorris T, Copeman R, Corcoran D, Saunders G, Potter S (1993) Anticipation of soccer goalkeepers facing penalty kicks. Sci footb II:250–253
40. Mori S, Ohtani Y, Imanaka K (2002) Reaction times and anticipatory skills of karate athletes. Hum Mov Sci 21(2):213–230
41. Multon F, Kulpa R, Bideau B (2008) Mkm: a global framework for animating humans in virtual reality applications. Presence: Teleoper. Virtual Environ. 17(1), 17–28
42. Noser H, Pandzic I, Capin T, Thalmann N, Thalmann D (1997) Playing games through the virtual life network. Artif life five 5:135
43. Ranganathan R, Carlton L (2007) Perception-action coupling and anticipatory performance in baseball batting. J mot behav 39(5):369–380
44. Ripoll H, Kerlirzin Y, Stein J, Reine B (1995) Analysis of information processing, decision making, and visual strategies in complex problem solving sport situations. Hum Mov Sci 14(3):325–349
45. Salmela J, Fiorito P (1979) Visual cues in ice hockey goaltending. Can J appl sport sci 4(1): 56–59
46. Savelsbergh G, Van der Kamp J, Williams A, Ward P (2005) Anticipation and visual search behaviour in expert soccer goalkeepers. Ergonomics 48(11–14):1686–1697
47. Savelsbergh G, der Williams A, kamp JV (2002) Visual search, anticipation and expertise in soccer goalkeepers. J Sports Sci 20(3):279–287
48. Sebanz N, Shiffrar M (2009) Detecting deception in a bluffing body: The role of expertise. Psychon bull rev 16(1):170–175
49. Shank M, Haywood K (1987) Eye movements while viewing a baseball pitch. Percept Mot Skills 64(3):1191–1197
50. Singer R, Cauraugh J, Chen D, Steinberg G, Frehlich SG (1996) Visual search, anticipation, and reactive comparisons between highly-skilled and beginning tennis players. J Appl Sport Psychol 8(1):9–26
51. Slater M (1999) Measuring Presence: A Response to the Witmer and Singer Presence Questionnaire. Presence 8(5):560–565
52. Slater M, Linakis V, Usoh M, Kooper R (1996) Immersion, presence and performance in virtual environments: an experiment with tri-dimensional chess. ACM Virtual Reality Software and Technology (VRST) pp 163–172
53. Slater M, Usoh M (1993) Presence in immersive virtual environments. In: I. Conference (ed.) Virtual Reality Annual International, Symposium, pp 90–96

54. Tyldesley D, Bootsma R, Bomhoff G (1982) Skill level and eye movement patterns in a sport oriented reaction time task. Motor learning and movement behavior: Contribution to learning in sports pp 290–296
55. Usoh M, Catena E, Arman S, Slater M (2000) Presence questionnaires in reality. Presence: Teleoperators and Virtual Environments 9(5), 497–503
56. Vignais N, Kulpa R, Craig C, Bideau B Virtual thrower versus real goalkeeper: the influence of different visual conditions on performance. Presence: Teleoper. Virtual Environ. 19, 281–290
57. Watson G, Brault S, Kulpa R, Bideau B, Butterfield J, Craig C (2010) Judging the 'passability' of dynamic gaps in a virtual rugby environment. Hum Mov Sci 30(5):952–956
58. Williams A, Burwitz L (1993) Advance cue utilization in soccer. Science and football II: 239–244
59. Williams A, Davids K (1998) Visual search strategy, selective attention, and expertise in soccer. Res Q Exerc Sport 69(2):111–28
60. Williams A, Davids K, Burwitz L, Williams J (1992) Perception and action in sport. Journal of Human Movement Studies 22:147–204
61. Williams A, Davids K (1999) Williams. Visual perception and action in sport
62. Williams A, Elliott D (1999) Anxiety, expertise, and visual search strategy in karate. J Sport Exerc Psychol
63. Williams AM, Davids K, Burwitz L, Williams J (1994) Visual search strategies of experienced and inexperienced soccer players. Research Quarterly for Exercise and Sport 65:127–135
64. Witmer B, Singer M (1998) Measuring Presence in Virtual Environments: A Presence Questionnaire. Presence 7(3):225–240
65. Zatsiorsky V, Seluyanov V, Chugunova L (1990) Methods of determining mass-inertial characteristics of human body segments. Contemporary Problems of, Biomechanics, pp 272–291
66. von Zitzewitz J, Wolf P, Novakovic V, Wellner M, Rauter G, Brunschweiler A, Riener R (2008) Real-time rowing simulator with multimodal feedback. Sports Technology 1(6):257–266

Chapter 14
Redirected Walking in Mixed Reality Training Applications

Evan A. Suma, David M. Krum and Mark Bolas

Abstract To create effective immersive training experiences, it is important to provide intuitive interfaces that allow users to move around and interact with virtual content in a manner that replicates real world experiences. However, natural locomotion remains an implementation challenge because the dimensions of the physical tracking space restrict the size of the virtual environment that users can walk through. To relax these limitations, redirected walking techniques may be employed to enable walking through immersive virtual environments that are substantially larger than the physical tracking area. In this chapter, we present practical design considerations for employing redirected walking in immersive training applications and recent research evaluating the impact on spatial orientation. Additionally, we also describe an alternative implementation of redirection that is more appropriate for mixed reality environments. Finally, we discuss challenges and future directions for research in redirected walking with the goal of transitioning these techniques into practical training simulators.

E. A. Suma (✉) · D. M. Krun
Institute for Creative Technologies, University of Southern California,
5318 McConnell Ave, Los Angeles, CA 90066, USA
e-mail: suma@ict.usc.edu

D. M. Krun
e-mail: krum@ict.usc.edu

M. Bolas
School of Cinematic Arts, Institute for Creative Technologies,
University of Southern California, 5318 McConnell Ave,
Los Angeles, CA 90066, USA
e-mail: bolas@ict.usc.edu

F. Steinicke et al. (eds.), *Human Walking in Virtual Environments*,
DOI: 10.1007/978-1-4419-8432-6_14, © Springer Science+Business Media New York 2013

14.1 Locomotion in Virtual Environments

To create effective immersive training experiences, it is important to engage users in simulated environments that convincingly replicate the mental, physical, and emotional aspects of a real world situation. Such applications seek to invoke a sense of presence—the feeling of "being there" in the virtual world, despite the fact that the user is aware that the environment is a simulation. For users to respond realistically in a virtual environment, the system must support the sensorimotor actions that allow them to walk around and perceive the simulated content as if it were real [18]. Indeed, experiments have demonstrated that walking yields benefits over less natural forms of locomotion such as joystick travel. When walking naturally, users experience a greater sense of presence [28] and less cognitive demand on working memory and attention [23, 30]. Additionally, studies have also shown that walking results in superior performance on travel and search tasks [16, 24].

The military has long been interested in providing immersive training experiences that attempt to replicate the energy and motions of natural locomotion. For example, the U.S. Naval Research Laboratory developed Gaiter, an immersive system that uses a harness and full-body tracking to allow users to locomote by walking-in-place [27]. Other examples include omni-directional treadmills (e.g. [4, 17]) and mechanical human-sized "hamster balls" (e.g. [9]). While recent research has shown that walking on an omnidirectional treadmill can be very close to real walking [19], these devices introduce translational vestibular conflicts when users alter their walking speed. Furthermore, omnidirectional treadmills are generally expensive, mechanically complicated, and only usable by a single person at a time. Thus, many immersive systems continue to rely on handheld devices to simulate walking through the virtual environment. For example, the U.S. Army recently awarded a $57 million contract to develop the Dismounted Soldier Training System, a multi-person immersive virtual reality training simulator using head-mounted displays and joysticks mounted on each soldier's weapon for locomotion [13].

In the past few years, advances in wide-area motion tracking technology have made it possible to realize natural walking through larger physical workspaces than were previously possible with traditional electromagnetic trackers. Additionally, portable rendering devices have made it possible to provide untethered movement through large-scale immersive spaces, such as the HIVE system developed at Miami University [29], and even outdoor environments with the integration of GPS tracking. Although it is now feasible to design virtual environments that can be explored entirely through natural walking, the finite dimensions of the physical tracking space will ultimately limit the size of the virtual environment that can be represented. To address this limitation, researchers developed *redirected walking*, a technique that subtly manipulates the mapping between physical and virtual motions, thereby enabling real walking through potentially expansive virtual worlds in comparably limited physical workspaces. In this chapter, we present design considerations for employing redirected walking techniques in immersive training applications. We also demonstrate an alternative redirection technique designed for mixed reality environments that is drastically different from previous implementations, and discuss the challenges and future work we have identified based on feedback from demonstrations and military personnel.

14.2 Redirected Walking

Redirected walking exploits imperfections in human perception of self-motion by either amplifying or diminishing a component of the user's physical movements [14]. For example, when the user rotates his head, the change in head orientation is measured by the system, and a scale factor is applied to the rotation in the virtual world. The net result is a gradual rotation of the entire virtual world around the user's head position, which in turn alters their walk direction in the real world. These *rotation gains* are most effective when applied during head turns while the user is standing still [6] or during body turns as the user is walking around [3]. However, the virtual world can also be rotated as the user walks in a straight line in the virtual world, a manipulation known as *curvature gains*. If these manipulations are applied slowly and gradually, the user will unknowingly compensate for the rotation, resulting in a walking path that is curved in the real world. Studies have shown that the magnitude of path curvature that can be applied without becoming perceptible is dependent on the user's velocity [11].

The illusion induced by redirected walking works perceptually because vision tends to dominate over vestibular sensations when these two modalities are in conflict [1]. In practice, redirection is useful in preventing the user from exiting the boundaries of the physical workspace. From the user's perspective, it will appear as though they are proceeding normally through the virtual environment, but in reality they will be unknowingly walking in circles. Thus, when applied properly, redirected walking can be leveraged to allow a user to physically walk through a large and expansive virtual world using a real world tracking area that is substantially smaller in size, an advantage that is especially useful for training environments.

Redirected walking is a promising method for enabling natural locomotion in virtual environments; however, there is a perceptual limit for the magnitude of conflict between visual and vestibular sensations. Excessive manipulation can become noticeable to the user, or at worst cause simulator sickness or disorientation. As a result, an important focus of research has been to quantify the detection thresholds for redirected walking techniques. Psychophysical studies have found that users can be physically turned approximately 49 % more or 20 % less than the perceived virtual rotation and can be curved along a circular arc with a radius of at least 22 m while believing they are walking in a straight line [21]. Because of these limitations, deploying redirected walking in arbitrary environment models remains a practical challenge—if the users' motions are not easily predictable, redirection may not be able to be applied quickly enough to prevent them from walking outside of the boundaries of the tracked space.

14.3 Practical Considerations for Training Environments

While redirected walking has been shown to work perceptually, these techniques have yet to be transitioned into active training simulators. In this section, we present practical considerations for using redirected walking techniques in training environments, and discuss recent research and design mechanics that have begun to address the challenges of deploying redirected walking in practical applications.

14.3.1 Impact of Redirection on Spatial Orientation

It is important to consider how soldiers will react when self-motion manipulation techniques are used in an immersive training simulator. Such applications often require users to rapidly assess the environment and respond quickly to perceived threats. However, when users are redirected, the mapping between the real and virtual worlds is manipulated. If a user reacts instinctively based on the spatial model of the real world, this could lead to confusion or poor performance in the virtual simulator, which may negatively impact the training value. Thus, we performed a study of the effect of redirection on spatial orientation relative to both the real and virtual worlds to determine which world model users would favor in spatial judgments [26].

Participants in the study performed a series of pointing tasks while wearing a Fakespace Wide 5 head-mounted display in our approximate $10 \times 10\,\text{m}$ tracking area. First, they were asked to remove the display and point at a real target using a pistol grip-mounted Wiimote that was also tracked using the motion capture system (see Fig. 14.1a). Next, they put the display back on and pointed to a different target that was visible in the virtual environment (see Fig. 14.1b). Participants were then asked to walk down a virtual road and visit a virtual room, and during this process, they were redirected by 90°. After leaving the room, and walking down to the end of the road, they then were asked to point back to the locations where they thought the original real and virtual targets were, even though they were no longer visible (see Fig. 14.1c–d).

The results from this experiment showed that participants pointed to the real targets *as if they had been rotated along with the virtual world.* In other words, their spatial models of the real world also appeared to be updated by the manipulations to the virtual world. In general, we speculate that users will often trust what they see in virtual environments, and will therefore tend to rely on visual information over vestibular sensation for spatial judgments. Our observations, both from this experiment and from our own informal tests, suggest that it is very difficult to hold on to both the real world and virtual world spatial models simultaneously. These findings suggest that redirected walking is highly promising for use in immersive training environments, as we can expect users to respond correctly to the virtual content with little or no spatial interference from residual memories of the real world.

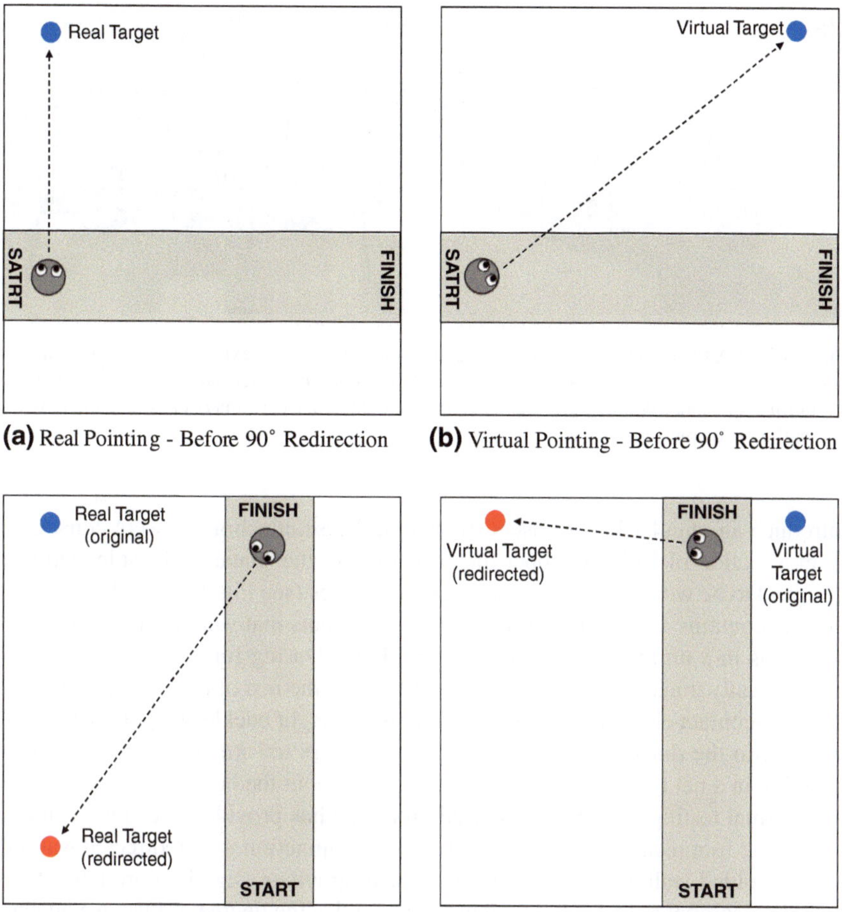

(a) Real Pointing - Before 90° Redirection **(b)** Virtual Pointing - Before 90° Redirection

(c) Real Pointing - After 90° Redirection **(d)** Virtual Pointing - After 90° Redirection

Fig. 14.1 Walkthrough of the pointing task in the spatial orientation experiment. **a** Participants pointed at a target in the real world. **b** Next, they put on the head-mounted display and pointed at a visible target in the virtual world. **c–d** After being redirected by 90°, participants pointed back to where the thought the real and virtual targets were. Participants generally pointed to target positions as if they had been rotated along with the virtual world, instead of their original locations

14.3.2 Augmenting Effectiveness of Redirected Walking

In order to increase the utility of redirected walking in immersive training scenarios, we have explored possibilities for augmenting the magnitude of redirection that can be achieved without becoming noticeable to the user. In particular, the interface between the ground and the user's feet is frequently taken for granted, but if floor contact can be manipulated, it may be possible to more strongly curve the user's walking path. To probe this effect, we constructed shoe attachment prototypes designed to

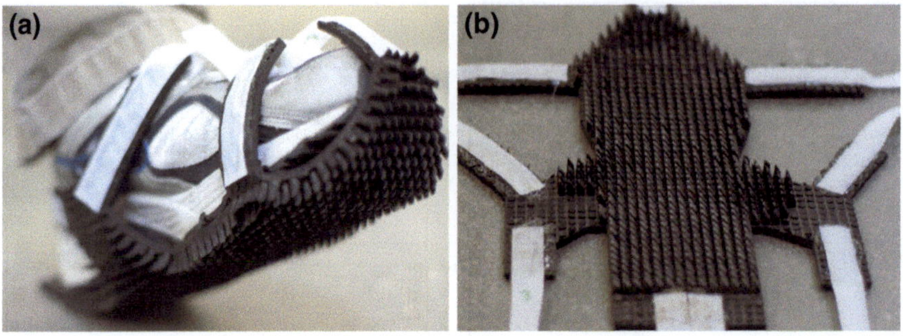

Fig. 14.2 **a** Attachments were designed to fit around the user's existing shoes and induce an angular rotation of the user's foot about the heel. **b** The *bottom* of each shoe attachment comprised a multitude of rubber floor contact elements that were diagonally trimmed, causing them to buckle and pull towards one side as weight is applied

introduce an angular bias to users' footsteps. These attachments were constructed from a section of NoTrax rubber finger mat cut to the shape of the sole, and were designed to be worn around the user's existing shoes (see Fig. 14.2a). The bottom of the mat contains a multitude of floor contact elements that were trimmed at 45–60° diagonals in a uniform direction (see Fig. 14.2b). During forward steps, the user's heel typically makes contact with the floor first. At the rest of the foot rolls forward, the floor-contact elements that begin to bear new weight buckle and pull towards one side due to the diagonal trim. This effect progresses towards the front of the foot, resulting in a net rotation of the shoe attachment about the heel.

Informal testing with our left-biased prototype has provided anecdotal evidence that haptic foot manipulation may be useful in conjunction with redirected walking. In blindfolded walking tests, users appeared to gravitate slightly more towards the left when compared to normal walking. To quantify the degree of bias, we attached LED markers to the user's foot and tracked the orientation before and after each footstep. Wearing the shoes resulted in an average foot rotation of approximately 4° per step when walking slowly. However, when walking quickly or running, the effect is reduced because the foot hits the ground with too much speed and force for the progressive buckling of floor contact elements to induce angular bias.

These prototypes have indicated that it is possible to introduce a bias to a user's walking path using an intermediary between ground and the user's shoes. However, formal study is required to determine if this effect can be used to augment redirected walking techniques. Our long-term vision would be to design a pair of "active shoes" that can dynamically alter the rotational bias as the user explores a virtual environment. Such devices, if proven effective, would be useful training simulators that needed to maximize the effectiveness of redirected walking techniques. Additionally, we also suggest that other often neglected sensory modalities may be useful for augmenting redirected walking. For example, previous studies have shown that spatialized audio can be used to influence the perception of self-motion [15], but it has yet to be explored in conjunction with redirection techniques.

14.3.3 Designing Experiences for Redirected Walking

The effectiveness of redirected walking depends largely upon the users' motions—the virtual world can be rotated faster to redirect the user's walking path during turns of the head or body. Unfortunately, user behavior in immersive simulators is often unpredictable, and if the user chooses to follow an unexpected route, the redirected walking technique may fail to prevent them from exiting the tracking space. However, the content of the virtual world can be designed to support redirected walking by subtly influencing and guiding user behavior. We have identified three design mechanics that can be leveraged to maximize the effectiveness of redirected walking: attractors, detractors, and distractors. We further suggest that in the context of immersive training, these mechanics should be ecologically valid and seamlessly merged into the content of the training scenario, so that the redirection will be as unintrusive as possible.

Attractors are designed to make the user walk towards a specific location or direction. For example, an environment to train soldiers how to search for improvised explosive devices may include suspicious objects or containers that attract users to come investigate them. Alternatively, avatars also have shown to be highly promising for supporting redirected walking (e.g. [11]). This may be especially useful in the context of training, since soldiers typically operate in squads. A virtual squadmate can move around the virtual environment and communicate with users to entice them to walk towards locations that would be advantageous for redirection.

Detractors are obstacles in the virtual environment that can be employed to prevent users from approaching inaccessible areas or force them to take a less direct route through the environment. For example, researchers have used partially opened doors to elicit body turns while navigating through doorways instead of walking straight through, thus allowing greater amounts of redirection to be applied [2]. Again, we suggest that an avatar may be ideally suited for such a role. Studies of proxemics have shown that people will attempt to maintain a comfortable distance from another person, and this social behavior has also been replicated when interacting with a virtual character [8]. If the user is approaching the boundaries of the tracking space, a virtual squadmate might walk in front, thereby forcing the user to turn and providing an opportunity for redirection.

Distractors were first introduced by Peck et al. to capture the user's attention and elicit head turns during reorientation [12]. This design mechanic has many potential uses in training environments. For example, explosions, gunfire, and vehicles are just a few events that might be encountered in a military combat simulator. These phenomena provide opportune moments for redirection, while the user's attention has been diverted away from the act of walking. We suggest that any practical training scenario seeking to use redirected walking techniques should be designed to make maximal use of these opportunities for distraction.

14.4 Redirection in Mixed Reality Environments

Mixed reality experiences that combine elements from the physical and virtual worlds have also been a focus for training applications, such as the Infantry Immersion Trainer at the Marine Corps Base Camp Pendleton [10]. Traditionally, mixed reality is often used to refer to the visual merging of real and virtual elements into a single scene. However, in the Mixed Reality Lab at the Institute for Creative Technologies, we are particularly interested in enhancing virtual experiences using modalities beyond just visuals, such as passive haptic feedback using real objects that are aligned with virtual counterparts. Experiments have shown that the ability to reach out and physically touch virtual objects substantially enhances the experience of the environment when using head-mounted displays [5].

Because redirected walking requires a continuous rotation of the virtual environment about the user, it disrupts the spatial alignment between virtual objects and their real world counterparts in mixed reality scenarios. While researchers have demonstrated that it is possible to combine redirected walking with passive haptic feedback, solutions have been limited in their applicability. For example, Kohli et al. presented an environment that redirects users to walk between multiple virtual cylindrical pedestals that are aligned with a single physical pedestal [7]. Unfortunately, this solution does not generalize to other types of geometry that would not be perceptually invariant to rotation (i.e. non-cylindical objects). Steinicke et al. extended this approach by showing that multiple virtual objects can be mapped to proxy props that need not match the haptic properties of the virtual object identically [20]. However, due to the gradual virtual world rotations required by redirected walking, synchronizing virtual objects with corresponding physical props remains a practical challenge.

Recent research has presented a drastically different approach to redirection that does not require gradual rotations of the virtual world. This technique, known as *change blindness redirection*, reorients the user by applying instantaneous alterations to the architecture of the virtual world behind the user's back. So long as the user does not directly observe the manipulation visually, minor structural changes to the virtual environment, such as the physical orientation of doorways (see Fig. 14.3), are difficult to detect and most often go completely unnoticed. Perceptual studies of this technique have shown it to provide a compelling illusion—out of 77 users tested across two experiments, only one person noticed that a scene change had occurred [22]. Furthermore, because this technique shifts between discrete environment states, it is much easier to deploy in mixed reality applications that provide passive haptic feedback [25].

Figure 14.4 demonstrates change blindness redirection being used in a mixed reality environment that combines synthetic visuals with a passive haptic gravel walking surface. In our example application, which was themed as an environment similar to those that might be used for training scenarios, users are instructed to search for a cache of weapons hidden in a desert village consisting of a gravel road connecting a series of buildings (see Fig. 14.4a). Inside of each building, the location of one of

(a) Before **(b)** After

Fig. 14.3 An example of a structural alteration to an environment that can be used to redirect the user. A door is swapped to the adjacent wall in the corner of a room, which results in users exiting in a different direction than when they entered

the doors is moved as they are looking inside strategically-placed containers (see Fig. 14.4b). Upon exiting the building, users will be located at the opposite end of the gravel road from where they entered, allowing them to walk repeatedly over the same surface to enter the next building (see Fig. 14.4c). Thus, the gravel road could be extended infinitely, so long as the user stops to explore each building along the way. Of course, if users choose to continue walking straight down the road without stopping, an intervention would be required to prevent them from exiting the physical workspace. Feedback from demonstrations and informal testing of this environment suggest that the physical sensation of walking on a gravel surface provides a compelling illusion that makes the virtual environment seem substantially more real.

Our vision is to provide a dynamic mixed reality experience with physical props that can be moved to a different location in the tracking space when the user is redirected. Dynamically moving props would be difficult and unpredictable with redirected walking techniques that use gradual, continuous rotations. However, because change blindness redirection shifts between predictable discrete environment states, it would be relatively simple to mark the correct locations for each prop on the floor, and have assistants move them to the appropriate place whenever a scene change is applied to the environment. Thus, we expect spatial manipulation techniques to prove useful for mixed reality training scenarios that incorporate a variety of passive haptic stimuli such as walking surfaces, furniture, and other physical props.

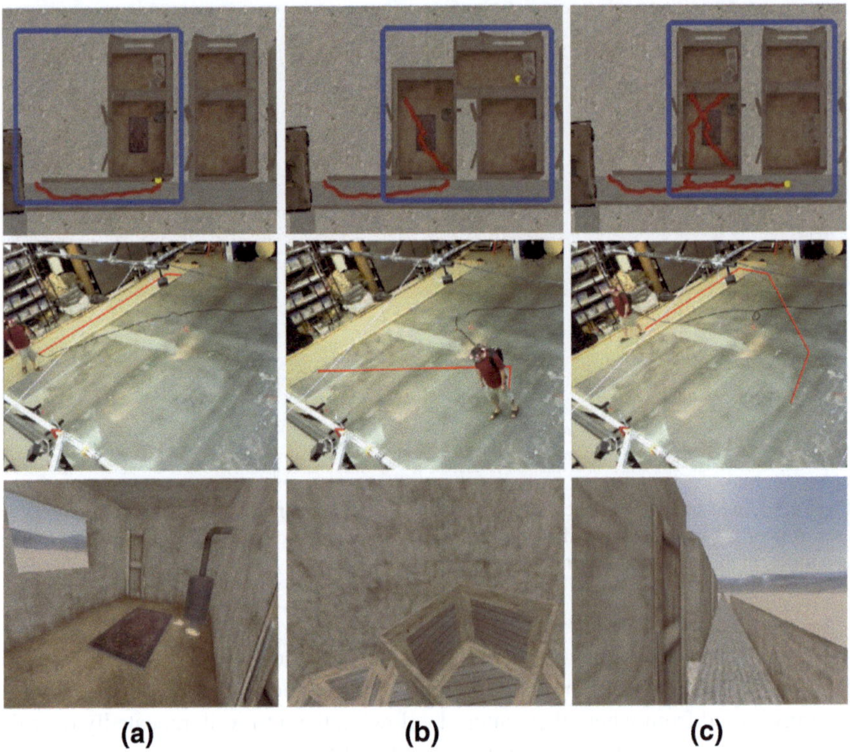

Fig. 14.4 (*top row*) Overhead map views of the virtual environment through the three stages of change blindness redirecton, with the user's location indicated by the *yellow* marker. The *blue rectangle* indicates the boundaries of the tracking area, and the user's path through the virtual world is plotted in red. (*middle row*) The view from an overhead camera mounted in the tracking space. (*bottom row*) The virtual environment as viewed through the head-mounted display. **a** The user walks down the gravel path and enters the first virtual building. **b** The user enters the back room of the building. When the user searches through the crates, the door behind him is moved. **c** The user exits and continues down the same gravel path. The original layout is restored as the user approaches the second virtual building

14.5 Challenges and Future Directions

In this section, we discuss the practical challenges and future directions towards transitioning redirected walking in active training environments, based on feedback we have received from public demonstrations and discussions with domain experts and military personnel.

Automated redirection in arbitrary environments. Much of the previous research has focused on purpose-built environments with constrained scenarios designed to test individual techniques. However, in a practical training setting, users will not always follow the designer's expectations. Such a system would need to be able to figure out how to automatically redirect in an optimal manner without imposing unnatural restrictions on freedom of movement. Automated approaches are a non-trivial problem, as they would require analyzing the scene structure, predicting the user's walking path within the virtual and physical spaces, selecting the most appropriate technique to employ based on the current state of the user and system, and gracefully intervening when failure cases are encountered.

Redirection with multiple users. Small squad training is an notable topic of interest for the U.S. Army, and the possibility of employing redirected walking with multiple users is one of the most frequent questions we have received from military personnel. To the best of our knowledge, research thus far in redirected walking has focused exclusively on the single user experience. We believe that it is possible to redirect multiple users within a single virtual environment, but this would also introduce new challenges when the two users need to interact and their virtual world coordinate systems do not align well with each other. Thus, redirected walking with multiple simultaneous users is an important area for future work.

Evaluating impact on training value. While research thus far has shown promising results for using redirected walking in virtual environments, the open question remains whether redirected walking techniques will be compelling and effective specifically in a training context. Many of the studies conducted by the virtual reality community tend to draw their subjects from either the general population or university students. However, soldiers are a self-selected population with specialized skills and training, and it seems logical to conclude that their experiences in an immersive training simulator could be drastically different from those of a randomly selected person drawn from the general population. Before redirected walking can be transitioned to these practical environments, there is a need to understand the impact that these techniques will have on learning gains and training outcomes. Thus, domain-specific evaluation remains an important focus for future studies.

References

1. Berthoz A (2000) The brain's sense of movement. Harvard University Press, Cambridge
2. Bruder G, Steinicke F, Hinrichs K, Lappe M (2009) Arch-explore: a natural user interface for immersive architectural walkthroughs. In: IEEE symposium on 3D user interfaces, pp 145–152
3. Bruder G, Steinicke F, Hinrichs K, Lappe M (2009) Reorientation uring body turns. In: Joint virtual reality conference of EGVE-ICAT-EuroVR, pp 145–152
4. Darken RP, Cockayne WR, Carmein D (1997) The omni-directional treadmill: a locomotion device for virtual worlds. In: ACM symposium on user interface software and technology, pp 213–221
5. Insko BE (2001) Passive haptics significantly enhances virtual environments. Ph.D. thesis, University of North Carolina, Chapel Hill
6. Jerald J, Peck TC, Steinicke F, Whitton MC (2008) Sensitivityto scenemotion for phases of head yaws. In: Symposium on applied perception in graphics and visualization, ACM Press, New York, pp 155–162. doi:10.1145/1394281.1394310

7. Kohli L, Burns E, Miller D, Fuchs H (2005) Combining passive haptics withredirected walking. In: International conference on artificial reality and telexistence, ACM Press, New York, pp 253–254. doi:10.1145/1152399.1152451
8. Llobera J, Spanlang B, Ruffini G, Slater M (2010) Proxemics withmultiple dynamic characters in an immersive virtual environment. ACM Trans Appl Percept 8(1):1–12. doi:10.1145/1857893.1857896
9. Medina E, Fruland R, Weghorst S (2008) Virtusphere: walking in a human size VR "hamster ball". In: Human factors and ergonomics society annual meeting, pp 2102–2106
10. Muller P, Schmorrow D, Buscemi T (2008) The infantry immersion trainer: todays holodeck. Marine Corps Gazette
11. Neth CT, Souman JL, Engel D, Kloos U, Buthoff HH, Mohler BJ (2011) Velocity-dependent dynamic curvature gain for redirected walking. In: Proceedings of the IEEE virtual reality, pp 151–158
12. Peck TC, Fuchs H, Whitton MC (2009) Evaluation of reorientation techniques and distractors for walking in large virtual environments. IEEE Trans Vis Comput Graph 15(3):383–394. doi:10.1109/TVCG.2008.191
13. Quinn K (2011) US Army to get dismounted soldier training system. Defense News Train Simul J
14. Razzaque S (2005) Redirected walking. Ph.D. thesis, University of North Carolina, Chapel Hill
15. Riecke BE, Feuereissen D, Rieser JJ (2010) Spatialized sound inuences biomechanical self-motion illusion ("Vection"). In: Symposium on applied perception in graphics and visualization, p 158
16. Ruddle RA, Lessels S (2009) The benefits of using a walkinginterface to navigate virtual environments. ACM Trans Comput Hum Interact 16(1):1–18. doi:10.1145/1502800.1502805
17. Schwaiger MC, Thummel T, Ulbrich H (2007) A 2D-motion platform: the cybercarpet. In: Joint EuroHaptics conference and symposium on haptic interfacesfor virtual environment and teleoperator systems. IEEE, pp 415–420. doi:10.1109/WHC.2007.1
18. Slater M (2009) Place illusion and plausibility can lead to realistic behaviour in immersive virtual environments. Philos Trans R soc London Ser B Biol Sci 364(1535):3549–3557. doi:10.1098/rstb.2009.0138
19. Souman JL, Giordano PR, Schwaiger MC, Frissen I, Thumel T, Ulbrich H, Luca AD, Bulthoff HH, Ernst MO (2011) CyberWalk: enabling unconstrained omnidirectional walking through virtual environments. ACM Trans Appl Percept 8(4):1–22. doi:10.1145/2043603.2043607
20. Steinicke F, Bruder G, Hinrichs K, Jerald J, Frenz H, Lappe M (2009) Real walking through virtual environments by redirection techniques. J Virtual Real Broadcast 6(2)
21. Steinicke F, Bruder G, Jerald J, Frenz H, Lappe M (2010) Estimation of detection thresholds for redirected walking techniques. IEEE Trans Vis Comput Graph 16(1):17–27. doi:10.1109/TVCG.2009.62
22. Suma EA, Clark S, Finkelstein SL, Wartell Z, Krum DM, Bolas M (2011) Leveraging change blindness for redirection in virtual environments. In: Proceedings of the IEEE virtual reality, pp 159–166
23. Suma EA, Finkelstein SL, Clark S, Goolkasian P, Hodges LF (2010) Effects of travel technique and gender on a divided attention task in a virtual environment. In: IEEE symposium on 3D user interfaces. IEEE, pp 27–34. doi:10.1109/3DUI.2010.5444726
24. Suma EA, Finkelstein SL, Reid M, V Babu S, Ulinski AC, Hodges LF (2010) Evaluation of the cognitive effects of travel technique in complex real and virtual environments. IEEE Trans Vis Comput Graph 16(4):690–702. doi:10.1109/TVCG.2009.93
25. Suma EA, Krum DM, Bolas M (2011) Redirection on mixed reality walking surfaces. In: IEEE VR workshop on perceptual illusions in virtual environments, pp 33–35
26. Suma EA, Krum DM, Finkelstein SL, Bolas M (2011) Effects of redirection on spatial drientation in real and virtual environments. In: IEEE symposium on 3D user interfaces, pp 35–38
27. Templeman JN, Denbrook PS, Sibert LE (1999) Virtual locomotion: walking in place through virtual environments. Presence Teleoper Virtual Environ 8(6):598–617. doi:10.1162/105474699566512

28. Usoh M, Arthur K, Whitton MC, Bastos R, Steed A, Slater M, Brooks FP (1999) Walking > walking-in-place > flying, in virtual environments. In: ACM conference on computer graphics and interactive techniques (SIGGRAPH). ACM Press, New York, pp 359–364. doi:10.1145/311535.311589
29. Waller D, Bachmann E, Hodgson E, Beall AC (2007) The HIVE: a huge immersive virtual environment for research in spatial cognition. Behav Res Methods 39(4):835–843
30. Zanbaka CA, Lok BC, Babu SV, Ulinski AC, Hodges LF (2005) Comparison of path visualizations and cognitive measures relative to travel technique in a virtual environment. IEEE Trans Vis Comput Graph 11(6):694–705. doi:10.1109/TVCG.2005.92

Chapter 15
VR-Based Assessment and Rehabilitation of Functional Mobility

Adam W. Kiefer, Christopher K. Rhea and William H. Warren

The advent of virtual reality (VR) as a tool for real-world training dates back to the mid-twentieth century and the early years of driving and flight simulators. These simulation environments, while far below the quality of today's visual displays, proved to be advantageous to the learner due to the safe training environments the simulations provided. More recently, these training environments have proven beneficial in the transfer of user-learned skills from the simulated environment to the real world (e.g., [5, 31, 48, 51, 57]). Of course the VR technology of today has come a long way. Contemporary displays boast high-resolution, wide-angle fields of view and increased portability. This has led to the evolution of new VR research and training applications in many different arenas, several of which are covered in other chapters of this book. This is true of clinical assessment and rehabilitation as well, as the field has recognized the potential advantages of incorporating VR technologies into patient training for almost 20 years (e.g., [7, 10, 18, 45, 78]).

Many of the early desktop VR clinical interventions unfortunately suffered from technological constraints that limited their value as training tools for clinical populations. In particular, they often required patients to remain stationary (seated or standing) and interact with displays on a computer monitor. Recently, however, new technological advances that allow the user to navigate virtual environments by walking (either over ground or on a treadmill), combined with the steady improvement of visual displays, serve to enhance the immersive nature of VR and introduce new behavioral measurement opportunities. As a result, we are in the midst of a paradigm shift in rehabilitation science; the field is beginning to move away from predominantly stationary interventions viewed on a computer monitor and toward dynamic,

A. W. Kiefer (✉) · W. H. Warren
Department of Cognitive, Linguistic and Psychological Sciences,
Brown University, Box 1821, Providence, RI 02912-1821, USA
e-mail: adam_kiefer@brown.edu

C. K. Rhea
Department of Kinesiology, University of North Carolina at Greensboro,
Greensboro, NC, USA

F. Steinicke et al. (eds.), *Human Walking in Virtual Environments*,
DOI: 10.1007/978-1-4419-8432-6_15,

interactive user-controlled virtual environments. The impact of this shift has been intensified by emergent technologies such as the *Nintendo Wii* (Nintendo Co. Ltd., Kyoto, Japan) and *Microsoft Kinect* (Microsoft Corp., Redmond, Washington) systems as well. These systems, and others like them, have led to the widespread cost-effective availability of interactive VR and may provide new opportunities for the application of VR interventions both inside the home and in local clinical settings. All of these technological enhancements have potentially far-reaching implications for clinical assessment and rehabilitation and, accordingly, serve as the impetus for this chapter.

One of the primary advantages of VR is that it provides a platform for the development of unique and customizable interventions that are not available or easily implemented in the real world. Specifically, VR enables the manipulation of training duration, intensity and feedback to satisfy clinical demands for intensive and repetitive patient training [9].[1] When developing VR interventions, it is important to consider both the construction of the virtual environment and the interfaces for measurement and feedback that accompany them. A useful framework to guide the development of VR-based rehabilitation was introduced by [74] in the form of a nested three-circle schema in Fig. 15.1. The schema represents the VR-based rehabilitation process as it relates to the patient, with the three circles illustrating each component of this process (listed in order from inner to outermost): (1) the interaction space, (2) the transfer phase, and (3) the real world.

The inner circle, or *interaction space*, signifies the interface between the user and the virtual environment. The user's characteristics (e.g., age and anthropometrics), function (e.g., sensory and mobility deficits) and the targeted anatomical structures engaged during the task all contribute to the user's interaction with the virtual world [75]. This allows for a VR intervention that is aligned with the user's real world experiences and results in a natural task environment with adequate visual and idiothetic information.[2] Further, the realism and ecological validity of VR environments is important to the enhancement of training efficiency in VR-based rehabilitation [17]. The middle circle, or band, represents the *transfer phase* and refers to the transferability of learned skills from the virtual environment (i.e., interaction space) to the real world. This phase requires varied levels of clinician support and training time depending on the severity and type of disability facing the patient [26]. It may even require combining virtual imagery with the real world (e.g., augmented reality)[3] to facilitate or catalyze skill transfer, in order to promote improved daily

[1] It is important to note that there are two different applications of VR in rehabilitation. When VR is used as an adjunct to rehabilitation, it is typically referred to as VR-augmented rehabilitation. Conversely, VR provided alone as a rehabilitation intervention is referred to as VR-based rehabilitation [8]. The latter is the predominant focus of this chapter.

[2] If the visual and idiothetic information are not aligned with the user's actions a disruption of the user's sense of realness, or *presence*, in the virtual environment can result, leading to feelings of physical disorientation and even nausea [55].

[3] Augmented reality is a tool in which the virtual world is superimposed over the real world, with the virtual information serving as a supplement to what is available in the real world alone [17].

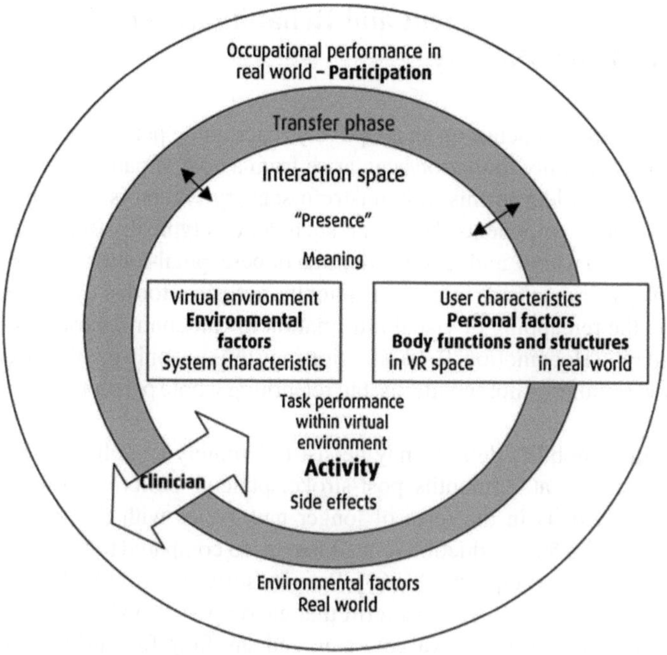

Fig. 15.1 The three-nested *circle* schema. Reproduced with permission (A request of permission for reproduction of this figure was submitted in the Fall of 2011.) from Weiss et al. [74]

function. The final, outermost circle in the schema refers to the *real world* and denotes changes in the affordances of the environment [15, 65] as a result of rehabilitation. For example, objects that previously prevented patients from interacting with the real world—the presence of low curbs or moving obstacles in a crowded environment (e.g., a busy airport or a busy intersection)—no longer present barriers but now afford passage for walking. This component is wholly dependent on the skills gained in the transfer phase and symbolizes the final rehabilitation goal of increasing the patient's participation in the real world, ultimately leading to an improved quality of life [75].

The nested three-circle schema introduces a useful heuristic for the development and implementation of immersive VR-based rehabilitation interventions. Moreover, each individual component highlights important considerations for researchers and clinicians as they seek to employ these techniques in patient populations. The rest of this chapter is devoted to reviewing different VR-based approaches to rehabilitation and assessment. Each of these approaches is distinguished by novel methodologies developed by clinicians and researchers alike. Despite these differences, each method shares a commitment (whether intentional or not) to the principles of the three-circle schema and offers a framework, in its own right, for the development of new and exciting VR-based interventions for improving functional mobility.

15.1 VR-Based Assessment and Rehabilitation to Promote Functional Mobility

A person's mobility depends on an adaptively functioning perception-action system. Consequently, mobility limitations can arise from a host of pathologies and injuries that affect various loci in this system, from sensory receptors to cortical areas to musculo-skeletal components. However, such deficits typically impact the function of the system as a whole, and require adaption of perceptual-motor control strategies. For example, a chronic knee injury may alter the actions afforded by the environment and require the remapping of visual information to gait control variables in order to generate adaptive locomotion. Rehabilitation may thus not only involve strengthening muscles and retraining motor patterns, but relearning whole perceptual-motor control relations.

The ensuing mobility deficits may persist indefinitely and often deteriorate over time. For instance, at 12 months post-stroke, patients suffering from hemiplegia exhibit motor deficits in the form of longer gait cycles with decreased cadences, and this results in a 50 % reduction in walking speed compared to the gait patterns of unaffected control participants [38, 63]. Patients suffering from Parkinson's disease frequently exhibit *freezing gait*—a term that encompasses both the inability of the patient to initiate or sustain a walking gait, and shuffling forward with small steps as their legs exhibit muscle trembling—and these symptoms worsen as the disease progresses [6, 44]. Mobility issues are also the typical sequelae of sensory deficits such as "tunnel vision" due to conditions like retinitis pigmentosa (RP)—a group of hereditary disorders characterized by retinal pigmentary degeneration that often leads to progressive visual field loss [13, 22, 27, 32, 58]. This spectrum of deficits detracts from a patient's functional mobility by reducing their ability to adapt (prospectively and/or reactively) to normally varying environmental conditions during locomotion. Moreover, their physiological basis influences the type and severity of the deficit, as well as the type of intervention that can be utilized to improve patient mobility. In direct response to these problems, researchers have started to employ VR training interventions that focus on increasing the walking speed and adaptability of patients with mobility deficits [11, 24, 36]. Others have developed unique VR assessment protocols that exploit the flexibility of VR and may have potential advantages over real-world clinical assessments [12, 28].

15.1.1 VR-Based Assessment and Rehabilitation Following Motor Dysfunction

One of the unique capabilities of VR is that optical information can be enhanced or manipulated during ongoing walking. For example, *optic flow*—the pattern of motion available from the ground and background surfaces during locomotion [14, 67]—provides information about one's speed and direction of travel (or *heading*). The rate of optic flow has been shown to influence the perceived speed of participants and to elicit changes in walking speed [37, 40, 43, 62]. Similarly, shifting the pattern of optic flow to the left or right influences the perceived heading direction [70], and elicits compensatory postural [3, 69] and steering adjustments when walking to a goal [59, 68]. Using VR to manipulate optic flow thus has the potential to alter the interaction space and provide salient information about locomotion speed and heading to the patient.

Lamontagne et al. [28] used such a manipulation to examine the perceptual-locomotor adaptability of patients suffering from post-stroke hemiplegia. During two experiments patients and unaffected control participants walked on a human-driven treadmill while virtual corridors provided optic flow information through a head-mounted display (HMD). The first experiment required participants to walk at a comfortable speed as the optic flow rate was varied continuously in an open-loop sinusoidal pattern at 0.017 Hz. This resulted in a compensatory out-of-phase relation between gait and optic flow speed for all participants (i.e., participants walked faster during slower optic flow conditions and vice versa),[4] although this was less pronounced for the patients and their phase relation was much more variable. In the second experiment the walking speed of participants during a baseline optic flow trial (1:1 mapping between walking pace and optic flow) was compared to their walking speed in a series of trials in which optic flow was discretely manipulated above or below the comparison trial. Again, walking speed was inversely related to rate of optic flow, but the patients were equal to the healthy controls in their gait response to optic flow. Taken together, the results of these two experiments provide evidence that patients with hemiplegia following stroke are influenced by optic flow in a similar way to healthy controls. This indicates, preliminarily, that virtual optic flow might be useful in training these patients to increase their walking speed over the course of a training intervention.

VR has also been used to manipulate visual cues to modulate the gait characteristics of patients with Parkinson's disease through both continuous optic flow (e.g., [49]) and continuous information paired with discrete visual stimuli (e.g., [60]). Similar to Lamontagne et al. [28], Schubert et al. required patients with Parkinson's disease and control participants to maintain a preferred walking speed on a human-driven treadmill while they viewed an optic flow pattern that varied at a constant speed perceived to be either faster or slower relative to the preferred walking speed of each participant. The results indicated that the patients with Parkinson's disease were more susceptible to changes in optic flow speed (i.e., their preferred walking speed was more variable) compared to control participants. The researchers concluded that the patients were more reliant on visual information, perhaps due to their

---

[4] The relation between optic flow and gait speed has been studied extensively (see text). While the findings of Prokop et al. [43] and Mohler et al. [37] parallel those of Lamontagne et al. [28], it is unclear why, exactly, the out-of-phase relation was observed. One possibility, as suggested by the authors, is that a sinusoidal change in optic flow speed may lead to a more pronounced time lag between the change in stimulus and the behavioral response. Another is that when the flow rate decreases, the participant walks faster to compensate for a perceived decrease in speed, in order to maintain a constant or preferred speed [37].

decreased ability to utilize proprioceptive information, which resulted in impaired adaptation to optic flow.

The work of van Wegen et al. [60] expanded on the optic flow approach by introducing various discrete stimuli into optic flow scenarios. Specifically, they required Parkinson's patients and control participants to walk on an automatic treadmill while they viewed a virtual corridor (synchronized with the treadmill speed) displayed on a screen in front of them. Two conditions required participants to walk in front of a blank screen in the presence or absence of a rhythmic temporal cue (i.e., a flashing light that patients viewed while wearing a pair of glasses). Three additional conditions consisted of the virtual corridor either by itself or in combination with either the temporal cue or a spatial cue (i.e., transverse lines overlaid on top of the virtual corridor). Both the spatial and temporal cues lowered the patients' stride frequency even as they were able to maintain their walking speed, but this may have been due to the visual cues drawing the attention of patients to the walking pattern [60]. Interestingly, the virtual corridor did not have an effect when compared to the non-VR conditions. Here the automaticity of the treadmill may have washed out any potential contributions the virtual corridor might have had on the patients' gait patterns, particularly given the effects observed on a human-driven treadmill by Lamontagne et al. [28].[5] Regardless, the results of van Wegen et al. provides preliminary evidence that the rigid gait patterns of patients with Parkinson's disease are not tightly coupled to walking speed and may be manipulated by visual cues. Thus, VR-based rehabilitation may hold promise for training these types of patient populations.

Experimenters have also utilized VR to simulate patient interactions with the real world to promote successful obstacle avoidance and circumvention. This is done through either the use of virtual obstacles during patient testing or in the evaluation of the transfer of VR training to real world obstacles. Moreover, these methods can be utilized in conjunction with modified perceptual information (e.g., optic flow), or separately. For example, Fung et al. [11] conducted a feasibility study of two patients post-stroke. Three separate virtual environments (i.e., a corridor, a park and a street crossing) were viewed on a screen while each patient walked on a feedback-driven motorized treadmill. As each patient successfully traveled these environments, the task difficulty increased by requiring faster walking speeds in order to successfully avoid virtual collisions. They were also forced to cope with increasing surface slope changes on the treadmill. Patients were able to increase their walking speed and maintain that speed in the face of slope changes. However, these mobility improvements did not translate to improved virtual obstacle avoidance by either patient. The nature of this feasibility study limits its generalizability, for the researchers did not test a control group or a comparison group that trained in a real environment. Therefore, it is difficult to separate general training effects from specific effects of the virtual interventions. Nonetheless, the results hold promise for the viability of VR as a training tool in comparable walking scenarios.

---

[5] This is based on a variation of the posture-first principle [79] in which participants would prioritize locomotion on the treadmill over attending to the perceptual information on the screen in front of them.

Jaffe et al. [24] examined a similar cohort of patients as they walked on a motorized treadmill while stepping over virtual obstacles, and vibrotactile sensations were used to provide feedback when contact was made with an obstacle. Patient performance in VR was then compared to the performance of a separate group of patients who trained on a 10 m walkway in the real world while stepping over actual obstacles. The patients that were trained in VR exhibited increases in walking speed (in a separate fast walk test) compared to the patients who trained in the real world. The researchers suggested that the visual augmentation of the virtual obstacles combined with the enhanced safety of the VR intervention were contributors to these improvements. It is also possible that the treadmill forced participants to maintain their walking speed leading up to, and following, obstacle clearance, and that the absence of this in the over-ground walking conditions influenced patient improvement as well.

More recently, Mirelman et al. [36] examined the influence of a similar VR intervention on the gait characteristics of patients with Parkinson's disease, compared to previously collected data from an historical active control group. The training required patients to walk on a virtual path (via a treadmill) as they coped with visual distracters (i.e., moving objects and changes in environmental lighting) while negotiating obstacles of varying size and frequency. Both treadmill speed and visual complexity of the environment were increased as patient performance improved over six weeks of training. Gait characteristics were assessed prior to and after the VR intervention by testing patients during three real-world walking conditions: (1) walking over ground, (2) walking while stepping over real world obstacles, and (3) walking while performing a concurrent mental task. The real-world tests revealed an increase in walking speed during all three of the evaluation conditions, with retention effects present up to a month after the final training session. These results are perhaps the most promising to date because they demonstrate that patient improvements, trained in certain VR contexts, are retained by the patients for a substantial period of time outside of VR.

15.1.2 VR-Based Assessment and Rehabilitation Following Visual Dysfunction

In some cases, mobility problems are consequences of local deficits in the early visual system. One of the consequences of visual disorders such as retinitis pigmentosa (RP) or choroideremia—the latter a degeneration of the choroid and retina—is that patients suffer peripheral visual field loss (PFL), or tunnel vision. This makes it hard to see stationary and moving obstacles and obstructions, including other pedestrians. The problem is magnified when patients are faced with an unfamiliar setting, so even simple locomotor tasks can become very challenging, and increase the risk of trips, collisions or falls.

Li et al. [29] found that tunnel-vision patients can judge their heading from optic flow as accurately as age-matched controls, under free fixation conditions. However, Turano et al. [59] reported that RP patients have more difficulty judging their direction of heading relative to objects in the scene. To compensate for this limitation, patients employ an active scanning strategy in which they make a rapid sequence of fixations between objects, the floor ahead, and other features of the layout (e.g., [58]). This is different from normally sighted individuals who tend to focus their gaze in the direction of heading or toward the current goal. While an active scanning strategy may improve the perception of heading with respect to a known object, its effect on the detection of stationary and moving obstacles and the likelihood of collisions is unknown. For this reason, different assessment and training interventions are needed to understand the cost-benefit tradeoff of such a strategy and to develop new or improved strategies for enhanced mobility safety.

Given the nature of VR as a safe testing and learning environment, a group of researchers at the Schepens Eye Research Institute (Boston, MA) have conducted a pair of experiments with two specific objectives: they assessed VR as a viable tool for studying the mobility of patients with PFL and they explored the viability of studying visual-motor learning in surrogate patients by simulating PFL in normally-sighted participants [1, 33]. Apfelbaum et al. [1] examined the influence of different approach angles to a virtual obstacle on perceptual judgments of whether their path would pass to the right or left of the obstacle. The experimental setup consisted of a human-driven treadmill facing a projection screen displaying a passive VR model of a local shopping mall (i.e., not coupled to participant's eye or head positions). Patients with PFL (the mean field of view was equal to 5.9° for the patient group) and control participants with an artificially reduced field of view (matched to the patient group) either passively viewed or actively walked while viewing the display (in passive viewing patients remained standing as the virtual environment moved). In this experiment all participants viewed the virtual environment monocularly while they approached the obstacle at different heading angles (ranging from 4° to 20°, with 0° representing a straight on approach). Both the control participants and the patients with PFL were equally accurate in their judgments and made judgments at similar distances from the obstacle. Additionally, when patients approached the obstacle at small angles while walking their accuracy increased, in contrast to an opposite pattern of results from the control participants. Both groups delayed their responses when walking until they were closer to the virtual obstacle than in passive viewing, suggesting that a walking-based VR interface might be important for evoking perceptually guided behavior that generalizes to the real world [1]. We are currently collaborating with the Schepens group to investigate the detection and avoidance of stationary and moving obstacles by PFL patients during overground walking in immersive VR [25].

Luo et al. [33] continued this line of research while employing the Multiplexing Vision Rehabilitation Device (cf. [41]).[6] Using the same experimental set-up as the previous experiment, participants interacted with the virtual environment through

[6] The Multiplexing Vision Rehabilitation Device is an augmented reality device in which the user wears a see-through head-mounted display (HMD) with a 25° field of view to which a small monochrome video camera has been attached. When wearing the device the user not only sees the real world in full resolution, but also sees real-time edge detection from a field of view between 75° and 100°, minified and displayed on the smaller field of view provided by the HMD [41].

either a minified view or a normal view across different conditions. The goal was to make sure the multiplexing device did not cause individuals to overestimate collision risks during active walking or passive viewing. The perceived passable space around the obstacle and variability of collision judgments were both greater for patients than for normally sighted participants during simulated walking (i.e., passive viewing), absent the minified device. The collision judgments were also more accurate for the normally sighted controls during the walking condition. Consequently, the minified device had no effect on the patients with PFL or the controls during either condition. These findings indicate that while the multiplexing device did not degrade performance in either population—an important finding given the increased attentional demands imposed by the device—it also did not improve perceptual judgments of collisions in the virtual environment.

These two experiments demonstrate the advantages of VR-based assessment of patients suffering from visual disorders. Specifically, important research questions about obstacle avoidance can be investigated without risk of injury to patients. In addition, VR enables simulation environments that mimic pathological deficits in healthy participants. This helps to ease the burden of participation by the clinical populations while researchers can draw from a large participant pool. While more research is necessary to ensure the viability of approaches such as these, these two experiments provide a solid foundation for exploring similar types of questions.

15.2 Dynamical Disease and VR-Based Assessment

Up to this point we have reviewed research associated with new developments in rehabilitation science sparked by interactive, immersive virtual environments. Over the last 30 years, clinical assessment has been undergoing another, equally important shift in thinking—the emergence of the concept of *dynamical disease* and techniques to measure it (see Van Orden [61], and West [76], for reviews). Dynamical disease, broadly defined, involves a physiological control system operating within parameter ranges that constrain the system's dynamics in such a way that it generates pathological behavior [16, 34]. This shift challenges the premise that behavioral variability is adverse to healthy functioning—a prominent assumption in clinical locomotor research (e.g., [4, 19, 39, 52, 64, 73]). A central tenet of this approach is that the system's dynamics, indexed by continuous measurement of locomotor patterns, are more revealing than classic summary statistics alone. For example, healthy adult gait exhibits a movement signature that is altered by neurological insult due to injury, aging, or disease [20, 54]; a difference that is not adequately captured by the mean and variance of behavior. The question of how one should measure the system dynamics, specifically how to quantify the patterns of variability in gait measures, is now at the forefront of clinical assessment research.

Virtual reality has the potential to play an important role in this transformation, for it enables the control of information that could influence the dynamics of movement [66]. This offers the flexibility to manipulate visual stimuli during walking in an attempt to alter the pattern of variability exhibited by the individual's gait cycle (e.g., [47]). VR can also be used to manipulate the locomotor trajectory of patients during over-ground walking (e.g., [12]). In other words, VR can be used to modify control parameters, thereby allowing researchers to test specific predictions about the role of those parameters in clinical assessment. These behaviors are a result of complex interactions at various control levels. Consequently, the examination of the various control parameters must take place at multiple scales of observation to fully understand the system dynamics. The remainder of this chapter will focus on several novel VR applications for the assessment of functional mobility at the level of the gait cycle and the level of the locomotor trajectory.

15.2.1 Dynamic Measures for Assessing Local Functional Mobility Using VR

Synchronizing to a stimulus is an experimental method commonly used to influence the timing properties of motor behavior. For example, much like the van Wegen et al. [60] study in which visual cues were employed to influence the step frequency, and consequently the mobility, of patients with Parkinson's disease, rhythmic auditory stimulation with a metronome has been used to improve the mobility of these patients as well [30, 35, 56, 77]. The perceptual-motor differences between synchronizing to an auditory versus a visual metronome have been described elsewhere [23], but it remains unclear which is optimal for purposes of assessment and rehabilitation. Recently, it has been demonstrated that more efficient adaptation to a perturbation (i.e., visual or auditory disruption of the stimulus rhythm) occurs when elderly participants synchronize to a visual stimulus [2]. This finding provides evidence for the privileged role visual information seems to play in the modification of acute changes to the gait cycle in healthy elderly adults. Given the biological nature of human gait, however, synchronizing to a metronome with fixed time intervals may not be effective in facilitating adaptive gait patterns and enhancing functional mobility.

Variability in the gait cycle, once thought to be a random by-product of biological noise, is now believed to reflect adaptive, functional gait (c.f., [20, 54]). Specifically, the variation in the stride-to-stride time intervals of healthy adults exhibits scale invariant (fractal) temporal correlations, as indexed by detrended fluctuation analysis (DFA[7]; [21]). Accordingly, asking a patient to synchronize to a metronome having fixed time intervals may actually reduce this natural variation, interfering with functional gait. Conversely, if humans can synchronize to a variable, or "noisy", visual metronome, this may enhance adaptive variation in their gait. A noisy metronome produces irregular intervals—some are shorter and some are longer

[7] DFA computes scaling exponents that relate a measure of variability, the detrended fluctuation function, to the time scale over which the function was computed. It is used to identify the presence or absence of persistence (i.e., a large value tends to follow a large value and a small value tends to follow a small value) in a time series. For full details, see Peng et al. [42].

than the previous one. A fractal pattern of dynamic variability can be generated to mimic those observed in healthy human gait. Instructing a patient to synchronize to a fractal metronome might induce desired patterns of dynamic variability in their gait cycle, enhancing adaptive functional mobility.

Evidence that participants can synchronize to a noisy visual metronome was first observed in finger tapping [53]. A flashing square on a computer screen prescribed the inter-tap intervals for the participant. The long-range correlations of the visual metronome intervals (indexed by DFA) were manipulated between conditions, and the participants' inter-tap intervals were shown to exhibit the same long-range correlations as the visual metronome. This provided evidence that the structure of variability of a movement task could be manipulated by altering the dynamic properties of a visual stimulus.

We recently extended this methodology to the gait domain to determine whether similar shifts in gait dynamics could be elicited in a desired direction [47]. Participants synchronized their steps to a flashing square on a computer screen while walking on a treadmill. The visual metronome generated intervals with a variety of long-range temporal correlations (indexed by DFA), yielding either a more "fractal" metronome (with a more correlated pink noise structure) or a more "random" metronome (with a more decorrelated white noise structure). The stochastic variability in participants' stride-to-stride intervals correspondingly shifted in the prescribed direction, from a normal pink noise pattern toward a more fractal pattern or a more random pattern, respectively. This result provides a proof-of-concept for the efficacy of using noisy visual metronomes to manipulate the nonlinear dynamics of the gait cycle. The exciting possibility is that this effect might be harnessed clinically to enhance adaptive gait and functional mobility.

It should be noted that visual stimuli can be presented continuously as well as discretely. A discrete visual stimulus (i.e., a classic visual metronome) only prescribes the time when an event should occur (e.g., a heel-strike during locomotion). A continuous visual stimulus, on the other hand, provides information that anticipates and specifies the timing of the upcoming event (e.g., motion of the foot and/or limb leading to and including a heel-strike). While a discrete stimulus has been shown to be useful, a continuous stimulus might enable a participant to more precisely synchronize to irregular events. VR has the potential to present novel classes of stimuli, such as virtual humans and avatars[8] that provide continuous information. It is therefore possible to imagine a number of ways that continuous information about the desired gait pattern could be presented to a patient. For example, footprints could appear discretely on the ground plane in a virtual environment, providing visual information about the timing leading up to heel-strike (see Fig. 15.2a). A stick figure could

[8] A distinction must be made about the origin of the continuous information. If a computer algorithm drives the character in virtual reality, then it is presenting continuous information about walking biomechanics that is non-biological and is termed a virtual human. Alternatively, the character can be driven by the actual motion of a human in either real-time or via a recording, which is deemed biological motion and termed an avatar. Current literature has not made a distinction about which type of motion is optimal for a gait synchronization task.

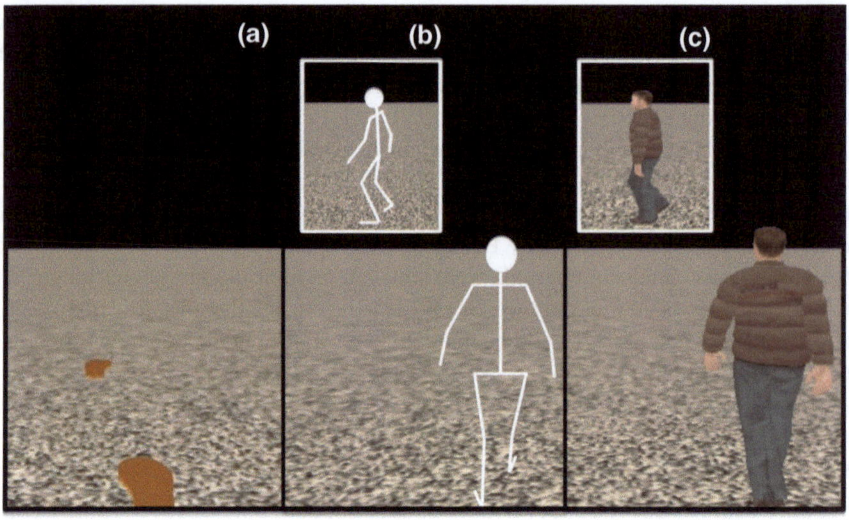

Fig. 15.2 Representation of a sample display of **(a)** virtual footprints, **(b)** a stick figure, and **(c)** a virtual human to be used as visual cues to modify the gait patterns of patients

walk through a virtual landscape, providing information to the patient about joint angles in the different phases of the gait cycle (see Fig. 15.2b). It is possible that a humanoid figure would be even more salient for synchronization purposes, so high-definition virtual humans or avatars may be an appropriate choice (see Fig. 15.2c). Finally, a third-person display of dual figures could be presented: the desired movement might be specified by a virtual human driven by a computational model or an avatar driven by motion capture data, while visual feedback is provided by an avatar yoked to the patient. This scenario would not only give the patient immediate feedback about their own performance, but would provide a model character for on-line movement comparison. Investigations into these types of stimuli are currently underway in the Virtual Environment for Assessment and Rehabilitation Laboratory (VEAR Lab) at the University of North Carolina at Greensboro.

15.2.2 Dynamic Measures for Assessing Global Functional Mobility Using VR

The previous examples illustrate the potential strength of VR applications for rehabilitation; namely, the opportunity to manipulate environmental information to probe the control parameters and index the dynamics of functional mobility at the local level of the step cycle. VR also lends itself to the flexible design of assessment protocols that yield continuous measures of behavior at a more global level, such as the locomotor trajectory.

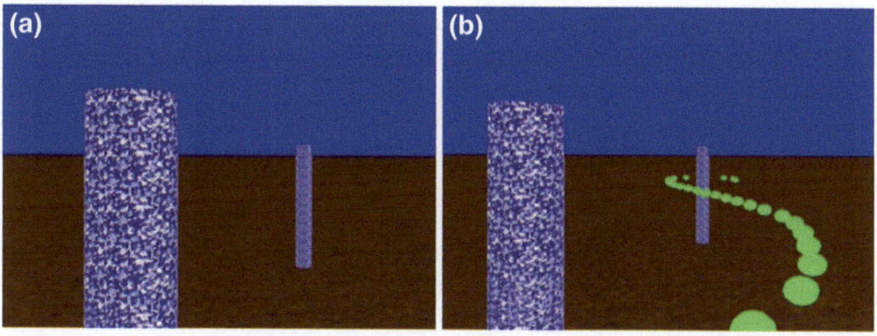

Fig. 15.3 Example displays from the baseline/controlled speed (**a**) and controlled path (**b**) conditions similar to those used in the figure-8 experiment conducted by Gerin-Lajoie et al. [12]

Consider the problem of evaluating the functional mobility of patients with a knee injury—a tear in their anterior cruciate ligament (ACL)—before and after surgery. Functional mobility in the real world subjects the knee joint to a wide range of forces and torques at various joint angles and velocities and with various patterns of muscular co-contraction, which are not currently measured in a clinical assessment. We are beginning to develop a battery of functional mobility tasks that exploit the flexibility of ambulatory VR to manipulate the affordances of the environment and capture the natural range of variability in an assessment context. Tests may include turns of varying curvature, quick stops and sharp reversals, stepping over gaps of varying widths, stepping up or down through various heights, and so on.

As a first step, Gérin-Lajoie et al. [12] developed an over-ground walking task in a virtual environment that varied the path's radius of curvature, to assess the impact of an emulated knee disability on the locomotor trajectory. Participants wearing an immobilizing knee splint walked in a figure-8 path around two virtual poles 6 m apart. There were three VR conditions: (1) natural walking at a self-selected pace, (2) speed-controlled walking, in which auditory feedback prompted participants to maintain a speed at or above their natural walking pace, and (3) path-controlled walking, in which participants followed virtual markers while receiving performance-based auditory feedback (Fig. 15.3). The participant's trajectory was then assessed to identify gait impairment indicators, and revealed a trade-off between path curvature and walking speed. Specifically, participants with an immobilized knee either decreased their speed to maintain path geometry, or increased their path radius to maintain walking speed, compared to the controls.

In addition, for the first time we exploited nonlinear dynamical methods to analyze the structure of variability at the level of the locomotor trajectory. Recurrence quantification analysis (RQA)[9] of the heading direction provided several measures of repeating temporal patterns in the trajectory as a participant walked the figure-8 path.

[9] RQA is a nonlinear measure that indexes repeating, or recurrent, patterns in a time series. For a review see Webber and Zbilut [71, 72].

These measures also differentiated the two groups, revealing that locomotor trajectories with an immobilized knee were less repeatable, less stable over time, and less mathematically complex than with a normal knee [46]. We are currently in the midst of a longitudinal study that uses the figure-8 task to assess the functional mobility of patients with actual ACL injuries pre- and post-surgery, with a follow-up after rehabilitation [50].

This research illustrates the possibilities offered to clinicians by VR-based assessment and rehabilitation. It takes advantage of perceptual manipulations that are unique to VR and allows for dynamic measurements of changes in functional locomotor behavior. Such work suggests the potential future of VR-based assessment and rehabilitation.

15.3 Conclusion

It should be clear by now, based on the numerous VR methodologies presented in this chapter, that one of the major challenges facing VR-based assessment and rehabilitation is determining the type of VR installation to employ. The visual display and head tracking devices available, as well as systems for kinematic and kinetic measurement of movement, strongly constrain the type of locomotor behavior permitted. For example, whether the user traverses the virtual environment by walking over-ground, walking on an omni-directional or linear treadmill, or via some other Wii or Kinect interface, and whether the treadmill is human- or motor-driven, have important implications for mobility assessment. Over-ground walking allows for the most natural interaction between the user and the virtual environment, implying good validity and generalizability, but such systems are expensive and space limitations often constrain them to a small room. Motorized or human-driven treadmill systems allow virtual environments of almost unlimited size, but at the price of less natural navigation (e.g., restricted turns, unnatural acceleration or deceleration) and possibly reduced validity and generalizability. Although such sophisticated technology may find a place in a regional hospital or research setting, simple Wii and Kinect-based applications have the potential for greatest impact on rehabilitation in the living room. Accordingly, researchers and clinicians must carefully consider their options when adopting these technologies and recognize the potential limitations for VR-based assessment and rehabilitation.

Regardless of these issues, the pursuit of VR-based assessment and rehabilitation is likely to increase in the coming years, as the potential benefits offered by these systems outweigh their shortcomings. It is still too early to tell whether the promise of VR will ultimately pay off for rehabilitation science, but with almost limitless possibilities awaiting implementation the future looks very bright.

References

1. Apfelbaum H, Pelah A, Peli E (2007) Heading assessment by "tunnel vision" patients and control subjects standing or walking in a virtual environment. ACM Trans Appl Percept 4:1–16
2. Bank PJM, Roerdink M, Peper CE (2011) Comparing the efficacy of metronome beeps and stepping stones to adjust gait: steps to follow. Exp Brain Res 209:159–169
3. Bardy BG, Warren WH, Kay BA (1996) Motion parallax is used to control postural sway during walking. Exp Brain Res 11:271–282
4. Blin O, Ferrandez A, Serratrice G (1990) Quantitative analysis of gait in Parkinson patients: increased variability of stride length. J Neurol Sci 98:79–91
5. Bliss JP, Tidwell PD, Guest MA (1997) The effectiveness of virtual reality for administering spatial navigation training to firefighters. Presence: Teleoperators Virtual Environ 6:73–86
6. Bloem BR, Hausdorff JM, Visser JE, Giladi N (2004) Falls and freezing of gait in Parkinson's disease: a review of two interconnected, episodic phenomena. Mov Disord 19:871–884
7. Brooks B, McNeil JE, Rose FD (1999) Route learning in a case of amnesia: a preliminary investigation into the efficacy of training in a virtual environment. Neuropsychol Rehabil 9:63–76
8. Burdea GC (2003) Virtual rehabilitation—benefits and challenges. Methods Inf Med 42:519–523
9. Deutsch JE, Mirelman A (2007) Virtual reality-based approaches to enable walking for people poststroke. Topics Stroke Rehabil 14:45–53
10. Emmett A (1994) Virtual reality helps steady the gait of Parkinson's patients. Comput Graph World 17:17–18
11. Fung J, Richards CL, Malouin F, McFadyen BJ, Lamontagne A (2006) A treadmill and motion coupled virtual reality system for gait training post-stroke. CyberPsychol Behav 9:157–162
12. Gerin-Lajoie M, Ciombor DM, Warren WH, Aaron RK (2010) Using ambulatory virtual environments for the assessment of functional gait impairment: a proof-of-concept study. Gait Posture 31:533–536
13. Geruschat DR, Turano KA, Stahl JW (1998) Traditional measures of mobility performance and retinitis pigmentosa. Optom Vis Sci 75:525–537
14. Gibson JJ (1950) The perception of the visual world. Houghton Mifflin, Boston
15. Gibson JJ (1979) The ecological approach to visual perception. Houghton Mifflin, Boston
16. Glass L, Mackey MC (1979) Pathological conditions resulting from instabilities in physiological control systems. Ann N Y Acad Sci 316:214–235
17. Gobbetti E, Scatenis R (1998) Virtual reality: past, present and future. In: Riva G, Wiederhold BK, Molinari E (eds) Virtual environments in clinical psychology and neuroscience. Ios Press, Amsterdam, pp 1–18
18. Goldberg S (1994) Training dismounted soldiers in a distributed interactive virtual environment. U.S. Army Res Inst Newslett 14:9–12
19. Grabiner PC, Biswas ST, Grabiner MD (2001) Age-related changes in spatial and temporal gait variables. Arch Phys Med Rehabil 82:31–35
20. Hausdorff JM (2007) Gait dynamics, fractals and falls: Finding meaning in the stride-to-stride fluctuations of human walking. Hum Mov Sci 26:555–589
21. Hausdorff JM, Peng C-K, Ladin Z, Wei JY, Goldberger AL (1995) Is walking a random walk? Evidence for long-range correlations in stride interval of human gait. Model Physiol 78:349–358
22. Haymes S, Guest DM, Heyes A, Johnston A (1996) Mobility of people with retinitis pigmentosa as a function of vision and psychological variables. Optom Vis Sci 73:621–637
23. Hove MJ, Spivey MJ, Krumhansl CL (2010) Compatibility of motion facilitates visuomotor synchronization. J Exp Psychol: Hum Percept Perform 36:1525–1534
24. Jaffe DL, Brown DA, Pierson-Carey CD, Buckley EL, Lew HL (2004) Stepping over obstacles to improve walking in individuals with poststroke hemiplegia. J Rehabil Res Dev 41:283–292

25. Kiefer AW, Bruggeman H, Woods R, Warren W (2012) Obstacle detection during walking by patients with tunnel vision. J Vis 12:183
26. Kizony R, Katz N, Weiss PL (2004) Virtual reality based intervention in rehabilitation: relationship between motor and cognitive abilities and performance within virtual environments for patients with stroke. In: Proceedings of the 5th international conference on disability, virtual reality and associated technology. Oxford, UK
27. Kuyk T, Elliott JL, Biehl J, Fuhr PS (1996) Environmental variables and mobility performance in adults with low vision. J Am Optom Assoc 67:403–409
28. Lamontagne A, Fung J, McFadyen BJ, Faubert J (2007) Modulation of walking speed by changing optic flow in persons with stroke. J NeuroEng Rehabil 4:22–30
29. Li L, Peli E, Warren WH (2002) Heading perception in patients with advanced retinitis pigmentosa. Optom Vis Sci 79:581–589
30. Lim I, Wegen EV, Goede CD, Deutekom M, Nieuwboer A, Willems A, Kwakkel G (2005) Effects of external rhythmical cueing on gait in patients with Parkinson's disease: a systematic review. Clin Rehabil 19:695–713
31. Lintern G, Roscoe JM, Koonce JM, Jefferson M, Segal LD (1990) Transfer of landing skills in beginning flight simulation. Human Factors 32:319–327
32. Lovie-Kitchin JE, Mainstone JC, Robinson J, Brown B (1990) What areas of the visual field are important for mobility in low vision patients? Clin Vis Sci 5:249–263
33. Luo G, Woods RL, Peli E (2009) Collision judgment when using an augmented-vision head-mounted display device. Invest Ophthalmol Vis Sci 50:4509–4515
34. Mackey MC, Glass L (1977) Oscillation and chaos in physiological control systems. Science 197:287–289
35. McIntosh GC, Brown SH, Rice RR, Thaut MH (1997) Rhythmic auditory-motor facilitation of gait patterns in patients with Parkinson's disease. J Neurol Neurosurg Psychiatry 62:22–26
36. Mirelman A, Maidan I, Herman T, Deutsch JE, Giladi N, Hausdorff JM (2011) Virtual reality for gait training: can it induce motor learning to enhance complex walking and reduce fall risk in patients with Parkinson's disease? J Gerontol A: Biol Sci Med Sci 66A:234–240
37. Mohler BJ, Thompson WB, Creem-Regehr SH, Willemsen P, Pick HL et al (2007) Calibration of locomotion resulting from visual motion in a treadmill=based virtual environment. ACM Trans Appl Percept 4:1–17
38. Olney SJ, Richards C (1996) Hemiparetic gait following stroke. Part I: characteristics. Gait Posture 4:136–148
39. Owings T, Grabiner M (2004) Step width variability, but not step length variability or step time variability, discriminates gait of healthy young and older adults during treadmill locomotion. J Biomech 37:935–938
40. Pailhous J, Ferrandez A, Fluckiger M, Baumberger B (1990) Unintentional modulations of human gait by optical flow. Behav Brain Res 38:275–281
41. Peli E (2001) Vision multiplexing: an engineering approach to vision rehabilitation device development. Optom Vis Sci 78:304–315
42. Peng CK, Havlin S, Stanley HE, Goldberger AL (1995) Quantification of scaling exponents and crossover phenomena in nonstationary heartbeat time series. Chaos 5:82–87
43. Prokop T, Schubert M, Berger W (1997) Visual influence on human locomotion: modulation to changes in optic flow. Exp Brain Res 114:63–70
44. Riess TJ (1998) Gait and Parkinson's disease: a conceptual model for an augmented-reality based therapeutic device. In: Riva G, Wiederhold BK, Molinari E (eds) Virtual environments in clinical psychology and neuroscience. Ios Press, Amsterdam, pp 200–208
45. Riess TWS et al (1995) Augmented reality in the treatment of Parkinson's disease. In: Morgan K, Satava M, Sieburg HB (eds) Interactive technology and the new paradigm for healthcare. IOS Press, Amsterdam, pp 298–302
46. Rhea CK, Gérin-Lajoie M, Ciombor DMcK, Warren WH, Aaron RK, Recurrence quantification analysis of walking path trajectories in a functional mobility task with imposed constraints. In: Paper presented at the North American Society for Psychology in Sport and Physical Activity, Tucson, AZ

47. Rhea C, Kiefer A, Warren W, D'Andrea S, Aaron R (2011) Synchronizing to a "noisy" metronome induces corresponding shifts in fractal gait dynamics. In: Paper presented at the North American Society for Psychology in Sport and Physical Activity, Burlington, VT
48. Rose FD, Attree EA, Brooks BM (2000) Training in virtual environments: transfer to real world tasks and equivalence to real task training. Ergonomics 43:494–511
49. Schubert M, Prokop T, Brocke F, Berger W (2005) Visual kinesthesia and locomotion in Parkinson's disease. Mov Disord 2:141–150
50. Shalvoy RM, Bruggeman H, D'Andrea S, Warren W, Aaron RK (2013) Virtual environmental navigation to quantify functional disability in ACL-deficient knees. In: Paper presented at the Orthopedic Research Society, San Antonio, TX
51. Seymour NE, Gallagher AG, Roman SA, O'Brien MK, Bansal VK, Andersen DK, Satava RM (2002) Virtual reality training improves operations room performance. Ann Surg 236:458–464
52. Spaulding SJ, Livingston LA, Hartsell HD (2003) The influence of external orthotic support on the adaptive gait characteristics of individuals with chronically unstable ankles. Gait Posture 17:152–153
53. Stephen DG, Stepp N, Dixon JA, Turvey MT (2008) Strong anticipation: sensitivity to long-range correlations in synchronization behavior. Phys A 387:5271–5278
54. Stergiou N, Decker LM (2011) Human movement variability, nonlinear dynamics, and pathology: is there a connection? Hum Mov Sci 30:869–888
55. Tarr MJ, Warren WH (2002) Virtual reality in behavioral neuroscience and beyond. Nat Neurosci 5:1089–1092
56. Thaut MH, McIntosh C, Rice R, Miller RA, Rathbun J, Brault JM (1996) Rhythmic auditory stimulation in gait training for Parkinson's disease patients. Mov Disord 11:193–200
57. Torkington J, Smith SGT, Rees BI, Darzi A (2001) Skill transfer from virtual reality to a real laparoscopic task. Surg Endosc 15:1076–1079
58. Turano KA, Geruschat DR, Baker FH, Stahl JW, Shapiro MD (2001) Direction of gaze while walking a simple route: persons with normal vision and persons with retinitis pigmentosa. Optom Vis Sci 78:667–675
59. Turano KA, Yu D, Hao L, Hicks JC (2005) Optic-flow and egocentric-direction strategies in walking: central vs. peripheral visual field. Vis Res 45:3117–3132
60. van Wegen E, de Goede C, Lim I, Rietberg M, Nieuwboer A, Willems A, Jones D (2006) The effect of rhythmic somatosensory cueing on gait in patients with Parkinson's disease. J Neurol Sci 248:210–214
61. Van Orden GC (2007) The fractal picture of health and wellbeing. Psychol Sci Agenda 21. http://www.apa.org/science/psa/vanorden.html. Accessed 21 Sept 2008
62. Varraine E, Bonnard M, Pailhous J (2002) Interaction between different sensory cues in the control of human gait. Exp Brain Res 142:374–384
63. von Schroeder HP, Coutts RD, Lyden PD, Billings E, Nickel VL (1995) Gait parameters following stroke: a practical assessment. J Rehabil Res Dev 32:25–31
64. von Porat A, Henriksson M, Holmstrom E, Thorstensson CA, Mattsson L, Roos EM (2006) Knee kinematics and kinetics during gait, step and hop in males with a 16 years old ACL injury compared with matched controls. Knee Surg Sports Traumatol Arthrosc 14:546–554
65. Warren WH (1995) Constructing an econiche. In: Flach J, Hancock P, Caird J, Vicente K (eds) The ecology of human-machine systems. Erlbaum, Hilldsale, pp 210–237
66. Warren WH (2006) The dynamics of perception and action. Psychol Rev 113:358–389
67. Warren WH (2008) Optic flow. In: Basbaum AI, Kaneko A, Shepherd GM, Westheimer G (eds) The senses—a comprehensive reference: vision II (vol 2, Albright TD, Masland R, eds). Academic Press, Oxford, pp 219–230
68. Warren WH, Kay BA, Duchon AP, Zosh W, Sahuc S (2001) Optic flow is used to control human walking. Nat Neurosci 4:213–216
69. Warren WH, Kay BA, Yilmaz E (1996) Visual control of posture during walking: functional specificity. J Exp Psychol: Hum Percept Perform 22:818–838
70. Warren WH, Morris M, Kalish M (1988) Perception of translational heading from optical flow. J Exp Psychol: Hum Percept Perform 14:646–660

71. Webber CL, Zbilut JP (1994) Dynamical assessment of physiological systems and states using recurrence plot strategies. J Appl Physiol 76:965–973
72. Webber CL, Zbilut JP (2005) Recurrence quantification analysis of nonlinear dynamical systems. In: Riley MA, Van Orden GC (eds). Tutorials in contemporary nonlinear methods for the behavioral sciences. National Science Foundation, Arlington, pp 353–400. http://www.nsf.gov/sbe/bcs/pac/nmbs/nmbs.jsp. Accessed 28 Sept 2006
73. Webster KE, Merory JR, Wittwer JE (2006) Gait variability in community dwelling adults with Alzheimer disease. Alzheimer Dis Assoc Disord 20:37–40
74. Weiss PL, Kizony R, Feintuch U, Katz N (2006) Virtual reality in neurorehabilitation. In: Selzer ME, Cohen L, Gage FH, Clarke S, Duncan PW (eds) Textbook of neural repair and neurorehabilitation, vol 2. University Press, Cambridge
75. Weiss PL, Rand D, Katz N, Kizony R (2004) Video capture virtual reality as a flexible and effective rehabilitation tool. J Rehabil Res Dev 1:12
76. West BJ (2006) Where medicine went wrong: rediscovering the path to complexity. World Scientific Publishing Co. Pte. Ltd., Hackensack
77. Willems AM, Nieuwboer A, Chavret F, Desloovere K, Dom R, Rochester L, Wegen EV (2006) The use of rhythmic auditory cues to influence gait in patients with Parkinson's disease, the differential effect for freezers and non-freezers, an explorative study. Disabil Rehabil 28:721–728
78. Wilson PN, Foreman N, Tlauka M (1996) Transfer of spatial information from a virtual to a real environment in physically disabled children. Disabil Rehabil 18:633–637
79. Woollacott M, Shumway-Cook A (2002) Attention and the control of posture and gait: a review of an emerging area of research. Gait Posture 16:1–14

Chapter 16
Full Body Locomotion with Video Game Motion Controllers

Brian Williamson, Chadwick Wingrave and Joseph J. LaViola Jr.

Abstract Sensing technologies of increasing fidelity are dropping in costs to the point that full body sensing hardware is commonplace in people's homes. This presents an opportunity for users to interact and move through environments with their body. Not just walking or running, but jumping, dodging, looking, dancing and exploring. Three current generation videogame devices, the Nintendo Wii Remote, Playstation Move and Microsoft Kinect, are discussed in terms of their sensors and data, in order to explore two questions. First, how do you deal with the data from the devices including error, uncertainty and volume? Second, how do you use the devices to create an interface that allows the user to interact as they wish? While these devices will change in time, understanding the sensing methods and approach to interface design will act as a basis for further improvements to full body locomotion.

16.1 Introduction

The ability to naturally interact in a virtual environment (VE), as the name implies, has been a long-standing goal of 3D user interface and virtual reality research for more than 20 years. In particular, the ability to travel or navigate through a VE is a critical component of most of these application's interfaces. One of the most natural travel techniques is full-body interaction (e.g., walking, running, jumping), as it represents seamless adaption of real-world movements to the virtual world.

B. Williamson · C. Wingrave · J. J. LaViola (✉)
Department of Electrical Engineering and Computer Science, University of Central Florida, Orlando, FL 32816-2362, USA
e-mail: bwilliam@eecs.ucf.edu

C. Wingrave
e-mail: cwingrav@eecs.ucf.edu

J. J. LaViola
e-mail: jjl@eecs.ucf.edu

F. Steinicke et al. (eds.), *Human Walking in Virtual Environments*,
DOI: 10.1007/978-1-4419-8432-6_16,

Until recently, most full-body locomotion tasks have required sophisticated and expensive hardware configurations to track various points on the body, making these types of interfaces only available in research laboratories or local entertainment centers. However, there has been an explosion of less expensive, commodity hardware devices, stemming from the video game industry that makes the ability to track either part or all of the user's body affordable for almost anyone.

Despite this potential, these devices remain challenging to work with. They fuse different sensing technologies together and are often not as robust in terms of accuracy and the types of data they provide as compared to their expensive research-quality counterparts. In this chapter, we go through the details of controllers currently at the forefront of this video game technology, we explore how to deal with the data from these devices including error, uncertainty and volume and how to use these devices to create robust interfaces that let the user interact as they wish. While these low cost commodity devices will change in time, understanding their methods of sensing and approach to interface design will act as a basis for further improvements to virtual environment locomotion .

In the next section, we provide details on the Nintendo Wii Remote (Wiimote), Playstation Move, and Microsoft Kinect video game devices in terms of how they work and the data they provide.

16.2 Video Game Motion Controllers

The notion of a video game motion controller is a relatively new term, since, until recently, video game controllers solely used buttons and digital or analog controllers that did not sense anything about the user's body [9]. One reason for not having motion sensing in the video game industry was a lack of low-cost sensing technologies. Many of these technologies, such as electromagnetic, inertial and vision-based tracking systems have been available for many years in the virtual and augmented reality communities [2, 18, 24] but their cost (e.g., tens of thousands of dollars) made it prohibitive for the majority of users. However, as computational power has increased and sensing technologies have become available at the commodity level, the video game motion controller became viable.

As of today, the three most common video game motion controllers are the Nintendo Wiimote, Playstation Move and Microsoft Kinect. Each of these devices has unique characteristics that make them interesting to examine because, although they use different sensing technology configurations, they all provide data which tells us something about where someone is or how they are moving in 3D physical space. Of course, in the future we expect that other input devices will be developed that will improve upon the video game motion controllers we discuss in this chapter. These improvements will likely come from two sources, current motion sensing technologies that are too expensive today for commodity use (e.g., sophisticated multi-camera motion capture systems, mechanical trackers associated with various haptic technologies, body sensor networks), and new research results that will take years to develop commercially (e.g., using the body as a human antenna for full-body interaction [4], using swept frequency capacitive sensing to detect interactions between a user's hands and ordinary physical objects [17]). However, regardless of what technology is used, the focus of these devices will be to extract 3D spatial locations and motions of users to support natural, body-based interfaces. Understanding how to use this data in locomotion interfaces is the main focus of the chapter. The current devices we discuss here are a representative sample of devices which provide motion and 3D spatial data that are applicable now and in the future.

Before going into the details of the devices, it is important to understand the timeline and design approaches taken by their manufacturers. The Nintendo Wiimote, the earliest device, uses IR sensors and internal accelerometers and gyroscopes to provide a general idea of where the controller is located and the motion it is going through. This was designed to feel like a remote control in the user's hands so that it had a general appeal to a wider audience, however no official SDK was planned and the data output was very raw with little pre-processing performed. The Playstation Move Controller arrived after the Wiimote and has similarities, in that it is a hand held controller with gyroscopes and accelerometers so the motion of the controller is known. The Move, however, makes use of the already existing Playstation Eye camera to more accurately track the position of the controller's glowing orb on top. Sony also designed the Move with a smoother surface to better fit in the user's hands and developed an SDK that fuses all of the raw data into an accurate position and orientation of the controller. The Kinect arrived after the Wiimote as well, and slightly after the Playstation Move, but introduced an interface that was controller-less. This system functions by watching the user's body and recreating a skeletal representation of them from both a colored image and depth image, all of which is available via a Microsoft provided SDK.

Now, let us go into the hardware details of each of these devices along with some design considerations to account for when using them, especially for full body locomotion solutions.

16.2.1 Wiimote

The Nintendo Wii Remote (Wiimote) incorporates many useful input and output features in an inexpensive, consumer-oriented, easy-to-replace and easy to repurchase package. The controller also incorporates several buttons (some in a gamepad and trigger configuration), has a speaker, programmable LEDs and a rumble device. Because this device is easy to set up, turn on, and maintain it allows game developers, researchers and homebrew engineers to use it as best serves their needs.

Importantly, the Wiimote changed console gaming due to its ability, albeit limited, to provide position data which altered the concept of how users interfaced with video games. It provides three axes of acceleration data in no particular frame of reference (FOR), with intermittent optical sensing via the infrared sensor bar. Additionally, the Wii MotionPlus gyroscope attachment can add three axes of orientation change, or angular velocity. Although this means the Wiimote's spatial data doesn't directly map to a real-world positions, the device can be employed for this under constrained use [17, 20]. For these reasons, it makes a good study of how incomplete sensor data can be fused together to make a useful controller.

16.2.1.1 Frames of Reference

The accelerometer and gyroscope sensors used in the Wiimote mean it has no single Frame of Reference (FOR). Instead, there are three that need to be considered when working with the device. Figure 16.1 shows the Wiimote's personal FOR where x, y and z axes are labeled, along with rotations pitch, roll and yaw, respectively. The Earth's gravity, detected by accelerometers, is a second FOR and a third FOR is the Wiimote's relationship to the sensor bar.

The following three examples clarify why these FOR are important and how all must be understood when working with the device. First, consider a user holding a Wiimote naturally, when +z is up in both the Wiimote's and the Earth's FOR and the Wiimote's front points away from the user and toward a sensor bar, which is usually on top of a display. In this first example, the user moves the Wiimote toward the sensor bar. This results in acceleration reported in the y-axis of both the Earth's and Wiimote's FOR and the sensor bar reporting decreased distance between the Wiimote and it. In the second example, the user rotates the Wiimote down to point toward the earth (a 90° pitch). When the user again moves the Wiimote directly toward the sensor bar, the controller reports that there is an acceleration in its z-axis. However, the Earth's FOR has acceleration in its y-axis because the 90° downward pitch didn't change the Earth's FOR. The sensor bar has no FOR because as the Wiimote rotates

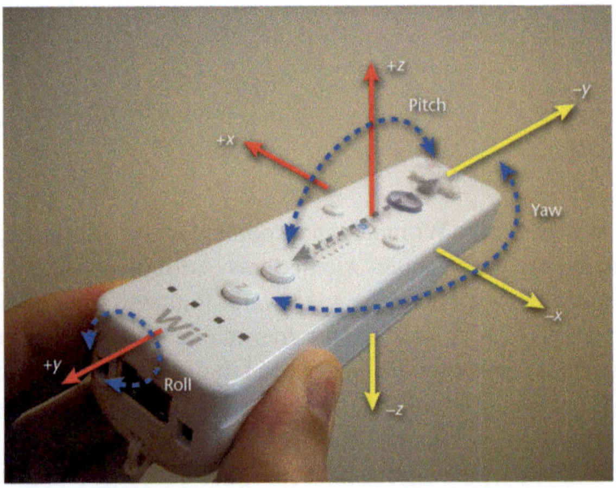

Fig. 16.1 The Wiimote, with labels indicating the Wiimote's coordinate system. Multiple coordinate systems and partial spatial data make the Wiimote difficult to design for

Fig. 16.2 The Wiimote sensor bar has two groups of IR LEDs at fixed widths

away, the controller's camera loses contact with it. The third example has the same configuration as the second example, but with the sensor bar on the ground directly below the Wiimote, at the user's feet. When the user repeats that forward motion, the sensor bar reports the Wiimote moving up in the sensor bar's z-axis, whereas the Earth and Wiimote data are the same as in the previous example. These examples show that each FOR captures important, different and incomplete information. The following sections explain in more detail what this incomplete data looks like, but it is important to keep in mind how the frame of reference may alter what the data really means.

16.2.1.2 The Sensor Bar Connection

The sensor bar connection (SBC) is one of the Wiimote's two primary spatial sensors. It occurs when the Wiimote's infrared optical camera points at a sensor bar and sees the infrared (IR) light emitted by its LEDs. A sensor bar has LEDs on each side (see Fig. 16.2), with a known width between them. This produces IR blobs that the Wiimote tracks and reports in x and y coordinates, along with the blob's width in pixels. To improve tracking distance (the Wiimote can sense blobs up to 16 ft away), the sensor bar LEDs are spread in a slight arc, with the outer LEDs angled out and the inner LEDs angled in. Interestingly, any IR source will work such as custom IR emitters, candles or multiple sensor bars, provided you have the means to differentiate between the sensor bars. More details on extracting distance information between Wiimote and Sensor Bar can be found in [29].

16.2.1.3 3-Axis Accelerometer

The second Wiimote input is the device's 3-axis accelerometer. The accelerometer reports acceleration data in the device's x, y and z directions, expressed conveniently in g's (approximately 9.8 m/s^2), which is a common unit of many devices employing 3-axis accelerometers such as cell phones, laptops and camcorders. With this information, the Wiimote is able to sense motion, reporting values that are a blend of accelerations exerted by the user and by gravity. As the gravity vector is constantly oriented towards the Earth ((0,0,1) in Earth's FOR), the gravity vector can be used to discover part of the Wiimote's orientation in terms of earth's frame of reference using:

$$pitch = \tan^{-1}\left(\frac{a_z}{a_y}\right)$$
$$roll = \tan^{-1}\left(\frac{a_z}{a_x}\right).$$

Unfortunately, determining yaw in the Earth's FOR isn't possible because the Earth's gravity vector aligns with its z-axis. Another unfortunate issue is that determining the actual acceleration of the Wiimote is problematic owing to the confounding gravity vector. To determine the actual acceleration, one of the following must take place:

- the Wiimote must be under no acceleration other than gravity so that you can accurately measure the gravity vector (in which case, you already know the actual acceleration is zero);
- you must make assumptions about the Wiimote's orientation, thus allowing room for errors;
- you must determine the orientation by other means, such as by the SBC or a gyroscope (we discuss this in more detail later).

The implications for orientation tracking by the accelerometers are that the Wiimote's orientation is only certain when it is under no acceleration. For this reason, many Wii games require that users either hold the Wiimote steady for a short period of time before using it in a game trial or have it point at the screen, oriented to the SBC.

16.2.1.4 Wii MotionPlus

This attachment uses two gyroscopes to report angular velocity along all three axes (one dual-axis gyro for x and y and a single-axis gyro for z). Mechanical gyroscopes would typically be too large and expensive for a Wiimote so it uses MEMS (micro-electromechanical system) gyroscopes, which operate using a vibrating structure, are inexpensive, use little power and are fairly accurate. A MotionPlus-augmented Wiimote provides information on changes to the Wiimote's orientation, alleviating many of the device's initial data limitations.

While the MotionPlus isn't yet fully reverse engineered, we know it reports orientation changes in two granularities, fast and slow, with fast being roughly four times the rate per bit. The gyroscope manufacturer reports that the two gyroscopes have a linear gain but that the different gyroscopes report values in two different scales, so there is no single scaling factor. Additionally, temperature and pressure changes can impact this scale factor and change the value associated with zero orientation change.

Merging the acceleration and gyroscopic data isn't simple as both sensors have accuracy and drift errors that, albeit small, amount to large errors over short time periods. When using the infrared sensor bar, you can compensate for these accumulating errors by providing an absolute orientation and position. Researchers have improved orientation by merging accelerometer and gyroscopic data but did not test a system under translational motion [9]. Other research has shown that you can combine accelerometers and gyroscopes for accurate position and orientation tracking [26]. In addition, researchers have successfully used Kalman filters to merge accelerometer and gyroscopic data [1, 27].

16.2.2 Playstation Move

The PlayStation Move system consists of a PlayStation Eye and one to four PlayStation Move motion controllers (see Fig. 16.3). The controller is used with one hand and has several buttons on the front and a long analog "T" button on the back. This hybrid device combines the advantages of camera tracking and motion sensing with traditional buttons, but achieves better results than the Wiimote due to the sensor differences.

Internally, it has several MEMS sensors like the Wiimote, including a three-axis gyroscope and three-axis accelerometer (see previous subsection for details on these sensors). The distinctive feature of the Playstation Move is the 44 mm-diameter sphere on the top that houses a RGB LED. The sphere color can be changed dynamically to enhance interaction, but the sphere's primary purpose is to track the controller's 3D position with the PlayStation Eye. Because the sphere generates its own light, tracking in a dark room works very well and even under non-optimal lighting it manages well. The spherical shape also makes the color tracking invariant to rotation, simplifying position recovery and improving precision. Deriving the Playstation Move state involves two major steps: image analysis and sensor fusion. Though the exact details of these steps are beyond the scope of this chapter, the following overview provides a qualitative understanding of each step.

16.2.2.1 Image Analysis

Conceptually, the image analysis can be broken up into two stages: finding the sphere in the image and then fitting a shape model to the sphere projection. Color segmentation is used to find the sphere and includes two steps; segmentation and pose recovery. Segmentation consists of labeling every video pixel that corresponds to the object being tracked. Pose recovery consists of converting 2D image data into 3D object pose (position and/or orientation). This can be accomplished robustly for certain shapes of known physical dimensions by measuring the statistical properties of the shape's 2D projection. The approximate size and location in the image are derived from the area and centroid of the segmented pixels (see Fig. 16.4).

It is well-known that the 2D perspective projection of a sphere is an ellipse [21], though many tracking systems introduce significant error by approximating the projection as a circle. In theory, fitting such a model to the image data is straight-

Fig. 16.3 The PlayStation Move motion controller

forward, but in practice many issues arise, such as motion causing blur or warping in a low-cost camera or partial occlusion of the object. With the Playstation Move, several of these common problems are addressed on the Playstation 3 console before the data is ever handed over to the developer.

16.2.2.2 Sensor Fusion

The results of the image analysis are combined with the inertial sensor data using a modified unscented Kalman filter. The details of this state estimation technique are beyond the scope of this chapter (though basic common Kalman Filters are presented later), but there are many excellent explanations available [4, 7, 23]. Though

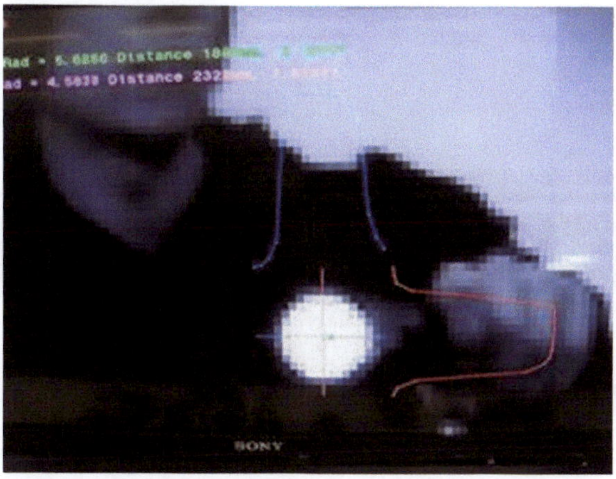

Fig. 16.4 Visualizing the 2D pixel intensity distribution of the Move's sphere is useful for debugging the model fitting stage

the sensors all contribute to the final state in a complex manner, each has a fundamental contribution that is necessary for the complete state computation. For example, the camera tracking provides an absolute measure of the 3D position and the accelerometers provides the direction of gravity when the controller is not moving, which gives an absolute measure of the pitch and roll angles. In addition, when the orientation is known, gravity can be "subtracted" from the accelerometer data to recover the controller acceleration. The acceleration is part of the state, and it can also be used to reduce noise in the 3D position and to derive the 3D velocity. The gyroscope data is also crucial because it directly provides angular velocity. When integrated, this provides a responsive measure of 3D rotation (relative orientation) and can be used to derive angular acceleration. The remaining unknown, absolute yaw, is the most tricky, but it can be inferred by comparing the motion direction computed from body-relative inertial measurements to the motion direction computed from camera-relative image measurements.

16.2.3 Microsoft Kinect

The Microsoft Kinect is an accessory for the XBOX 360 console that turns the user's body into the controller. It is able to detect multiple bodies simultaneously and use their movements and voices as input.

The hardware for the Kinect (see Fig. 16.5) is comprised of a color camera, a depth sensor, a multi-array microphone and a motorized tilt platform. The camera is used to determine different features of the user and space by detecting RGB colors and is mainly used for facial recognition of the user, an advanced feature the XBOX uses for

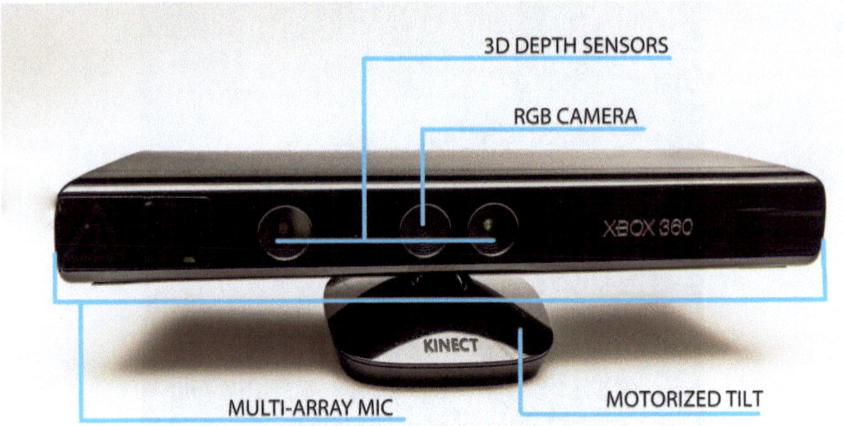

Fig. 16.5 The Microsoft Kinect sensor

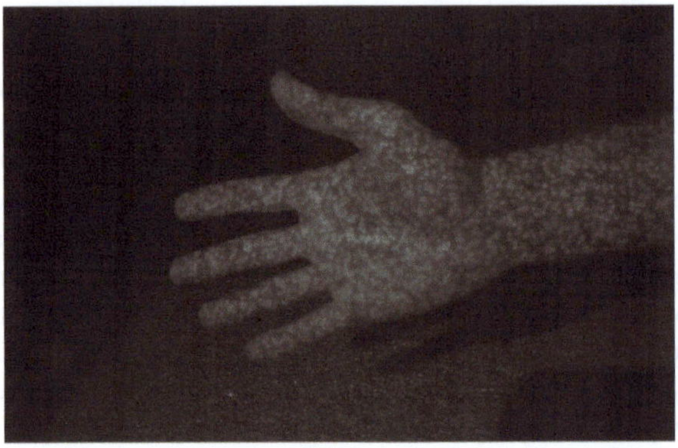

Fig. 16.6 The structured light pattern generated by the Kinect's infrared laser projector is shown. Through stereo triangulation between two images, a depth map is constructed

automatic login of players. The multi-array microphone is a set of four microphones that are able to isolate the voices of multiple users from the ambient noises in the room. It makes use of acoustic source localization and ambient noise suppression allowing users to be a few feet away from the device but still be able to use the voice controls in a headset-free manner. The third component of the hardware, the depth sensor (generally referred to as the 3D or depth camera), combines an infrared laser projector and a CMOS (complimentary metal-oxide semiconductor) sensor. The infrared projector casts out a myriad of infrared dots (see Fig. 16.6) that the CMOS sensor is able to "see" regardless of the lighting in the room and is the most differentiating feature of the Kinect.

To acquire 3D depth information, a software component of the Kinect interprets the data from the CMOS sensor. Rays are cast out via the infrared projector in a

Fig. 16.7 A depth image from the Microsoft Kinect clearly shows a human waving at the camera. Here we show nearer object in warmer colors and more distant colors in cooler colors

pseudo-random array across a large area. The CMOS sensor is able to then read the depth of all of the pixels at 30 frames per second. It is able to do this because it is an active pixel sensor, which is comprised of a two-dimensional array of pixel sensors. Each pixel sensor has a photo detector and an active amplifier. This camera is used to detect the location of the infrared dots. Following this, depth calculations are performed in the scene using stereo triangulation, which adds a requirement of two cameras. The depth measurement requires that corresponding points in one image need to be found in the second image. Once those corresponding points are found, we can then find the disparity (the number of pixels between a point in the right image and the corresponding point in the left image) between the two images. If the images are rectified (along the same parallel axis), then, once we have the disparity, we can then use triangulation to calculate the depth of that point in the scene. The depth data is then interpreted in and used in the system. To visualize the depth information, a depth image can be generated by assigning a color coding to the data as shown in Fig. 16.7.

The Kinect software is able to track users' skeletons by combining the depth information with knowledge about human body kinematics that was obtained by gathering and labeling data from special rigs which captured user motions in everyday life. These images and labels were then used for training a machine learning algorithm to create probabilities and statistics about the human form and movement. In real time, when the user steps in front of the Kinect, a 3D surface is generated using the depth information, creating a point cloud of the user. The Kinect then creates a starting guess at the user's skeleton and using the kinematic data, the Kinect makes attempts to determine the different parts of the body. A level of confidence is also assigned to each guess based on how confident the algorithm is about guessing the correct parts. Once this is done, the Kinect finds the most probable skeleton (an

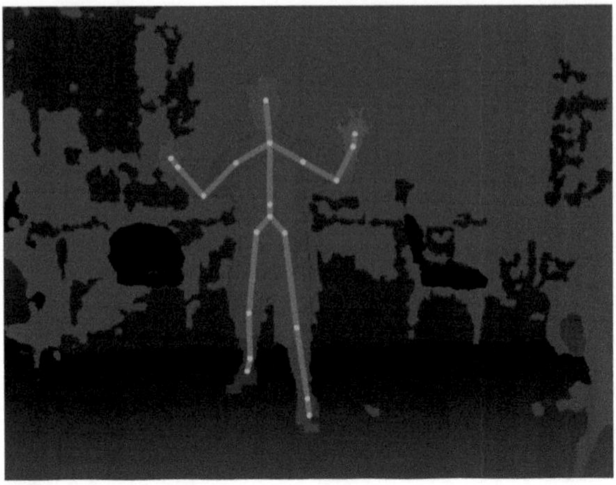

Fig. 16.8 An example skeleton representation of a user acquired from the Microsoft Kinect which can be used to track the major joints of the body frame

example available in Fig. 16.8) that would fit these body parts and their confidence levels assigned to them. All of these steps are performed real time at 30 frames per second. For more detail on the Kinect skeleton tracking algorithm, see Shotton et al. [22].

16.3 Dealing with the Data

As we can see from our discussion above, while the sensor data from these devices can potentially lead to some innovative interaction, the interpretation of the sensor data can present problems of its own. Remember that data is only an opportunity to understand the user and in many cases, more data can often be overwhelming.

In this section, we answer questions on how low-level device data is interpreted to achieve full body locomotion (see Fig. 16.9). For example, how does accelerometer data indicate the user is moving their legs? What type of machine learning algorithms adapt to the user? What levels of certainty do we have regarding a user action? We answer these questions by understanding what the data represents, followed by a short review of some common machine learning algorithms. From there, we apply these algorithms to gesture recognition to handle uncertainty. Lastly, we provide advice in applying the results from the algorithms toward locomotion solutions.

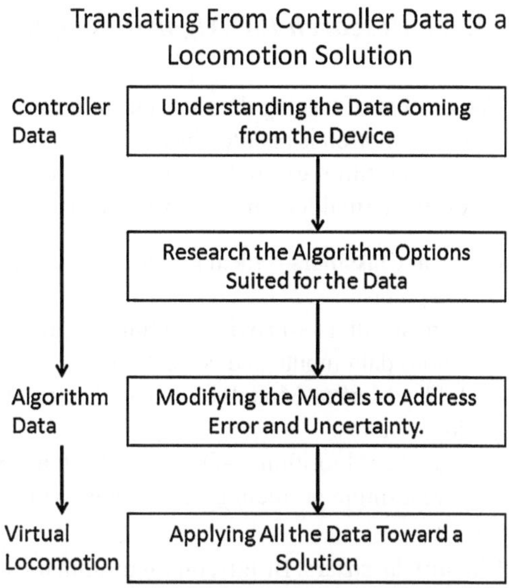

Fig. 16.9 The process we will be going through to translate from the data a controller provides to a virtual locomotion solution

16.3.1 Understanding the Data Coming from the Device

The first step in working with a new device is accessing its data and understanding what the reported values mean. Each device will have a different software mechanism for obtaining the data, specific to that device, but once you obtain your data stream, you need to understand what it means, its limitations and decide how best to make use of it. The game controller discussions above demonstrate how their creators cleverly synthesize useful data for a developer. In many cases, such as with the Wiimote, you may never have access to how the device actually does this, only having the raw data as input. Or, you may be able to find a toolkit that does much of this synthesis for you and it may be sufficient for your needs. In the case of the Kinect, Microsoft provides a software development kit (SDK) to provide the raw depth and color image as well as skeletal data interpreted by Microsoft's own algorithms. In this situation, you can choose whether to perform an analysis on the raw images or use the skeletal data for gesture recognition. Choosing to perform your own analysis should not be undertaken lightly, as it is complex and time consuming. However, many good reasons exist for this such as needing a classifier for a specific gesture, needing a new feature in the data stream recognized or needing functionality in the current toolkit that does not exist. In these cases, you will need to dig deeper into the data and understand it. Once you know the data, you can make a better choice in the algorithm to use. This is the topic of the next section, where we show how some algorithms work better with raw data while others may rely on a higher-level interpretation to already be known.

16.3.2 Research the Algorithm Options Suited for the Data

The process of turning data streams into useful data for gesture recognition and full body locomotion is handled by a range of algorithms. In this sub-section, we provide several algorithm examples that have either been used or have potential to be used to resolve virtual locomotion problems [6, 27, 28]. The examples shall include:

- Linear Classifiers—Ideal for recognizing discrete gestures with definitive start and stop points.
- Kalman Filters—Provide continuous predictions of an unknown state from continuous data inputs and corrections.
- Hidden Markov Models—Take in discrete or continuous inputs to predict discrete hidden states.
- Heuristic Algorithms—Quick, understandable and easy algorithms that normally solve a single problem given the context of the project.

Linear Classifiers. Briefly, this algorithm uses a feature vector, which represents key features of the input data, and has weights per gesture that indicate the importance of each feature to the gesture. In Hoffman et al. [6] a linear classification algorithm was used on Wiimote accelerometer data to recognize discrete gestures useful to game developers such as chopping, circles, whipping motions, etc. While the classifier's weights could be hand tuned, collected sample data can be used to set the weights to maximally differentiate between gestures. Then when a user performs an action, the classifier computes the feature vector values, weights them per gesture and determines which gesture was likely performed. A strength of this approach is the explainability of the classifier as one can look at the weights and readily determine if new features need to be created to differentiate between gestures.

An example of a linear classifier would be the Rubine algorithm [16], which evaluates the probability of a gesture by

$$g_c = w_{c0} + \sum_{i=1}^{F} w_{ci} f_i$$

where g_c is the probability of gesture c, of w_{c0} is a starting weight, a feature set of size F, w_{ci} is the weight of feature f_i for gesture c. In this algorithm training data is taken in to compute the weights each feature provides to a gesture. This process begins with a mean feature vector calculated by

$$\bar{f}_{ci} = \frac{1}{E_c} \sum_{e=0}^{E_c - 1} f_{cei}$$

where $\bar{f}_{ci}$ is the mean vector of feature i for gesture c, E_c is the number of training samples for gesture c, and f_{cei} is feature i, for gesture c, sample e.

Furthermore, Rubine's algorithm involves a covariance matrix for gesture c and all features i, j defined in

$$\sum\nolimits_{cij} = \sum_{e=0}^{E_c-1} \left(f_{cei} - \bar{f}_{ci}\right)\left(f_{cej} - \bar{f}_{cj}\right)$$

where f_{cei} is feature i, for gesture c, of the e sample (ranging from 0 to total samples E_c), f_{cej} is feature j for gesture c and sample e, and $\bar{f}_{ci}$, $\bar{f}_{cj}$ are the mean feature vectors for both features being compared. The covariance matrix is then averaged with

$$\sum\nolimits_{ij} = \frac{\sum_{c=0}^{C-1} \sum_{cij}}{-C + \sum_{c=0}^{C-1} E_c}$$

where $\sum_{cij}$ is the covariance matrix for features i, j and gesture c, divided by total gestures C plus the summation of all sample totals E_c for each gesture c.

With all of these calculations done on the training data, the weights for the main Rubine algorithm can be calculated with

$$w_{cj} = \sum_{i=1}^{F} \left(\sum\nolimits^{-1}\right)_{ij} \bar{f}_{ci}$$

where $\left(\sum^{-1}\right)_{ij}$ is the inverse covariance matrix for gestures i, j and $\bar{f}_{ci}$ is the mean feature, i for gesture c, and F is the total number of features in the feature vector. Finally w_{c0} is calculated with

$$w_{c0} = -\frac{1}{2} \sum_{i=1}^{F} w_{ci} \bar{f}_{ci}$$

where w_{ci} is the weight of feature i for gesture c and $\bar{f}_{ci}$ is the mean vector of feature i for gesture c.

The algorithms show the linear classifier is heavily dependent upon the feature set, which is purposely left to you to define depending on your device and gestures to recognize. For example, in Hoffman et al. [6] the accelerometer and gyroscope data had features such as the length of time for the gesture, their centroid, average accelerometer data, etc. Being able to create an application-specific feature vector allows you to tailor the classifier to selections between similar gestures. With good training data, a classifier can be quite accurate and work well with raw data. However, it does not perform well on streaming data, which requires identifying when a gesture starts and stops (data segmentation) as well as performing recognition. It is also, in the simplest form, limited to discrete actions. For example, it is capable of determining "the user jumped" but not necessarily how high or where to.

Kalman Filter. The Kalman Filter is computationally simple, relying on matrix multiplication to fuse data to reduce error and uncertainty, however the setup of the filter can be complicated to master. It has many uses but is of interest to us in how it can determine a user's position in real time for locomotion [27]. The Kalman filter is ideal for controllers like the Wiimote and Sony Move where several sensors provide raw data that can be fused together.

The basic algorithm works in two stages, the first performs prediction of hidden states based on what is known about the model. For example, in determining position, this first step takes accelerometer data and uses kinematics to predict from the previous position what the new position is. This is done via two equations

$$\hat{x}_t = X_{t-1} * A_{\Delta t}$$
$$\hat{p}_t = A_{\Delta t} * P_{t-1} * A_{\Delta t}^T + E$$

where $\hat{x}_t$ is the predicted state, $X_{t-1}$is the previous state that the Kalman Filter determined and $A_{\Delta t}$ is a time dependent matrix representing the model of the system. $\hat{p}_t$ is the probability matrix of the prediction, P_{t-1} is the previous probability/confident matrix determined by the Kalman Filter, and E is the inherent error matrix of the prediction model.

The second step of a Kalman Filter goes through a measurement step, where an observation concerning the state is used to both compute the optimal Kalman gain of the model, and to correct errors inherent in the prediction step. This is done with

$$K_t = \hat{p}_t * H^T * \left(H * \hat{p}_t * H^T + R\right)^{-1}$$
$$X_t = \hat{x}_t + K_t * \left(Z_t - H * \hat{x}_t\right)$$
$$P_t = (I - K_t * H) * \hat{p}_t$$

where K_t is the optimal Kalman gain, H is the observation model that correlates observation states to the prediction model, R is the error matrix for the observation data, Z_t is the observation, X_t is the corrected predicted state and P_t is the corrected probability matrix.

As can be seen in these equations, the prediction model based on the first set of data is blended with the observation data, while at the same time learning and adjusting the probability matrix.

The Kalman Filter is a method that can work well with raw data or interpreted data depending on how the models above are developed. Also, it can provide real time continuous data updates; for example this filter is commonly used in GPS systems to fill in the gaps while waiting for the next satellite update. While that isn't directly related to motion controllers, it does show utility of Kalman Filters.

Hidden Markov Models. Another useful algorithm is the Hidden Markov Model (HHM). Like the linear classifier algorithm above, HMMs are great for determining discrete actions but unlike linear classifiers, can do so on streaming data (e.g., providing gesture segmentation and recognition). The Hidden Markov Model Tutorial by Rabiner [15] contains several uses of the model and is a well-known tutorial for further details on HMM use. The basic Hidden Markov Model is composed of five matrices (labels come from the tutorial):

- N—Discrete set of hidden states the HMM is predicting.
- M—Set of possible observation symbols (the data input that we know).
- A—The state transition matrix, the probability of going from one state to another.
- B—The observation matrix, the probability of seeing a specific observation while in a specific state.
- Π—The initial state matrix, which contains the probability of starting in a specific state.

The HMM matrices are then setup to represent a particular problem set and trained on existing data (see [15] for a full discussion). Once complete, current observational data can be fed into the HMM to provide the probability that the system is in each state. From this, you determine how to use the outputs of an HMM for your system. For example, if you have states representing different gestures, you might take the highest state probability above a threshold to determine a gesture was completed such as walking, jumping or ducking. In a well-trained system, gestures may be predicted before the user has even finished the jump, giving the feel of real time locomotion.

There are complications to overcome when using an HMM with most modern video game controllers. For example, all of the controllers we have discussed provide you with continuous data and the traditionally designed HMM observation matrix has a discrete set of symbols. Suggested in the HMM tutorial [15], you may consider the simple solution of using codebooks to translate the continuous data to discrete numbers, though this provides a loss in resolution of data. An example of a simple codebook would be if you know your data can only go from 0 to 100, you may have 100 discrete states, each being the whole number of the data (resolution of floating point is lost). A better solution is to create an observation density function to replace matrix B in the Hidden Markov Model. This means rather than having a B matrix, you may have a B function that maps the continuous data to some probability distribution (typically Gaussian). Once this is figured out and adapted the HMM provides an excellent way to determine discrete gestures in real time as the Kalman Filter is able to provide continuous data in real time. Furthermore it is a system capable of adapting to the user with training data, and being updated as it is used.

Heuristic Recognizers. When the gesture set to be recognized is simple enough, a viable approach is to use heuristics to differentiate between gestures. This is best when the gesture set is small and clearly different enough from each other. In Williamson et al. [28] a heuristic recognition method was used instead of a linear classifier because it provided adequate accuracy with no training data required by the user. This was only possible because the gesture set being recognized was only focused on jumping, crouching and a turning gesture. An example heuristic recognizer for jumping would be to assume a jump was made when the head is a certain height above its normal position, explained as

$$J = H_y - \bar{H}_y > C$$

where J is a true or false value based on if a jump has occurred, H_y is the height of the head position, $\bar{H}_y$ is the calibrated normal height of the head joint and C is some constant. C would then be set to a height that a person would only get to by jumping from the ground. Such recognition is very specialized but simple, explainable and can determine in an instant whether a jump has occurred.

Selecting between algorithms for your recognition will be highly dependent on your devices and the needs of your application. When performed properly, these choices can result in recognition with high accuracy [6, 27, 28]. However, errors and uncertainty are always going to remain, even with your best efforts. No translation from raw data to real world actions is going to be perfect but you can minimize the impact of these issues, as discussed in the next sub-section.

16.3.3 Modifying the Models to Address Error and Uncertainty

With any device or solution to a difficult problem, there are chances for error and uncertainty. From the data itself, to the results from the algorithms, or even the user themselves, there are several possibilities and variables to consider. While all error and uncertainty is not likely to be completely removed, by properly addressing them, they can be minimized. In this section we start with issues that arise at the device level and present solutions following the process all the way to the end user.

Errors from the Device. Most raw sensor data, such as the accelerometers and gyroscopes, have the potential to fluctuate slightly as a result of their mechanical design. This can be observed often while the device is just sitting steady on the table (in fact this is a good technique to measure the error ranges), and in some cases it is exasperated as the device is moved around. While these fluctuations are small, over time they can begin to add up creating monumental drift in a matter of seconds from the user's perspective. In some cases the algorithms themselves have built in options to reduce these errors. For the Kalman Filter, error matrices are present which can tell the system how much to trust raw data, or for a linear classifier as long as the error variance is present in the training data the algorithm is relatively unaffected in its predictions, though confidence values might be slightly lower. Thus it is important to look into the details of the hardware being used to gather exact variance numbers, even if supplied by the manufacturer. Perform experiments with the device under known conditions to see what levels of drift are present and map these into the algorithms that you decide to use or apply filters to the data before it enters the algorithm to minimize variances.

Errors in the Algorithms. Most of the algorithms we discussed above are designed for general purpose and need to be modified to match a specific project. Even so, they may encounter situations that were unplanned and will produce incorrect data. In the case of the linear classifier, often better training data will resolve many issues. For the Kalman Filter and Hidden Markov Models, the initial states can greatly change the overall performance and prediction capability of the model. To reduce these errors, review your algorithm design and modify them to match the system and expectations from the user as best as possible. If an initial state is needed, do not just provide all zeros, actually research what it should be and define this for the model. Often this is done by running the model repeatedly and finding the optimal starting point for the experiment. Even for the linear classifier, one set of training data may be too obscure and must be removed for optimal recognition accuracy.

Uncertainty with the Users. Perhaps the most difficult variables to predict is what users not used to the system may try to do. For example, they may change their mind and switch the gesture they are trying to perform midway. In the case of the Kinect, say you have a heuristic solution that waits for the height of the user's head to pass a predefined point to consider it a jump and the first user is much taller than expected, so much so that their natural skeleton representation is over this point. These situations must be considered for an accurate system. Often times a pilot study, which is a small sampling of the potential users, can be used to find the types of actions and variations in users you can expect. From there, the project needs to be modified to address these issues. For example, maybe training the linear classifier to recognize an incomplete gesture and throw it out or to normalize all skeletal representations from the Kinect.

Any device or system is going to have to address these issues, however by planning ahead, they will be minimized and can cause less heartache to you as you perform your work. For more example of reducing error and uncertainty from a recognizer, check out other research and tutorials on the matter [11]. Now that we have gone through the process of understanding the data, choosing algorithms and planning for errors and uncertainty, we conclude with the final step to the problem: translating the results from the algorithms into actual solutions for full body locomotion.

16.3.4 Applying All the Data Toward a Solution

It may seem you are done once the algorithms are working correctly, however there is a final step in applying the results from the algorithm into a real world solution, and sometimes making use of the context of the situation to achieve even better results. For example, a Hidden Markov Model supplies a list of probabilities to the chances of being in a hidden state which you may have chosen to be possible gestures. A simple setup may just take the highest probable gesture and react to it, but a better system would check to make sure the probabilities of all gestures aren't very low so choose to do nothing (this would be an example of the user being in an unknown or idle state). Another example would be accelerometer readings showing the user is actually moving rather than just walking in place, which would determine how important the data coming from the Kalman Filter may be. Those were just a couple of examples, and are very specialized to the interface that you are designing, discussed in the next section, but should be considered to get the best from the system as possible.

After you've taken the steps to understand the data from the controller, and designed the proper algorithms and system models to receive the best results, it is essential to remember the context of your situation and apply these algorithms as is appropriate. Now that you have learned the basics of the devices and how to use their data to solve real world problems, the next important step is to consider how you are resolving these problems. In the next section we cover tips on creating an interface for a user so that they find your solution to virtual locomotion both fun and natural.

16.4 Creating an Interface

Humans are masters at working in a space about their body and have senses developed for this. However, not all of this mastery can be transferred to a full body interface (see challenges below) and this leads to problems for the user. It makes sense then to incorporate as much of our human ability as possible, the naturalness that we have in working in the real world, to allow for natural locomotion in virtual spaces. However, a video camera staring at a user does not tell them what to do (i.e. give an affordance) like a button or joystick would. This is because people do not always know how they would move or react to achieve some effect, even for a natural interface. Consider then how "magic" interaction, i.e. a non-natural gesture that is easier to do, can be more effective and less fatiguing. Know the tradeoffs you can make in your interface... how *do* people want to interact?

An interface begins with a solid understanding of devices and data, but the goal is to allow the user to achieve the application's requirements quickly, effortlessly, accurately and sometimes enjoyably. Incorporating video game devices properly can provide more engaging experiences, more freedom and a more natural, healthy and immersive experience. Good design is about understanding all requirements and their importance in order to create a holistic interface; gracefully handing situations when recognition fails. Additionally, heed the practices of game developers that include a warm-up or tutorial in their applications, to train the user.

In this section, we present challenges in designing an interface, learning how to make your design tradeoffs, discovering how your users like to interact and finally some ideas on compensating for technology limitations.

16.4.1 Challenges

Consider the following challenges when designing your full-body interface:

- Virtual locomotion is limited by real space and physics: walking runs out of room, walls are tangible and virtual ladders have no rungs.
- Virtual worlds present new experiences where players do not always know the correct full-body motions, such as skating a half-pipe or snow boarding.

- Virtual worlds do not always have real-world counterparts; without affordances, how do you cast spells or work alien technology?
- Full body interaction is not always fun: exhaustion, injury and boredom need to be avoided (or incorporated!).
- Technology limitations may not be able to detect player actions to the levels of accuracy you desire: occlusion, accuracy, delay, etc.

Therefore, to meet these challenges, important questions must be addressed. These include: What full body actions are needed? What full body actions are usable? How can we build natural and compelling full body video game interfaces when we don't always have the data we desire?

16.4.2 Controlling Travel

Whenever users need to travel through a virtual space, whether with expensive high-end equipment or common video game controllers, the strategy the user uses to move through the environment must be addressed. The user needs to be able to indicate the beginning of travel, the continuation of travel, the cessation of travel, steering and possibly even define the rate of travel. Traditional controllers simply gave the user a joystick, where these actions are easily mapped and easily recognized. Natural gestures complicate the interface design, as there is not a simple mapping between gesture and result. However, there is a benefit to using natural gestures, such as real walking or walking in place, which Whitton et al. showed [25] was better in several areas over joystick movement. Real walking in movement are the most preferred, but can only cover the relatively small ranges of sensors, with walking in place an accepted substitute for navigating very large areas.

There are a range of options available to you in how you achieve natural travel. In [27], a single Wiimote attached to a hat provided sufficient data to give real movement within the range of the infrared SBC and to recognize a walking in place gesture. This approach used Kalman Filters to merge accelerometer, infrared and gyroscope data into a position vector and orientation quaternion while heuristic recognition determined if the user was stepping up and down with their feet. In [28], a Kinect provided skeletal data capable of giving real time movement representations in the sensor's field of view. For walking past this area, a walking in place gesture was used and determined by user's leg joints. Full-body interaction is a widely studied subject and while simple game controllers are limited in their data, they are still entirely capable of providing a natural interface.

16.4.3 Understand Your Design Tradeoffs and Users

Design tradeoffs, i.e. making changes to one part of an interface so the interface as a whole improves, are especially important for full-body interaction as limitations will be encountered in what you can sense of the user and recognize in the data. You will have to know what is key to your interface and what can be changed. The best way to understand this is to perform a requirement and task analysis (a full discussion of this is beyond this chapter's scope). Knowing this required functionality informs the design process when problems are encountered so a designer can make informed decisions between: (1) spending time on specific algorithmic recognition improvements, (2) adding new hardware, or (3) making changes to the user interface based on which action is more important (i.e. differentiation between similar gestures is problematic so swap in a new gesture).

In our experience, a few general assumptions can be made about how users will want, or expect, to move in your interface, based upon *naturalness*, the "*weight*" and *fatigue* of gestures. Expect players will start with *natural* movements; movements directly matched to real-world counterparts, and then they change their actions as these natural movements fail due to your design, real-world constraints or even your recognition algorithms. For example, they walk a few feet forward but then realize, when they hit a wall, they need to run in place for longer distance travel. Next, consider the *weight* of a user's motion (i.e. how much cognitive and physical effort is required), as users will tend to match task and motion energy. Consider a user staring at a fixed large screen display, who needs to both look around and turn: "lighter" head glances can be used for quickly looks and the "heavier" torso rotations can change their in-world orientation. Lastly, expect *fatigue* to play a role in the long-term use of an interface. While a simple running in place gesture might satisfy some design considerations, it quickly tires a user. Even low fatigue motions become fatiguing over time, especially if it requires odd motions that can lead to repetitive stress. In these cases, create non-realistic gestures to compensate. One alternative is to use fatiguing gestures for actions that give extra rewards, such as running in place for sprinting, coupled with a light-weight gesture for typical walking speeds. Keep in mind too that sometimes fatigue is wanted, such as exertion games [12] or military trainers where fatigue is a part of the simulation.

Full body locomotion and interaction design is difficult and if the first design is potentially not possible, know what design tradeoffs can be made. For example, if people do not need to move their arms, then map leg movements to the arms. People are going to fatigue quickly so know if it is important to be realistic or if less fatiguing "magic" gestures can replace fatiguing real gestures. For instance, leaning forward could replace running in place. Or, remove the need for the user to move vast distances by moving them automatically or provide a game pad. Even small error rates result in problems for the user so make sure that there is a real reason to use full-body interaction. In these situations, keep it novel and reserve full-body interaction for only those cases where there is a real benefit or purpose to the application, as dictated by your requirements. Design is about tradeoffs. Know which ones can be made.

16.4.4 Find How People Want to Interact

A developer may follow their intuition to rapidly iterate on prototype systems, but this method is costly and leads to ambiguous results: *Did the prototype fail because of the design or because the recognition was poorly implemented*? A method to avoid this ambiguity while exploring the design space is the Wizard of Oz protocol [3] where a human operator interprets a user's actions (if an operator observing the user cannot recognize their intent, a computer probably can't either). In this way, several "prototypes" can be tested without time spent implementing recognizers. As well, cleaner results can be obtained as the operator can adapt to a user as they explore how they would move for all the actions needed in an environment and as a design solidifies, the protocol allows for capturing sample data. This method allows users complete freedom to: choose their movements for their intent, create new actions on-the-fly and adapt as new requirements of the system are given that would conflict. The operator also sees how fatigue affects the user. Video recording analysis, post-experiment interviews or even data returned from devices you gave the user during the Wizard of Oz session can inform the development of the recognition algorithms.

This approach was used on a commercial video game Mirror's Edge [13]. Mirror's Edge is a first person action-adventure video game incorporating Parkour-like methods of travel as well as basic fighting motions. Mirror's Edge requires the player to run, jump, climb and duck as they travel between buildings and across rooftops in an urban landscape. Using the Wizard of Oz protocol, the operator watched a real-time video feed, interpreted the player's actions and controlled the game accordingly with a keyboard and mouse. The result was seamless: not all players even realized there was a human in the loop. From this, a generalized gesture set for locomotion was created [13]. These same actions are common to many locomotion systems and can form the starting point for most systems, having already been incorporated into trainers [27, 28] (Table 16.1).

16.4.5 Compensate For Technology Limitations

Once you have your devices, recognizers and interface design, which works to minimize errors, you still can expect recognition errors and the impact of this can be minimized through design. One approach is to focus improvements on the most commonly used and most critical actions. For instance, a military trainer should have walking be responsive, shooting triggered by a button and the grenade throwing recognizer minimize false positives (i.e. not randomly toss a grenade at your feet). The following three approaches offer non-technical compensation strategies.

Table 16.1 A proposed set of gestures for natural locomotion

Task	Technique
Translation (local)	Normal walking
Translation (extended)	Running in place
Orientation	Shoulder steering
Combat	Punch and kick
Sliding	Duck
Balance	Arms out and leaning
Climb pole	Repeated placement of fists over the other
Climb up (from hang)	Arms up and pull down
Climb up (ladder)	Alternating arm pull-down
Jump (low)	Hop, no arms
Jump (medium)	Hop, arms out front
Jump (high)	Hop, arms up

These can form the starting points for larger or smaller gesture sets

Context. Understand the role that context plays in the interface. A gesture for jump may indicate climb when in front of a ladder. At a minimum, understand how context can play a role in mutually disambiguating [14] between multiple outcomes. While gestures may look similar, their use may not be and this becomes a dimension of differentiation that can be incorporated.

"Waggling" Motions. "Waggling" or "cheating" motions are commonly used in Wiimote games due to the devices not properly recognizing the absolute motions of the user and accepting more chaotic or less "weighty" versions of the gesture. So, instead of performing a realistic punching gesture, a user might tap the Wiimote forward which the game recognizes as a punch. While this might seem a limitation, these types of waggling motions have a place in some interfaces when the purpose is not for exercise or training, but fun or utility. The result are games winnable by simply moving the device randomly about, but that are still fun and can be played for longer periods of time. When building recognizers, make sure to support the purpose of the gesture and not just the absolute form.

Compensation by "Story". Whenever possible, the easiest means of compensating for input hardware limitations is through the use of story. By "use of story," we mean that by careful manipulation of the user's tasks and their goals, the shortcomings of the technology can be avoided. For example in augmented reality, where perfect registration to the real world is very difficult, use NPC ghosts because we expect ghosts to float around, be translucent and pass through walls. For a hard to recognize gesture, create a story element that requires the user to perform some easily recognized gesture first to create context or allow the recognizer to make an assumption. This is common in Wii gaming where accuracy is second to enjoyment and playability. To compensate for the Wiimote's drift, games require the user to return to a known position where the gravity vector or position can be assumed. So, Wii Sports Resort's Frisbee requires you to point at a disk right before you throw it, We Cheer instructs players to hold the Wiimote as if they're holding pom-poms and Wii Sports Boxing makes players raise their hands when preparing for a fight. Hardware and recognition limitations will persist as technology improves, so make users believe there is a valid reason in the story for this or guide them so that they never notice.

16.5 Conclusion

It may seem unlikely at first that a low cost video game controller could ever solve a complex locomotion problem, however if you use the ideas and techniques discussed above, success with these devices are easily plausible. In this chapter we have shown you the details to three different popular game controllers, the Nintendo Wiimote, Sony Move and Microsoft Kinect. You have seen how these devices work, some considerations made during their design and what data the SDKs might provide from them. You've also seen how to take that data, understand what you are dealing with and use that to select the best algorithm for your problem. Tips were provided on modifying these algorithms to reduce uncertainty and taking that final step toward a real world solution. In our last section of the chapter, advice and considerations were given toward developing an interface appropriate to the user and task. After all, just because you think a gesture is natural does not mean the average person will think so; or if they do, they might not like it anyway. Run tests to see what the users themselves prefer before you build up training data that may end up being useless. While using low cost hardware to recognize such interactions may be difficult, it can be done, with the added benefit of using hardware users are familiar with and that exists in their homes.

References

1. Azuma R, Bishop G (1994) Improving static and dynamic registration in an optical see-through HMD. In: SIGGRAPH '94: Proceedings of the 21st annual conference on computer graphics and interactive techniques. ACM, New York, pp 197–204
2. Bowman D, Kruijff E, LaViola J, Poupyrev I (2005) 3D user interfaces: theory and practice. Addison-Wesley, Boston
3. Buxton B (2007) Sketching user experiences: getting the design right and the right design. Morgan Kaufmann, San Francisco
4. Cohn G, Morris D, Patel S, Tan D (2012) Humantenna: using the body as an antenna for real-time whole-body interaction. In: Proceedings of the 2012 ACM annual conference on human factors in computing systems (CHI '12), pp 1901–1910
5. Crassidis JL, Markley FL (2003) Unscented filtering for spacecraft attitude estimation. J Guid Control Dyn 26(4):536–542
6. Hoffman M, Varcholik P, LaViola J (2010) Breaking the status quo: improving 3D gesture recognition with spatially convenient input devices. IEEE VR 2010
7. Julier S, Uhlmann J (1997) A new extension of the Kalman filter to nonlinear systems. In: International symposium: aerospace/defense sensing, simulation and controls.

8. Kalman RE (1960) A new approach to linear filtering and prediction problems. J Basic Eng 82 (Series D) 35–45
9. LaViola J (2008) Bringing VR and spatial 3D interaction to the masses through video games. IEEE Comput Graph Appl 28(5):10–15
10. Luinge H, Veltink P, Baten C (1999) Estimating orientation with gyroscopes and accelerometers. Technol Health Care 7(6):455–459
11. Mankoff J, Hudson S, Abowd GD (2000) Interaction techniques for ambiguity resolution in recognition-based interfaces. In: Proceedings of the 13th annual ACM symposium UIST, pp 11–20
12. Mueller F, Edge D, Vetere F, Gibbs M, Agamanolis S, Bongers B, Sheridan J (2011) Designing sports: a framework for exertion games. In: Proceedings of the 2011 annual conference on human factors in computing systems (CHI '11). ACM, New York, pp 2651–2660
13. Norton J, Wingrave C, LaViola J (2011) Exploring strategies and guidelines for developing full body video game interfaces. In: Foundations of digital games 2011
14. Oviatt S (1999) Mutual disambiguation of recognition errors in a multimodel architecture. In: Proceedings of the SIGCHI conference on human factors in computing systems: the CHI is the limit (CHI '99). ACM, New York, pp 576–583
15. Rabiner L (1989) A tutorial on hidden Markov models and selected applications in speech recognition. In: Proceedings of the IEEE, 1989
16. Rubine D (1991) Specifying gestures by example. Comput Graph 25(4):329–337
17. Sato M, Poupyrev I, Harrison C (2012) Touché: enhancing touch interaction on humans, screens, liquids, and everyday objects. In: Proceedings of the 2012 ACM annual conference on human factors in computing systems (CHI '12), pp 483–492
18. Sherman B, Craig A (2002) Understanding virtual reality: interface, application, and design. Morgan Kauffman, San Francisco
19. Shirai A, Geslin E, Richir S (2007) Wiimedia: motion analysis methods and applications using a consumer video game controller. In: Sandbox '07: Proceedings of the (2007) ACM SIGGRAPH symposium on video games. ACM, New York, pp 133–140
20. Shiratori T, Hodgins JK (2008) Accelerometer-based user interfaces for the control of a physically simulated character. ACM Trans. Graph. 27(5):1–9
21. Shivaram G, Seetharaman G (1998) A new technique for finding the optical center of cameras. In: ICIP 98: Proceedings of 1998 international conference on image processing, vol 2, pp 167–171
22. Shotton J, Fizgibbon A, Cook M, Sharp T, Finocchio M, Moore R, Kipman A, Blake A (2011) Real-time human pose recognition in parts from single depth images. In: IEEE computer vision and pattern recognition (CVPR) 2011. IEEE
23. Wan EA, van der Merwe R (2002) The unscented Kalman filter for nonlinear estimation, pp 153–158
24. Welch G, Foxlin E (2002) Motion tracking: no silver bullet, but a respectable arsenal. IEEE Comput Graph Appl 22(6):24–38
25. Whitton M, Cohn J, Feasel J, Zimmons P, Razzaque S, Poulton S et al (2005) Comparing VE locomotion interfaces. In: IEEE virtual reality, pp 123–130
26. Williamson R, Andrews B (2001) Detecting absolute human knee angle and angular velocity using accelerometers and rate gyroscopes. Med Biol Eng Comput 39(3):294–302
27. Williamson B, Wingrave C, LaViola J (2010) Realnav: exploring natural user interfaces for locomotion in video games. In: Proceedings of IEEE symposium on 3D user interfaces 2010. IEEE Computer Society, pp 3–10
28. Williamson B, Wingrave C, LaViola J, Roberts T, Garrity P (2011) Natural full body interaction for navigation in dismounted soldier training. In: Proceedings of the interservice/industry training, simulation, and education conference (I/ITSEC) 2011, pp 2103–2110, Dec 2011
29. Wingrave C, Williamson B, Varcholik P, Rose J, Miller A, Charbonneau E, Bott J, LaViola J (2010) Wii remote and beyond: using spatially convenient devices for 3DUIs. IEEE Comput Graph Appl 30(2):71–85

Chapter 17
Interacting with Augmented Floor Surfaces

Yon Visell, Severin Smith and Jeremy R. Cooperstock

Abstract This chapter reviews techniques and technologies for interaction via the feet with touch-sensitive floor surfaces that are augmented with multimodal (visual, auditory, and/or haptic) feedback. We discuss aspects of human-computer interaction with such interfaces, including potential applications in virtual and augmented reality for floor based user interfaces and immersive walking simulations. Several realizations of augmented floor surfaces are discussed, and we review one case example that has been extensively investigated by the authors, along with evaluations that have been reported in prior literature. Potential applications in the domains of human-computer interaction and virtual reality are also reviewed.

17.1 Introduction

Despite growing interest in foot-based interaction for computationally augmented environments, there has been limited research to date on the design of interactions for such interfaces, or their usability. Two factors that may have contributed to this state of affairs are the lack of standard displays and techniques for installing computationally augmented floors (see Fig. 17.1 for an example), and the limited range and realism of the feedback that they are able to provide. In the present chapter, we review techniques for interacting with computationally augmented floor surfaces, including

Y. Visell (✉)
Department of Electrical Engineering and Computer Engineering, Drexel University, Philadelphia, PA19104, USA
e-mail: yon@zero-th.org

S. Smith · J. R. Cooperstock
McGill University, Montreal, Canada
e-mail: severin.smith@gmail.com

J. R. Cooperstock
e-mail: jer@cim.mcgill.ca

F. Steinicke et al. (eds.), *Human Walking in Virtual Environments*,

DOI: 10.1007/978-1-4419-8432-6_17,

Fig. 17.1 A floor interface situated within an immersive, rear projected virtual environment simulator

strategies that can be used for their technical implementation—notably for sensing and multimodal feedback—and interaction techniques that can be employed with them. Our discussion draws on known results from the research literature and from the commercial domain. We also survey current and potential applications of such floor-based displays, including a range of examples from prior literature, and discuss in detail one set of case studies based on prior research of the authors, associated with an interface based on a distributed network of low-cost, rigid floor tile components, with integrated sensing and actuation.

17.2 Background

Foot-based human-computer interaction has attracted increasing attention as a paradigm for interacting in virtual reality or in ambient computing environments. Some potential application domains include architectural visualization, immersive mission training, locomotor rehabilitation, and entertainment. Traditional techniques based on traditional mouse click and scrolling, or finger-based multitouch gestures, may be adapted for use with some of these displays. Interaction in immersive environments, or with other datasets distributed in the ambient space of the environment in which users are situated, could also benefit from the naturalness and intuitiveness with which we are accustomed to interacting on foot via unique walking or stepping behaviors. Additional benefits are possible by taking advantage of the relatively greater number of degrees of freedom that underly bodily control of movement.

However, methods of interaction must also be chosen according to constraints of computational task and context. Users of such environments can be presumed to need to interact in order to support activities that they value. Thus, a floor-based touch surface might be most appropriate in areas of human-machine interaction where foot-operated controls or interfaces are already commonplace. Many examples exist in manufacturing, transportation, or medicine. A virtualized display in such settings might be used to provide access to instrumentation or machine controls in ways that can adapt to a range of different tasks (e.g., to different dental procedures). Such an approach may, for example, be able to overcome problems with physically embodied foot controllers, such as the overabundance of physical foot pedal controls in medical interactions [41].

Another potentially relevant application of computational interfaces integrated in floor surfaces may be for the purpose of enhancing pedestrian navigation, or for providing interactive maps or other geospatial data visualizations related to a space or dataset that is being navigated immersively. In such cases, movement on foot may be seamlessly integrated as a means of virtual navigation. Previous researchers have utilized context-based interactive maps or ambiently represented menus in applications such as architectural walkthroughs or training simulations [19]. Other relevant application fields could include entertainment, music performance, gaming, or advertising, where companies such as Gesturetek (Fig. 17.2) and Reactrix have successfully commercialized interactive, floor-based visual displays for marketing purposes.

Although most potential applications, like those noted above, are either emerging, or remain to be defined, a few conceptual scenarios may be helpful toward further motivating the discussion in the sections that follow:

- A walkway provides ecologically based tactile, acoustic, and visual feedback simulating the response that is felt and observed when walking on ice, earth, or snow in order to aid patients with gait disorders in readapting to walking in challenging conditions. The associated virtual environment may also serve as a tool that is used to assess the extent to which different sensory cues can affect locomotion strategies, aiding clinical researchers in identifying potentially beneficial rehabilitation programs.
- A multi-function, reconfigurable operating room in a future medical facility is equipped with a virtual floor controller that allows doctors or their assistants to access the controls of instruments needed during a certain medical procedure without requiring the physical rearrangement of foot pedals or other physical interface devices.
- An immersive virtual environment that is presented on several walls surrounding its users displays data corresponding to the allocation of geospatial resources. This data can be navigated via a floor-based map interface that is operated via the feet. In another scenario, students at a remote location might use such an interface to navigate the streets of a site of special historical interest, such as Pompeii (captured through photographic data), or its virtual reconstruction, presented as a three-dimensional virtual world.

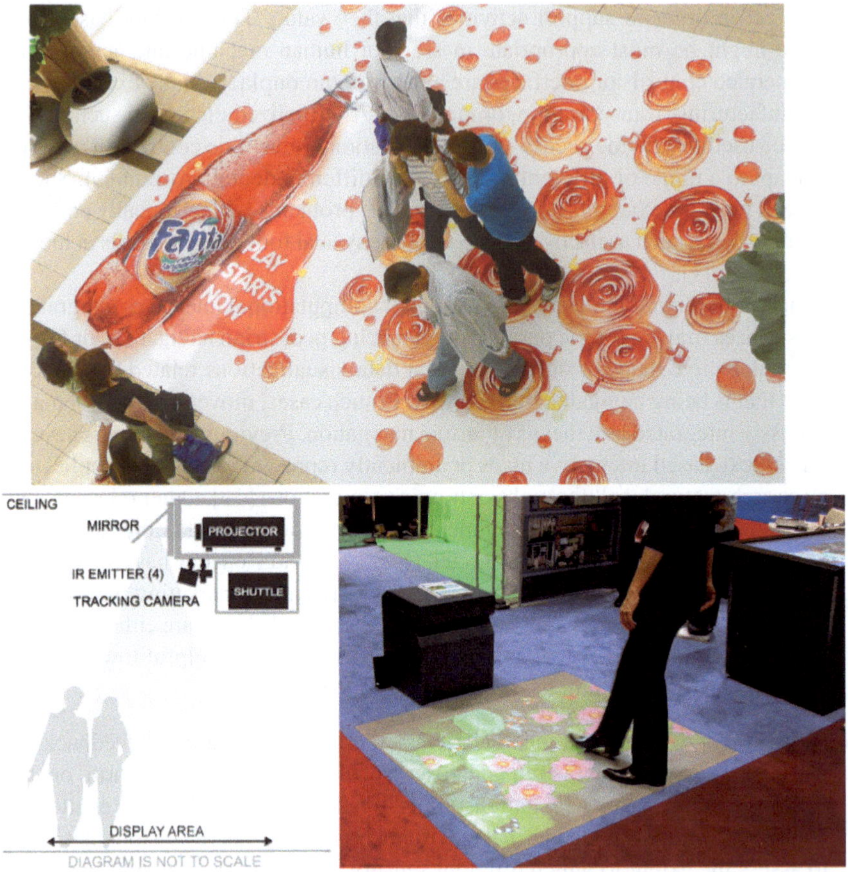

Fig. 17.2 The ground FX video-based interactive floor display system offered by the company Gesturetek can be used for purposes such as entertainment or product promotion

17.2.1 Input from the Foot in Human-Computer Interaction

Examples of the use of foot-controlled input in HCI, interactive arts and video games date at least to the early 1980s, with Amiga's Joyboard (1982) being one widely known early example [33]. The current Nintendo Wii balance board controller has achieved international commercial success by building on the success of such past video game control interfaces. Human-computer interaction researchers have investigated related issues, arguably beginning in the mid 1980s, when Pearson and Weiser studied foot input devices for desktop PCs, and invented a pedal-like device called the Mole [25]. However, despite the high-level of ongoing interest in touch screen-based interactions for the hands, less research and development has targeted touch-sensitive interfaces for the feet.

17.2.2 Relevance to Virtual Reality

Virtual reality (VR) aims to digitally simulate man-made or natural environments that users can experience perceptually, via one or more sensory modalities, and interact in, by moving or otherwise acting in real time. In order to improve the degree of realism of the virtual environment, and the presence of users within it, it is often desired to preserve as many features of the environment through as many modalities as possible. When users are permitted to navigate by walking within the virtual environment, as illustrated in the other chapters of this book, it can be desirable to represent features of the virtual ground surface, and, particularly in simulated natural environments, to represent dynamical and material-dependent aspects of interactions between the foot and ground surface, such as the sense of soft materials like snow or sand deforming underfoot. Sensations accompanying walking on natural ground surfaces in real world environments (sand in the desert, or snow in winter) are multimodal and can greatly reinforce users' sense of presence in the virtual settings in which they occur [38]. Limited previous research has addressed foot-based interaction with deformable ground surfaces in virtual reality [38]. This may be due to a lack of efficient techniques for capturing foot-floor contact interactions and rendering them over distributed floor areas, and to the emerging nature of the applications involved. Some newly developed methods for accomplishing these tasks are presented in this chapter and that of Marchal et al. in this book.

Related research on virtual and augmented reality environments has focused on the problem of natural navigation in virtual reality environments. Solutions such as walking in place [34] and redirected walking [30] map kinematically tracked movements onto a user's coordinates in a virtual environment (VE); see Chap. 11 for a review. A number of haptic interfaces for enabling omnidirectional in-place locomotion in VEs, based on treadmills or other mechanisms, have been developed, and can serve some of the same purposes as augmented floor surfaces in permitting their users to navigate virtual reality environments. The design of such locomotion interfaces and several examples of them are reviewed in Chap. 9 of this book, and in earlier literature [16], and consequently they are not discussed here. Instead, in this chapter we focus on effectively stationary ground surfaces that are augmented with audio, visual, and/or tactile feedback.

One example is the shoe-based Step WIM interface of LaViola et al. [19] (Fig. 17.3), which introduced foot gestures performed through a pair of electronically instrumented shoes for controlling navigation in a larger virtual environment. The system operated with reference to a visual map display representing the surrounding virtual environment, and did not provide auditory or haptic feedback.

Fig. 17.3 The step WIM interface of LaViola et al. introduced an interactive floor display for navigating within a virtual environment via foot gestures performed with instrumented shoes [19]

17.3 Techniques and Technologies

Augmented floor surfaces that can respond to actions of the feet of persons moving on them can, as envisioned here, be said to consist of interfaces comprising interactive visual, audio, and/or tactile displays located on the walking surface together with sensing technologies capable of capturing movements of the feet, contact or forces between the feet and ground, or other movements of the body. Existing devices can be distinguished according to the choices of sensing and display technologies that they employ.

17.3.1 Indirect Optical Sensing

One sensing method that is frequently used to implement interactive floor displays involves the inference of body position or movement from video capture in a region above the floor. This is the technique used in systems that have been commercialized by several companies (e.g., Gesturetek and Reactrix). The disadvantage of such an approach is that it normally provides no direct information about foot-floor contact forces, contact onset, or contact area, and is thus unable to distinguish between near-contact and the actual instant of contact between foot and ground. Arguably, this distinction is vital for the convincing rendering of interactions with virtual objects or controls, or for simulating highly contact-dependent interactions with virtual materials [37, 40].

Video-based sensing of the kinematics of walking interactions has been a mainstay in fields including gait analysis and biometrics [26], for example, in the identification of pathological gait patterns with machine learning algorithms. Begg et al. [5], among others, have used such algorithms to identify pathological gait using kinematic features acquired through optical motion capture, such as the minimum foot clearance (a local minimum of the vertical distance between the shoe and ground that occurs after toe off in the gait cycle).

Although most of these systems involve the capture and analysis of frames of video of the whole body, or optical motion capture systems for tracking discrete markers on the body, the Lightfoot interface adopted a somewhat different approach, using an array of interruptible lasers to capture interactions between the feet of users. In interactive applications, the foot movements of dancers were used to control the real-time synthesis of musical audio feedback [11].

17.3.2 Contact Sensing

Contact sensing via floor surfaces requires the measurement of forces, areas, or occurrences of physical contact between the body and the floor. This can be accomplished using surface mounted force sensing arrays, via force sensors embedded within the structure of elements of the floor itself (in the manner of a force measurement plate or scale), via optical measurement of the foot-floor contact region(s), or via other surface-based contact sensing techniques, such as those based on capacitance, acoustic waves, or piezoelectric effects. Direct tactile sensing for interaction with floor surfaces is often accomplished with surface-mounted force sensing arrays—as, for example, in the Z-Tiles project, Magic Carpet project, ORL Active Floor project, and others [1, 24, 28, 31, 32]. Such interfaces have been employed in applications including person tracking, activity tracking, or musical performance. Floor-mounted tactile sensing arrays are commercially available, but for large surfaces areas, costs are high and support for real-time interaction is often not offered commercially, since the predominant application areas involve offline gait and posture measurement that do not require such features.

A wide range of ambient computing tasks have served to motivate the development of several of these systems. Orr, Abowd, and Addlesee [1] developed an activity-aware smart home environment based, in part, on a floor surface (the ORL Active Floor) that captured foot-floor forces during normal activities of daily living. The Ubi-Floor allowed users to access context-aware multimedia applications selected via a footstep-activated menuing system [9]. Headon developed a system for interacting with games via full-body movements or gestures sensed via a floor surface. Input gestures were recognized using statistical classification of temporal patterns of force measurements acquired through force-sensing load cells embedded in the floor structure [14]. Commercially available sensing pads for video games have been used to implement novel human-computer interactive scenarios, such as the navigation of heritage sites presented in a virtual environment [10].

Steinicke et al. have investigated several scenarios associated with the use of combinations of floor-sensed body posture and hand gestures to navigate virtual environments [8, 35]. In one scenario, they employed the Nintendo Wii Balance Board interface in tandem with a manually operated touch-sensitive interface to allow users to navigate within a 3D model of a city presented in a video-projected virtual environment simulator.

Several research groups have also studied the acquisition and analysis of inertially sensed movements and foot-ground force profiles (ground reaction forces) for applications such as biometric authentication. Mostayed et al. studied the recognition of walker identity from ground-reaction force profiles acquired through a force plate [22], as did Headon and Curwen [13]. A further promising application in this area is the early identification of at-risk gait in aging populations from foot-ground force measurements. Holtzreiter and Kohle [17] were among the first to employ machine learning techniques to identify normal and pathological gait patterns from force platform recordings.

17.3.2.1 Under-Floor Camera-Projector Systems

Another method that has been used to enable foot-floor interactions with visual feedback employs under-floor cameras and projectors to render interactions via translucent floor plates. Groenboek et al. implemented an interactive floor surface called the iGameFloor using body gestures sensed optically through translucent floor plates, which were also used to display visual feedback via rear-projected video [12]. Four projectors and cameras were installed in a cavity beneath floor level, creating an interactive floor area 3×4 m in size. The locations of limbs near to the interactive surface were tracked, and were used to mediate interactions with multi-person floor-based video games.

Augsten et al. adapted the method of optical sensing via frustrated total internal reflection to enable the dynamic capture of foot-floor contact areas through back-projected translucent floor plates [4]. This method provides direct imaging of contact area, although it does not directly reveal forces. The main drawback to this approach is that it requires the installation of cameras and projectors within a potentially large recessed space that must be available beneath the floor. The authors implemented and evaluated the usability of foot-floor touch-surface interfaces using this approach (Fig. 17.4). Similar to the methods described in Sect. 17.4.1, they identified salient selection points within the foot-floor contact interface in order to enable precision pointing to targets or performance of other actions through a floor-based touch-screen interface.

17.3.3 Usability

A number of special considerations can be identified in relation to the usability of touch surfaces designed for the feet, including the appropriate size, distance, and arrangement of control areas, the type of gestures suitable for use with the feet, and the respective role of each foot during interaction, among others. Similar to the case of hand-based touch-surface interaction, the appropriate size of controls for use with the feet likely depends on a range of factors, including limitations in sensing resolution, user motor abilities, and feedback modalities. Such factors have been studied in the literature on task performance in human-computer interaction generally [18, 21] and on touch-screen usability in particular [3, 6], where the appropriate size of controls

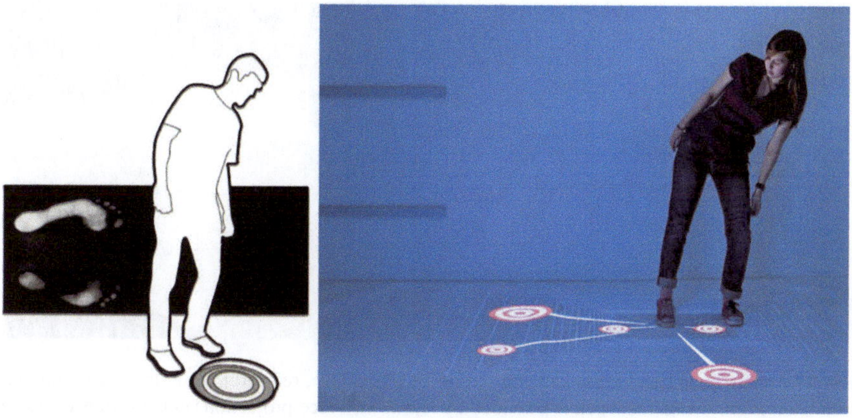

Fig. 17.4 The multitoe interface of Augsten et al. [4] is based on the optical sensing of foot-floor contact regions captured through back-projected translucent floor plates using the frustrated total internal reflection method (image reproduced from [4])

Fig. 17.5 The iGameFloor interface of Gronboek et al. [12] uses vision-based limb tracking on a rear-projected interactive floor comprised of back-projected translucent floor plates (image reproduced from [12])

has been found to depend on the interaction technique adopted or other factors. Using precision control strategies, single-pixel accuracy in finger-based touch-screen interaction has been demonstrated [3, 27], and similar techniques may be effective for enhancing the precision of control with the feet (Figs. 17.5, 17.6).

Fig. 17.6 The floor interface is situated within an immersive, rear-projected virtual environment simulator. *Right* visual feedback is provided by *top-down* video projection (in the instance shown, this corresponds to a virtual, multimodal sand scenario—see Chap. 12)

Limited research to date has addressed the usability of floor-based touch interfaces. However, since the feet play very different roles in human movement than the hands do, the extent to which the kinds of techniques that have been adopted for use with the fingers may be useful for interaction via the feet is questionable and requires further investigation. For example, when users are standing, there are strong constraints on the placement of the user's feet, due to the need to maintain stability. In addition, there are obvious anatomical differences between the hands and feet, with toes rarely used for prehensile tasks. Nonetheless, human movement research has studied foot movement control in diverse settings. Visually guided targeting with the foot has been found to be effectively modeled by a similar version of Fitts' Law as is employed for modeling hand movements, with an execution time about twice as long for a similar hand movement [15]. However, for many interfaces, usability is manifestly co-determined by both operator and device limitations (e.g., sensor noise or inaccuracy), imposing a window on both.

Augsten et al. studied users' performance in selecting keys using a touch-screen keyboard projected on the floor using precision optical motion capture tracking of the foot location, a visual crosshair display to indicate the location that was being pointed to, and a pressure threshold to determine pressing. They found that users were able to select target buttons (keyboard keys) having one of the three dimensions 1.1×1.7, 3.1×3.5, or 5.3×5.8 cm with respective error rates of 28.6, 9.5, or 3.0 %. Sensing accuracy was not the limiting factor in this study, since the motion capture input provided sub-millimeter accuracy. The authors undertook further studies to determine users' preferences with respect to the part of the foot to be used as a pointer for selection, and to determine the extent to which non-intentional stepping actions could be discriminated from volitional selection operations.

Visell et al. investigated the usability of a floor interface consisting of an array of instrumented tiles, by assessing users' abilities to select on-screen virtual buttons of different sizes presented at different distances and directions [39]. This example is described further in the following sections.

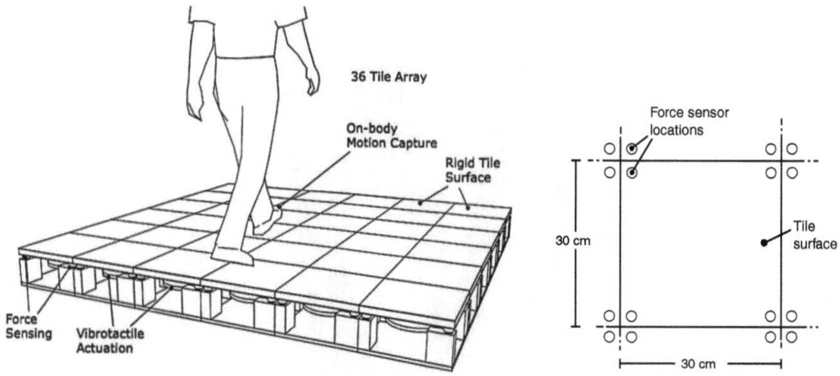

Fig. 17.7 *Left* diagrammatic view of the interface. Sensing and actuating components are integrated beneath the floor. *Right* view from above showing sensor locations

17.4 Case Study: A Distributed, Multimodal Floor Interface

In this section, we review a case example, consisting of an interface developed by the authors [39]. The interface is a modular collection of rigid floor components, each of which is instrumented with force sensors and a large vibrotactile actuator. The prototype shown in Fig. 17.7 comprises a square array of 36 tiles, with an area of approximately four square meters. The floor is coated in reflective projection paint, and a redundant pair of overhead video projectors is used for visual display, the visual redundancy making it possible to reduce the effect of shadows cast by users.

The individual tile interfaces are rigid plates with dimensions $30.5 \times 30.5 \times 2$ cm, supported by elastic vibration mounts, and coupled to a vibrotactile actuator (Clark Synthesis, model TST229) beneath each plate [36]. Actuator signals are generated on personal computers, output via digital audio interfaces, and amplified.

Normal forces are sensed at locations below the corner vibration supports of each tile using a total of four resistive force sensors (Interlink model 402 FSR). Analog data from the force sensors is conditioned, amplified, and digitized via a set of 32-channel, 16-bit data acquisition boards based on an Altera FPGA. Each sensor is sampled at a rate of up to 1 kHz, and the data is transmitted in aggregate over a low-latency Ethernet link. An array of six small-form-factor computers is used for force data processing and audio and vibrotactile rendering. A separate, networked server is responsible for rendering visual feedback and managing user input (Fig. 17.8).

17.4.1 Contact Localization

For processing sensor data acquired through the distributed floor interface, the authors applied intrinsic contact sensing techniques previously developed in the robotics

Fig. 17.8 Example of a portable, 9-tile version of the distributed floor interface of Visell et al. [39], shown without video feedback, as developed by Kemper, Franinovic, Willle, and Visell

community. These make it possible to resolve the locations of contact, forces at the interface, and the moment about the contact normals using internal force and torque measurements [7] using far fewer sensors than are needed in surface-based tactile sensing techniques. For simplification, we assume that there is frictionless contact between the foot and floor and we neglect relative displacement of the suspension elements in the tile (which is, by design, less than 3 mm [36]). We then resolve the location of the centroid $\mathbf{x}_c$ associated with the pressure distribution $p_R(\mathbf{x})$ exerted by the foot, such that a normal force F_c at $\mathbf{x}_c$ gives rise to the same measurements as $p_R(\mathbf{x})$ does [7]. For a floor tile with sensor locations $\mathbf{x}_j$ where measurements f_j are taken (j indexes the tile sensors), $\mathbf{x}_c$ and the normal force $\mathbf{F}_c = (0, 0, F_c)$ can be recovered from scalar measurements $\mathbf{F}_j = (0, 0, f_j)$ via force and torque equilibrium equations,

$$\sum_{j=1}^{4} f_j + F_c + f_p = 0, \quad \sum_{j=1}^{4} \mathbf{x}_j \times \mathbf{F}_j + \mathbf{x}_c \times \mathbf{F}_c + \mathbf{x}_p \times \mathbf{F}_p = 0. \qquad (17.1)$$

$\mathbf{F}_p = (0, 0, f_p)$ is the weight of the plate and actuator at the tile's center $\mathbf{x}_p$. The three nontrivial scalar equalities in (17.1) yield:

$$F_c = \sum_{i=1}^{4} f_i - f_p, \quad \mathbf{x}_c = \frac{1}{F_c}\left(\sum_{i=1}^{4} (\mathbf{x}_i - \mathbf{x}_p) f_i + f_c \mathbf{x}_p\right) \qquad (17.2)$$

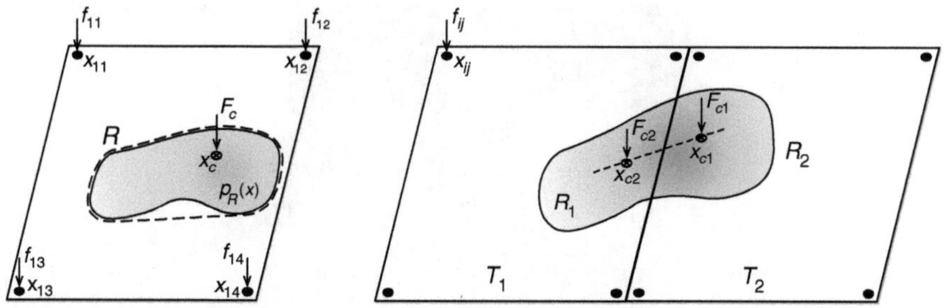

Fig. 17.9 *Left* a normal force distribution $p_R(\mathbf{x})$ during foot-floor contact the and associated contact centroid position $\mathbf{x}_c$, which is used to mediate interaction with an augmented floor surface. *Right* a pressure distribution $p_R(\mathbf{x})$ on a region R spanning adjacent tiles. The weighted sum of centroids $\mathbf{x}_c$ is the centroid location for the distribution with support $R = R_1 \bigcup R_2$. It lies on the line segment connecting $\mathbf{x}_{c1}$ and x_{c2}. The difference $\delta\mathbf{x} = \mathbf{x}_{c1} - \mathbf{x}_{c2}$ provides information about contact shape

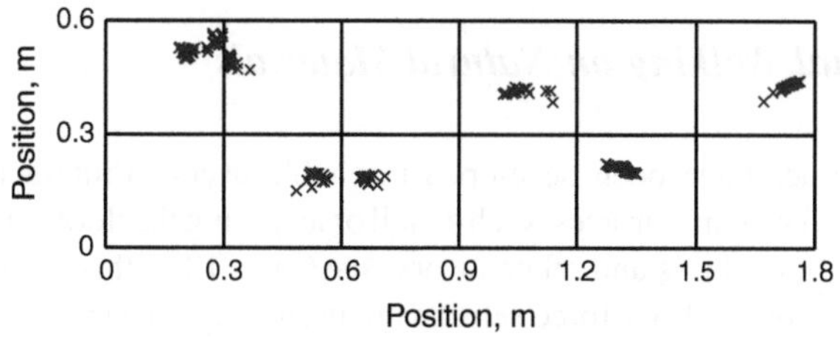

Fig. 17.10 A recorded sequence of contact centroids produced by an individual walking across the floor. Data was sampled each 100 ms to produce the figure. Each *square* corresponds to one floor tile. When the foot lies on a single tile, as weight shifts from heel to toe, an array of centroids is produced, moving in the direction of travel. At inter-tile boundaries, at each instant one centroid is produced on each tile with which there is contact

The contact centroid lies within the convex hull of the contact area (dashed line, Fig. 17.9), at the centroid of the pressure distribution [7], and thus provides a concise summary of the foot-floor contact locus, but not about shape or orientation. When the foot-floor contact area R overlaps multiple tiles, a pressure centroid $\mathbf{x}_c$ can be computed by combining those ($\mathbf{x}_{ck}$) for each tile, via $\mathbf{x}_c = w_1\mathbf{x}_{c1} + w_2\mathbf{x}_{c2}$, where $w_k = F_i/F$. This makes it possible to continuously track contact across tile boundaries. The difference vector $\delta\mathbf{x} = \mathbf{x}_{c1} - \mathbf{x}_{c2}$ is indicative of the orientation and extent of the contact distribution at the boundary (Fig. 17.9).

A sequence of contact centroid locations produced by an individual walking across the floor is shown in Fig. 17.10. When there is multi-tile foot-floor contact, as illustrated in the sequence shown, we use a simple clustering algorithm to associate nearby contact centroids that are assumed to belong to the same foot, combining those from nearby tiles.

Contacts were measured to be localized with a typical accuracy of 2 cm, with worst-case errors of ≈3 cm, considerably smaller than the linear dimensions of the tile (30 cm) or the typical width of an adult shoe. Distortion was observed to be highest, and accuracy lowest, near the edges of the tile.

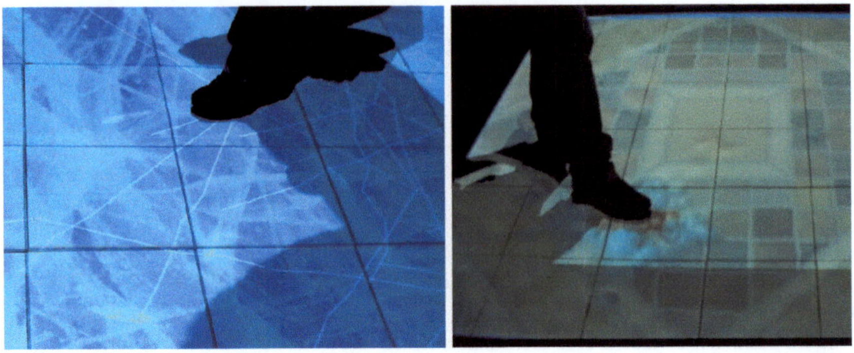

Fig. 17.11 *Left* still image of user interacting with the simulated frozen pond. *Right* a user engaged with a simulation of water in a shallow pool

17.4.2 Virtual Walking on Natural Materials

Several techniques have been developed to enable users to interact on foot with simulated natural ground surfaces, such as soil or ice, using the floor surface described above. For a review of this and related work, see Chap. 12 of this book. Position and force estimates from in-floor force sensors are used to synthesize plausible auditory and vibrotactile feedback in response. Sensations accompanying walking on natural ground surfaces in real world environments (sand in the desert, or snow in winter) are multimodal and can be highly evocative of the settings in which they occur [38]. By reproducing these sensations through auditory, visual, and vibrotactile channels, it is possible to recreate highly evocative scenarios of walking on virtual ground surfaces.

Figure 17.11 illustrates two interactive scenarios that were realized based on the interface and interaction techniques presented above. The first represents a virtual frozen pond demonstration that users may walk on, producing patterns of surface cracks that are rendered and displayed via audio, visual, and vibrotactile channels (Fig. 17.11). A second simulation uses the same feedback modalities to realize the experience of walking on a shallow pool of water. For further discussion of the multimodal rendering algorithms used in these simulations, see Chap. 12.

17.4.3 Floor Touch-Surface Interaction Techniques

The floor interface described above was used to implement virtual floor-based touch interfaces, using either foot based gestures or standard UI widgets controlled with the feet (Fig. 17.12). Input is derived from a multi-touch-screen metaphor relying on a set of interaction points, which are given by the contact centroid locations $\mathbf{x}_c$ associated with the largest forces. Force thresholds assigned to a control are used to determine selection. The controls can also provide positive tactile feedback via

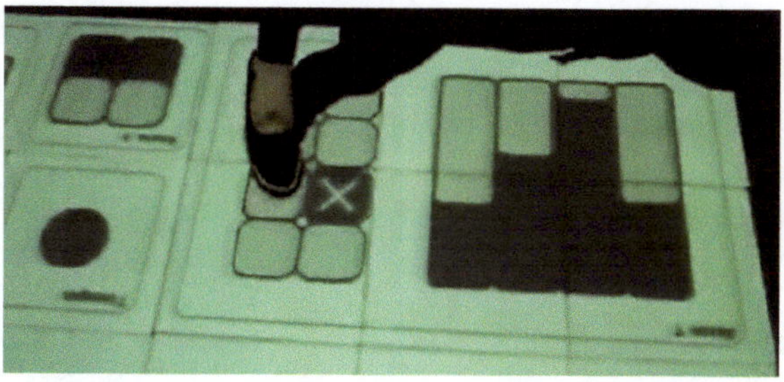

Fig. 17.12 A user interacts with floor-based interface widgets implemented using the interface design toolkit described in the text

the vibrotactile actuators, in the form of click-like transient vibrations or sliding frictional vibrations.

A software layer and network protocol is used to facilitate the design of interactive applications using these techniques. It abstracts the hardware systems, which are accessed over a local Ethernet network, connects them to the user interface, and allows an interface designer to instantiate an array of standard user interface controls, such as sliders, buttons, or toggle switches.

The software processes the sensor data to extract foot-floor contact points that are used for interaction, provides them with unique IDs that persist while contact is sustained, and allows to remotely cue localized vibrotactile feedback. Figure 17.12 illustrates a virtual floor-based touch interface.

17.4.4 Usability of Foot-Floor Touch-Surface Interfaces

The authors investigated users' abilities to select on-screen virtual buttons of different sizes presented at different distances and directions using the floor tile interface described in the preceding section [39]. In this task, the limited resolution of the force sensors (Fig. 17.7) was one factor that could influence performance. Users could activate a button by pressing it within the area of the button with a force exceeding a threshold of about 35 N. Round, virtual buttons ranging in diameter from 4.5 to 16.5 cm, and presented at four distances, on lines radiating at two angles from between their feet. Upon selection, the buttons provided visual feedback in the form of a 20 cm white disc centered in place of the original graphic. Results are summarized in Fig. 17.13. Users were found to have selected larger targets within the allotted two-second interval at a significantly higher rate of success than smaller ones. Performance with the largest was very high (98 %), and that for the smallest was low (44 %). Small targets can pose difficulties due to their tendency to be occluded by the foot during selection, which is mitigated during top-down projection, since they are projected on top of the foot, and can be more affected by limitations arising from factors such as shoe width, human motor abilities, and sensor positioning errors.

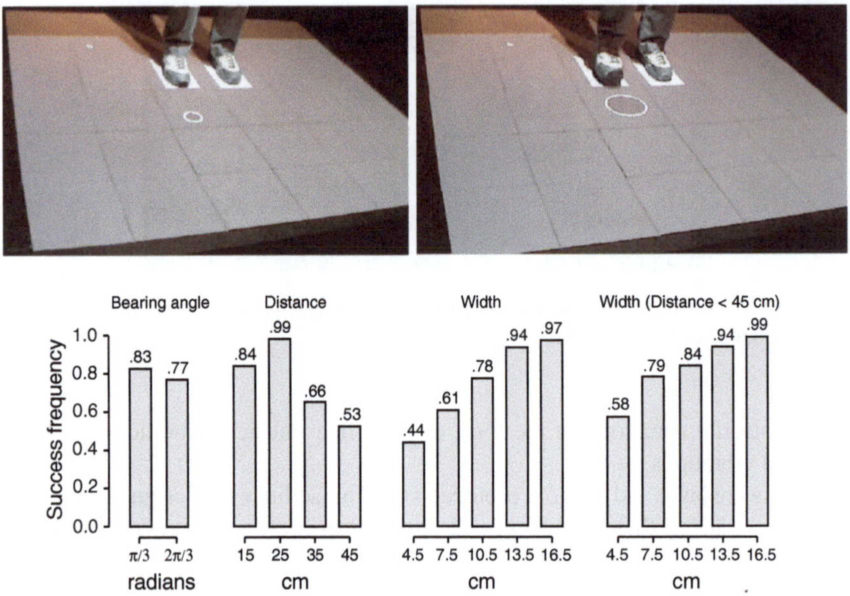

Fig. 17.13 *Top row* a user in the study of Visell et al. [39], in which circular targets were selected via stepping actions. *Bottom* successful target selection rate versus distance, angle of presentation (measured away from preferred foot), and button width, inclusive and exclusive of the farthest targets

Six out of eight participants in this study reported finding a strategy to activate the small buttons by using a feature of the shoe or changing the applied force. Nearby targets (distances of $D = 15$–25 cm) were selected at a higher rate. However, performance was better at 25 cm than at the nearest distance of 15 cm (98.5 vs. 84 %). One apparent reason was that when an interface element was too close, it could be occluded from view by the body, or could present a difficult viewing angle. Due to such effects selection time T might not be expected to follow a Fitts' Law relation, $T = a + b \log_2(D/W)$, but this was not tested. A mobile user might be able to avoid visibility problems, but they seem to be an important consideration. Neck fatigue was a frequently cited source of discomfort, suggesting that displacing the visual feedback relative to the foot-based interaction point might be beneficial.

Further work is needed on the usability of floor surfaces in order to characterize the usability aspects of floor-based touch-screen displays. A greater understanding of factors such as control element size, display scale, motor abilities, modalities, and other aspects salient to the use of such a device will certainly be needed. In addition, it would be valuable to know to what extent usability might be improved through the use of auditory or vibrotactile feedback. Although Augsten et al. were able to identify some strategies that could be used to avoid selecting controls that are walked across, there remain open questions concerning the interplay between users' movements on foot and their interactions with the touch surface. A novel aspect of interacting on foot is that, implicitly, both feet are involved in touching the interface, due to requirements of movement and of maintaining balance. In everyday actions, like striking a soccer ball, weight is often shifted onto one foot, which specifies an anchored location, while the opposite is used to perform an action within the reachable area. Thus, floor interfaces that involve movement may share similarities to bimanual interaction in HCI, a connection that provides another potential avenue of future investigation.

17.4.5 Application: Geospatial Data Navigation

The ubiquity of multi-touch mobile devices and computers has popularized the use of two-finger gestures to navigate and control zoom level in the display of data via a touch screen. Such an approach focuses on the fingers as a means of providing input and may not be appropriate for settings in which the display surface is not amenable to finger-based interaction, when the hands are occupied, or when the data visualization application occupies a larger volume of space. We argue that this focus has limited the possibilities for carrying out complex exploration tasks in ways that leverage the capabilities we exploit naturally in the physical world.

For target acquisition tasks, e.g., menu and object selection, the accuracy and speed of the input primitives are important. In contrast, for spatial navigation tasks, e.g., panning and zooming, using dragging, resizing, and scrolling operations, less accuracy is often acceptable. Consequently, spatial navigation is likely an appropriate candidate for foot-based input, leaving the hands free, in parallel, to specialize on the more time- and accuracy-critical operations [23]. For instance, in the context of collaborative design or decision making, e.g., urban and architectural planning, or emergency and crisis management [2, 20, 29], different roles are appropriate for foot input and hand input. Specifically, the feet might be better suited to controlling the location or region of interest, while the hands are used to perform other more critical or complex input tasks such as target selection, annotation, or drawing in the virtual design space, or dispatching tasks in emergency situations. Moreover, this represents a compelling alternative for scenarios where use of the hands is inconvenient. For example, in the context of airport information kiosks, where users might be carrying luggage, it may be desirable and appropriate to obtain directions to ones gate through a foot-based interface.

17.4.6 Foot-Based Gestures for Geospatial Navigation

The authors of this chapter implemented an application and several foot-based interaction methods for navigating geospatial data sets presented through the multitouch-sensitive floor surface described in the foregoing. In contrast to prior work, which has focused on foot-based interaction at a fixed location, we investigated the benefit of such interaction in a distributed floorspace. This approach can be seen as an alternative to desktop computer interfaces for navigating geospatial data, and may be particularly suitable for situations in which an immersive display is involved or in which users' hands are occupied with other tasks.

We developed these interactive techniques for exploring geospatial information by navigating with a floor-based map interface, presented via the multitouch floor surface. The application was developed with the user interface framework presented above. As participants navigated to specified locations using the map, via one of the foot-floor interaction techniques, the images of the streets for the locations visited were presented in real time via the wall surfaces whenever applicable. The available data was acquired from existing internet-based mapping applications from Google and Microsoft.

The interaction techniques investigated included a "pivot" interface, which uses relative foot position as a navigational input, a "magic tape" interface that uses absolute foot position within the workspace, a "sliding" interface that allowed users to virtually push themselves within the mapping interface, and a "classic" approach using virtual buttons and sliders to provide a comparison with a more conventional user interface paradigm. Additional gestures allowed participants to control the zoom level of the map.

Classic Interface

The Classic interface transposes the basic design used for spatial navigation in mouse-based applications to the setting of floor-based interaction. Four buttons in a cross arrangement control position and a discrete-valued slider provides control over the zoom level. The discrete slider levels match those used for the whole body gestures in the subsequent interfaces. In the Pivot mode, users establish a pivot point by standing still for a short period. Placing one foot outside of the pivot area, indicated by a circle around the feet, pans in the direction specified as the vector from the pivot center to the outside foot. The participant can, at any time, exit the pivot area and establish a new one elsewhere. In the Sliding interface, a user first establishes a pivot point by standing still for a short period. Then, by placing one foot outside the pivot area and using sliding or dragging gestures, akin to touch-screen scroll on an iPhone, the user can pan the map (Figs. 17.14, 17.15, 17.16, 17.17).

The Magic Tape interface was inspired by the work of Cirio et al. and takes advantage of the larger floor surface by employing an interaction paradigm based on absolute foot position. This metaphor allows users to navigate freely in the center of the floor space, without altering the displayed map contents. However, when participants walk past the boundary region of the floor surface, the map pans in a direction designated by the user's position. The farther the user strays from the center, the greater the panning speed.

Gestures: Crouch or Jump to Zoom

Unlike the direct-manipulation inspired control actions described above, gestures are recognized to allow temporally extended body movements, such as jumping,

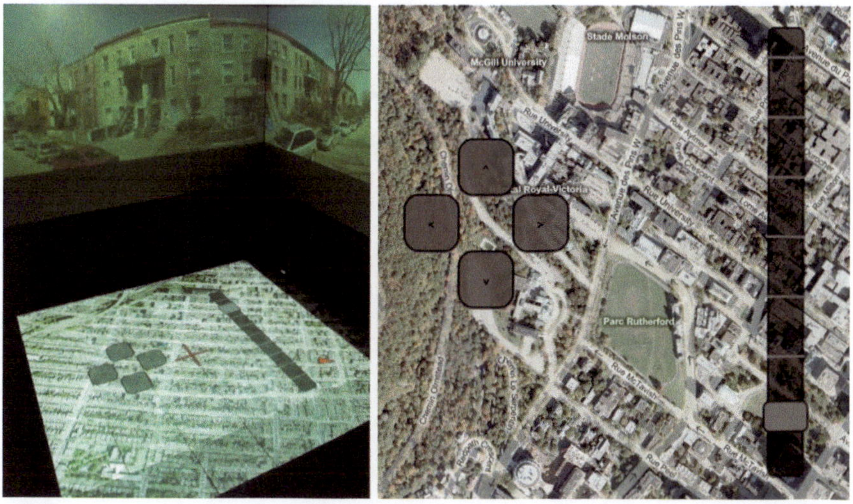

Fig. 17.14 *Four button arrows* corresponding to panning in the cardinal directions and a slider for zoom

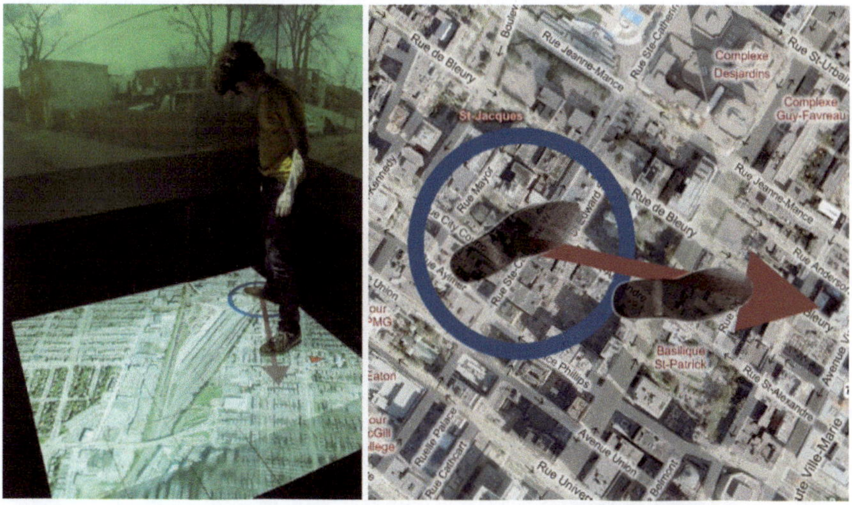

Fig. 17.15 The *first foot* to remain still establishes the pivot point, surrounded by a *blue circle*. The *second foot* then specifies the vector, relative to the pivot, for panning

to affect the application state in a way that is not directly or instantaneously linked to changes in the user interface. In the geospatial data navigation application, short body gestures can be used in conjunction with the control actions described above to control zooming. A curt "crouching" gesture zooms the map in, while a curt "jumping" gesture (raised onto the toes) zooms out. Since such gestures have distinctive normal-force profiles trough time they can be matched to a reference signal.

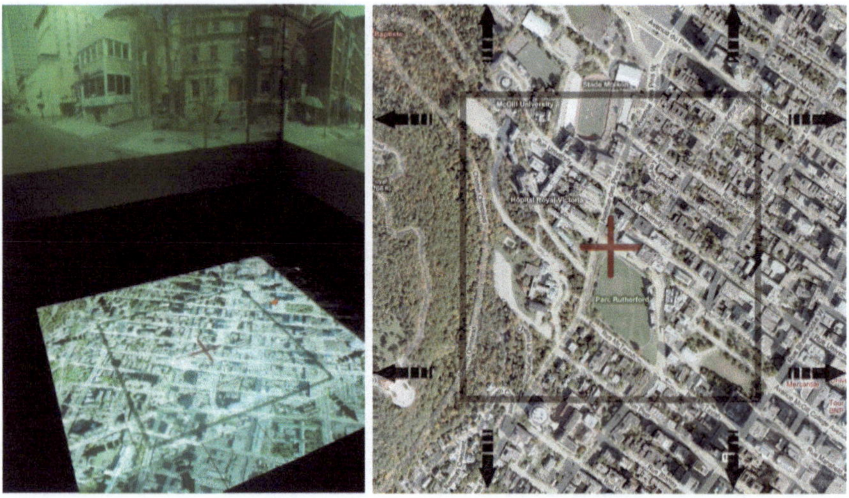

Fig. 17.16 The "magic tape" interface: A *rectangular* outline indicates the magic tape boundary, beyond which, the user's footstep will result in panning in the direction formed by a vector from the *center crosshair* and the user's foot. (It is assumed that one of the user's feet remains within the boundary at all times.)

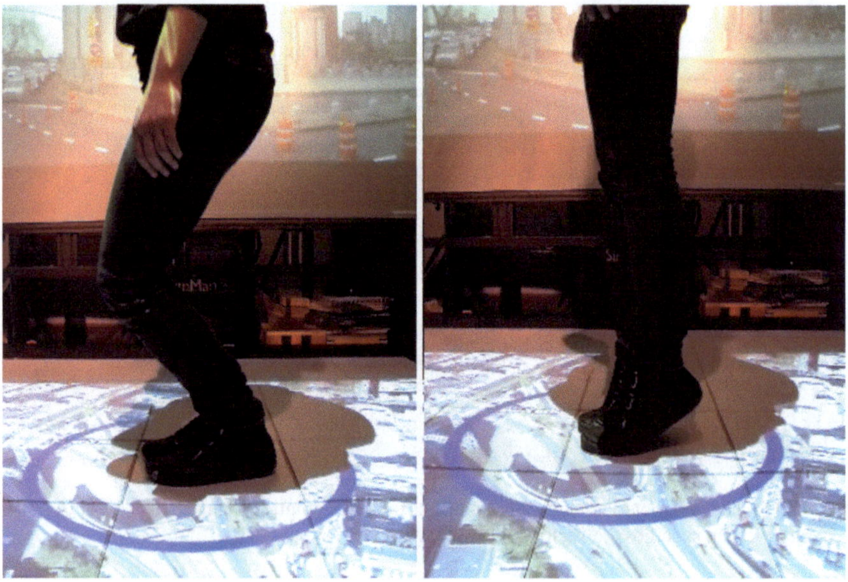

Fig. 17.17 Crouching and jumping gestures for zoom control

The analysis of these gestures, which relate to shifts in the weight and posture of users, is based on recognizing distinct sequences of whole body movements based on dynamic force signals sensed during weight shifts or posture changes. Many such movements are characterized by distinctive transient force signals from the floor sensors, due to the reaction forces the user generates when moving. A larger gestural vocabulary based on such movements could include crouching, jumping, leaning and tapping. Such movements are accompanied by force impulses as the user lifts and moves his or her body weight.

In order to compensate for variations in the timing, manner, and intensity of gesture execution, we use standard techniques from time series recognition, based on the dynamic time warping family of algorithms. Since the starting time of gestures is unknown, we dynamically spot gestures on a running window of data acquired continuously from the sensors, so that users may execute gestures to control the interface in a manner that is unconstrained with respect to the timing of execution.

17.5 Conclusions

This chapter reviewed approaches to interacting with computationally augmented ground surfaces, for purposes such as those of realizing virtual ground material simulations or foot-based touch surface interfaces. Several technical techniques for enabling such simulations were reviewed, and we described in detail one approach based on a distributed, multimodal floor interface that is capable of directly capturing foot-floor contact information via intrinsic force sensors. It presented interaction methods that are low in cost and complexity, and that can be made accessible to multiple users without requiring body-worn markers or equipment. In addition, this chapter presented examples in which these interaction techniques are used to realize generic virtual control surfaces or navigation interfaces for virtual reality or for immersive geospatial data navigation. Further issues related to the interactive rendering of multisensory simulations of virtual ground surfaces are reviewed in detail in Chap. 12 of this book. We also discussed guidelines for the use of such a display, including several factors that have been studied in the context of empirical usability evaluations from foot-based human-computer interaction and ergonomics. It is therefore hoped that this contribution succeeds in demonstrating a range of potential uses and design considerations for floor-based touch surfaces in virtual reality and human-computer interaction.

References

1. Addlesee MD, Jones AH, Livesey F, Samaria FS (1997) The ORL active floor. IEEE Pers Commun 4(5):35–51
2. Agrawal P, Rauschert I, Inochanon K, Bolelli L, Fuhrmann S, Brewer I, Cai G, MacEachren A, Sharma R (2004) Multimodal interface platform for geographical information systems (GEOMIP) in crisis management. In: ICMI '04: Proceedings of the 6th international conference on multimodal interfaces. ACM, New York, NY, USA, pp 339–340

3. Albinsson PA, Zhai S (2003) High precision touch screen interaction. In: Proceedings of the SIGCHI conference on human factors in computing systems. ACM, pp 105–112
4. Augsten T, Kaefer K, Meusel R, Fetzer C, Kanitz D, Stoff T, Becker T, Holz C, Baudisch P (2010) Multitoe: high-precision interaction with back-projected floors based on high-resolution multi-touch input. In: Proceedings of the 23nd annual ACM symposium on user interface software and technology. ACM, pp 209–218
5. Begg RK, Palaniswami M, Owen B (2005) Support vector machines for automated gait classification. IEEE Trans Biomed Eng 52(5):828–838
6. Benko H, Wilson AD, Baudisch P (2006) Precise selection techniques for multi-touch screens. In: Proceedings of the SIGCHI conference on human factors in computing systems. ACM, pp 1263–1272
7. Bicchi A, Salisbury JK, Brock DL (1993) Contact sensing from force measurements. Int J of Robot Res 12(3):249
8. Bruder G, Steinicke F, Hinrichs KH (2009) Arch-explore: a natural user interface for immersive architectural walkthroughs. In: Proceedings of IEEE symposium on 3D user interfaces, 3DUI 2009. IEEE, pp 75–82
9. Chang S, Ham S, Kim S, Suh D, Kim H (2010) Ubi-floor: design and pilot implementation of an interactive floor system. In: Proceedings of 2nd international conference on intelligent human-machine systems and cybernetics (IHMSC), 2010, vol 2. IEEE, pp 290–293
10. Fassbender E, Richards D (2008) Using a dance pad to navigate through the virtual heritage environment of macquarie lighthouse, Sydney. In: Virtual systems and multimedia, Springer, pp 1–12
11. Fernström M, Griffith N (1998) Litefoot-auditory display of footwork. In: Proceeding of ICAD, vol 98
12. Gronboek K, Iversen OS, Kortbek KJ, Nielsen KR, Aagaard L (2007) Interactive floor support for kinesthetic interaction in children learning environments. Lect Notes Comput Sci 4663:361
13. Headon R, Curwen R (2001) Recognizing movements from the ground reaction force. In: Proceedings of the workshop on perceptive user, interfaces
14. Headon R, Curwen R (2002) Movement awareness for ubiquitous game control. Pers Ubiquitous Comput 6(5):407–415
15. Hoffmann ER (1991) A comparison of hand and foot movement times. Ergonomics 34(4):397
16. Hollerbach J (2008) Locomotion interfaces and rendering. In: Lin M, Otaduy M (eds) Haptic rendering: foundations, algorithms and applications. A K Peters, Ltd., 2008
17. Holzreiter SH, Köhle ME (1993) Assessment of gait patterns using neural networks. J Biomech 26(6):645–651
18. Jacko JA, Sears A (2003) The human-computer interaction handbook: fundamentals, evolving technologies, and emerging applications. Lawrence Erlbaum Assoc Inc, Mahwah
19. LaViola JJ Jr, Feliz DA, Keefe DF, Zeleznik RC (2001) Hands-free multi-scale navigation in virtual environments. In: Proceedings of the 2001 symposium on interactive 3D graphics. ACM, New York, NY, USA, pp 9–15
20. MacEachren AM, Cai G, Sharma R, Rauschert I, Brewer I, Bolelli L, Shaparenko B, Fuhrmann S, Wang H (2005) Enabling collaborative geoinformation access and decision-making through a natural, multimodal interface. Int J Geogr Inf Sci 19(3):293–317
21. MacKenzie IS (1992) Fitts' law as a research and design tool in human-computer interaction. Hum Comput Interact 7(1):91–139
22. Mostayed A, Mynuddin M, Mazumder G, Kim S, Park S (2008) Abnormal gait detection using Discrete Fourier Transform. In: MUE, pp 36–40
23. Pakkanen T, Raisamo R (2004) Appropriateness of foot interaction for non-accurate spatial tasks. In: CHI '04: CHI '04 extended abstracts on human factors in computing systems. ACM, New York, NY, USA, pp 1123–1126
24. Paradiso J, Abler C, Hsiao KY, Reynolds M (1997) The magic carpet: physical sensing for immersive environments. In: ACM CHI extended abstracts. ACM, New York, NY, USA, pp 277–278

25. Pearson G, Weiser M (1986) Of moles and men: the design of foot controls for workstations. In: Proceedings of the ACM SIGCHI conference on human factors in computing systems. ACM, New York, NY, USA, pp 333–339
26. Perry J (1992) Gait analysis: normal and pathological function. SLACK Inc., Thorofare
27. Potter RL, Weldon LJ, Shneiderman B (1988) Improving the accuracy of touch screens: an experimental evaluation of three strategies. In: CHI '88: Proceedings of the SIGCHI conference on human factors in computing systems. ACM, New York, NY, USA, pp 27–32
28. Rangarajan S, Kidane A, Qian G, Rajko S, Birchfield D (2007) The design of a pressure sensing floor for movement-based human computer interaction. Smart Sensing and, Context, pp 46–61
29. Rauschert I, Agrawal P, Sharma R, Fuhrmann S, Brewer I, MacEachren A (2002) Designing a human-centered, multimodal gis interface to support emergency management. In: GIS '02: Proceedings of the 10th ACM international symposium on advances in geographic information systems. ACM, New York, NY, USA, pp 119–124
30. Razzaque S, Kohn Z, Whitton MC (2001) Redirected walking. In: Proceedings of EUROGRAPHICS, pp 289–294
31. Richardson B, Leydon K, Fernstrom M, Paradiso JA (2004) Z-tiles: building blocks for modular, pressure-sensing floorspaces. In: CHI'04 extended abstracts on Human factors in computing systems. ACM, pp 1529–1532
32. Schmidt A, Strohbach M, van Laerhoven K, Friday A, Gellersen H-W (2002) Context acquisition based on load sensing. In: UbiComp '02: Proceedings of the 4th international conference on Ubiquitous computing. Springer-Verlag, London, UK, pp 333–350
33. Sinclair J, Hingston P, Masek M (2007) Considerations for the design of exergames. In: Proceedings of the 5th international ACM conference on computer graphics and interactive techniques, Southeast Asia, p 295
34. Templeman JN, Denbrook PS, Sibert LE (1999) Virtual locomotion: walking in place through virtual environments. Presence 8(6):598–617
35. Valkov D, Steinicke F, Bruder G, Hinrichs KH (2010) Traveling in 3d virtual environments with foot gestures and a multi-touch enabled WIM. In: Proceedings of virtual reality international conference (VRIC 2010), pp 171–180
36. Visell Y, Cooperstock JR (2010) Design of a vibrotactile display via a rigid surface. In: Proceedings of IEEE haptics symposium, 2010
37. Visell Y, Cooperstock J, Giordano BL, Franinovic K, Law A, McAdams S, Jathal K, Fontana F (2008) A vibrotactile device for display of virtual ground materials in walking. In: Proceedings of eurohaptics symposium 2008
38. Visell Y, Fontana F, Giordano BL, Nordahl R, Serafin S, Bresin R (2009) Sound design and perception in walking interactions. Int J Hum Comput Stud 67:(11)947–959
39. Visell Y, Smith S, Law A, Rajalingham R, Cooperstock JR (2010) Contact sensing and interaction techniques for a distributed, multimodal floor display. In: IEEE symposium on 3D user interfaces (3DUI), 2010. IEEE, pp 75–78
40. Visell Y, Law A, Ip J, Smith S, Cooperstock JR (2010) Interaction capture in immersive virtual environments via an intelligent floor surface. In: Proceedings of IEEE virtual reality 2010
41. Wauben LSGL, van Veelen MA, Gossot D, Goossens RHM (2006) Application of ergonomic guidelines during minimally invasive surgery: a questionnaire survey of 284 surgeons. Surg Endoscopy 20(8):1268–1274

Index

F. Steinicke et al. (eds.), *Human Walking in Virtual Environments*,
DOI: 10.1007/978-1-4419-8432-6, © Springer Science+Business Media New York 2013

MIX
Papier aus verantwortungsvollen Quellen
Paper from responsible sources
FSC® C105338

If you have any concerns about our products,
you can contact us on
ProductSafety@springernature.com

In case Publisher is established outside the EU,
the EU authorized representative is:
Springer Nature Customer Service Center GmbH
Europaplatz 3, 69115 Heidelberg, Germany

Printed by Libri Plureos GmbH
in Hamburg, Germany